NUCLEAR WASTE CLEANUP TECHNOLOGY AND OPPORTUNITIES

NUCLEAR WASTE CLEANUP TECHNOLOGY AND OPPORTUNITIES

by

Robert Noyes

Library of Congress Catalog Card Number: 95-22674
ISBN: 0-8155-1381-X

Published in the United States of America by
Noyes Publications
Mill Road, Park Ridge, New Jersey 07656

Library of Congress Cataloging-in-Publication Data

Noyes, Robert.
Nuclear waste cleanup technology and opportunities / by Robert Noyes.
p. cm.
Includes index.
ISBN 0-8155-1381-X
1. Cleanup of radioactive waste sites--United States. 2. Nuclear weapons plants--Waste disposal--Environmental aspects--United States. 3. United States. Dept. of Energy--Buildings. I. Title.
TD898.118.N69 1995
621.48'38--dc20 95-22674
CIP

Transferred to Digital Printing 2009

About the Author

Robert Noyes is a graduate chemical engineer, with graduate studies in nuclear engineering. Prior to founding Noyes Data Corporation/Noyes Publications, he was nuclear sales manager for Burns & Roe, Inc., and Curtiss-Wright International; responsible for marketing research reactors.

He is the author of four previous books: *Handbook of Pollution Control Processes; Handbook of Leak, Spill and Accidental Release Prevention Techniques; Pollution Prevention Technology Handbook;* and *Unit Operations in Environmental Engineering.*

Preface

One of the largest, most complicated and expensive environmental problems in the United States is the cleanup of nuclear wastes. The U.S. Department of Energy (DOE) has approximately 4,000 contaminated sites covering tens of thousands of acres and replete with contaminated hazardous or radioactive waste, soil, or structures. In addition to high-level waste, it has more than 250,000 cubic meters of transuranic waste and millions of cubic meters of low-level radioactive waste. In addition, DOE is responsible for thousands of facilities awaiting decontamination, decommissioning, and dismantling.

DOE and its predecessors have been involved in the management of radioactive wastes since 1943, when such wastes were first generated in significant quantities as by-products of nuclear weapons production. Waste connected with DOE's nuclear weapons complex has been accumulating as a result of various operations spanning over five decades.

The cost estimates for nuclear waste cleanup in the United States have been rapidly rising. It has recently been estimated in a range from $200 billion to $350 billion. Costs could vary considerably based on future philosophies as to whether to isolate certain sites (the "iron fence" philosophy), or clean them up to pristine conditions (the "green fields" philosophy). Funding will also be based on congressional action that may reduce environmental cleanup, based on budget considerations.

The technologies discussed in this book include the following:

1. Processes currently being utilized
2. Technology in the demonstration phase
3. Processes being developed
4. Research needs

There is a vast amount of technical information on current, in demonstration, and potential cleanup technology, that would require an encylopaedic work to fully describe. Therefore, this book can only describe very briefly the technology involved.

Information contained in this book was obtained from published material issued by various government agencies including: Department of Energy (DOE), Department of Defense (DOD), Environmental Protection Agency (EPA), General Accounting Office (GAO), Office of Technology Assessment (OTA), and Energy Information Administration (EIA). Published material was also supplied by contractors working at DOE sites.

The Introduction presents information that would be helpful to those firms wishing to participate in DOE programs. In Appendix II, addresses, and telephone numbers of the important sites are indicated. Also, foreign nuclear contacts (with a description of their activities) are presented in Appendix III, for 24 countries.

The substantial amount of money that will be spent on nuclear waste cleanup offers an excellent opportunity for engineering, equipment, chemical, instrument, and other firms.

Contents

1

Introduction

1.1 OVERVIEW

Any activity that produces or uses radioactive materials generates radioactive waste. Mining, nuclear power generation, and various processes in industry, defense, medicine, and scientific research produce by-products that include radioactive waste. Radioactive waste can be in gas, liquid, or solid form, and its level of radioactivity can vary. The waste can remain radioactive for a few hours or several months or even hundreds of thousands of years.

Broadly defined, the management of radioactive wastes encompasses the handling, storage, treatment, transportation, and permanent disposal of all radioactive wastes.

Currently, a minimum of over 45,300 sites handle radioactive material or contain potential radioactive contamination. Of these, approximately half are in operation today. In some cases, a single complex may have as many as 1,500 contaminated sites, in other cases there may be 1 site per complex. However, for all practical purposes, a much smaller number of sites are responsible for most of the radioactive waste, and will involve a very high percentage of the cleanup costs. These important sites are under the control of the U.S. Department of Energy (DOE).

DOE and its predecessors have been involved in the management of radioactive wastes since 1943, when such wastes were first generated in significant quantities as by-products of nuclear weapons production.

Waste connected with DOE's nuclear complex has been accumulating as a result of various operations spanning nearly 5 decades: first in connection with DOE's Defense Programs, more recently in connection with programs in Nuclear Energy and Energy Research. When the nuclear age dawned in the 1940s with the Manhattan Project, there was little knowledge of the degree to which nuclear and hazardous waste materials posed a danger to human health and safety or to the environment. Furthermore, during the earlier part of this period, the demands of World War II, Korea, and the cold war days of the 1950s and 1960s placed higher priority on nuclear operations and lower priority on the wastes generated from such operations.

DOE's mission and priorities have changed dramatically over time so that the Department is now very different from what it was in 1977 when it was created in response to the nation's energy crisis. While energy research, conservation and policy-making dominated early DOE priorities, weapons production and now environmental cleanup overshadow its budget.

When DOE was created in 1977, it inherited the national laboratories with a management structure that had evolved from the World War II "Manhattan Project," whose mission was to design and build the world's first atomic bombs. From this national security mission, the laboratories generated expertise that initially developed nuclear power as an energy source. The laboratories' missions broadened in 1967, when the Congress recognized their role in conducting environmental as well as public health and safety-related research and development. In 1971, the Congress again expanded the laboratories' role, permitting them to conduct non-nuclear energy research and development. During the 1980s, the Congress enacted laws to stimulate the transfer of technology from the laboratories to U.S. industry. DOE estimates that over the past 20 years, the nation has invested more than $100 billion in the laboratories.

DOE is responsible for some of the nation's largest and most impressive scientific facilities. The agency's 9 national multiprogram laboratories employ more than 50,000 people and have annual operating budgets that exceed $6 billion. DOE estimates that more than $100 billion has been invested in the laboratories over the past 20 years. The laboratories' work covers many scientific areas—from high-energy physics to advanced computing—at facilities located throughout the nation.

Although DOE owns the laboratories, it contracts with universities and private-sector organizations for their management and operation—a practice that has made the laboratories more attractive to scientists and engineers. These contracts generally run for 5 years; however, some of the laboratories have been run by the same contractor for decades, even since their inception in the early 1940s. The laboratory contractors and DOE form partnerships at each site, but the Department remains responsible for providing the laboratories with their missions and overall direction, as well as for giving them specific direction to meet both program and administrative goals. There are more than 145,000 contractor employees at DOE nuclear sites. The original concept of nuclear reconfiguration was described in the January 1991 *Nuclear Weapons Complex Reconfiguration Study*. This study outlined potential configurations of the future Complex and charted the course necessary to achieve the goal of modernization. In February 1991, the Secretary of Energy announced DOE's intent to prepare a Reconfiguration Programmatic Environmental Impact Statement (PEIS) to analyze the environmental impacts of the alternatives presented in this study. However, there have been significant changes in the world since January 1991, especially with regard to projected future requirements of the United States' nuclear weapons stockpile. As a result, the *Nuclear Weapons Complex Reconfiguration Study* no longer provides a suitable framework for determining the appropriate configuration of the future Nuclear Weapons Complex. The framework for a new proposal is now being developed. Therefore, DOE decided to separate the Reconfiguration PEIS into two PEISs: a Tritium Supply and Recycling PEIS and a Stockpile Stewardship and Management PEIS.

The environmental task facing DOE is enormous and continues to expand. DOE has approximately 4,000 contaminated sites covering tens of thousands of acres and replete with contaminated hazardous or radioactive waste, soil, or structures. It has more than 250,000 cubic meters of transuranic waste and millions of cubic meters of low-level radioactive waste. In addition, DOE is responsible for thousands of facilities awaiting decontamination, decommissioning, and dismantling. Consequently, DOE faces major technical, planning, and institutional challenges in meeting its expanding environmental responsibilities while controlling cost growth.

Work performed at the DOE Weapons Complex has traditionally been divided into 4 categories:

1. Weapons research and development at 3 national laboratories, Los Alamos and Sandia in New Mexico and Lawrence Livermore in California;
2. Nuclear materials (plutonium and tritium) production and processing at the Hanford Plant in Washington State and the Savannah River Site in South Carolina, along with uranium processing at the Feed Materials Production Center in Ohio and the Idaho National Engineering Laboratory;
3. Warhead component production at the Rocky Flats Plant in Colorado, the Y-12 Plant in Tennessee, the Mound Plant in Ohio, the Pinellas Plant in Florida, the Kansas City Plant in Missouri, and the Pantex Plant (final assembly) in Texas; and
4. Warhead testing at the Nevada Test Site.

Although the Weapons Complex was developed in World War II as part of the Manhattan Project, a major expansion occurred in the early 1950s. Today, most operating facilities are more than 30 years old. Operations are in various stages of transition because of safety and environmental problems that have diverted attention from production and because of the uncertain future of the entire enterprise.

Although facilities in the DOE complex have much in common, there is no "typical" facility. Each site has a unique combination of characteristics that shapes its particular waste and contamination problems and affects the way those problems are addressed. Relevant facility characteristics include its functions and management; its size, location, and proximity to populated areas;

and its relationships with Federal and State regulators, neighboring communities, and the general public.

DOE is responsible for environmental cleanup and waste management at 15 major contaminated facilities and more than 100 smaller facilities in thirty-four states and territories. These facilities encompass a wide range of waste sites, including tanks or other storage facilities containing radioactive waste from nuclear weapons production, production facilities that are now idled and in need of cleanup, and locations where hazardous chemicals were dumped into the ground. Cleaning up these sites is an enormous task. Examples of some of these sites are indicated below.

DOE Weapons Complex facilities—both large and small—are spread across the Nation, from South Carolina to Washington State, and are located in both remote and populated regions. The Feed Materials Production Center (Fernald), which has produced uranium metal for weapons, is a 1,450-acre site, a relatively small facility located 20 miles northwest of Cincinnati, OH, in a rural area with a number of farms. The Rocky Flats Plant in Colorado, which has been producing plutonium "triggers" for weapons, is also a small facility situated close to densely populated suburbs of Denver.

Other sites are much larger than Fernald or Rocky Flats. The Hanford Reservation encompasses approximately 360,000 acres in the Columbia River Basin of southeastern Washington State. Hanford's primary mission has been to produce weapons-grade plutonium; it produced plutonium for the atom bomb dropped on Nagasaki during World War II. The Savannah River Site, built in the 1950s, produces tritium and plutonium. It consists of 192,000 acres on the north bank of the Savannah River. Most of the immediate plant environs are rural, and the surrounding area, which is heavily wooded, ranges from dry hilltops to swampland. More than 20,000 people are employed at Savannah River, making it the largest plant (in terms of employment) in the DOE Weapons Complex.

The Oak Ridge Reservation covers approximately 58,000 acres in eastern Tennessee. Oak Ridge carries out several activities including the production of weapons components. The area immediately around the reservation is predominately rural except for the City of Oak Ridge. The City of Knoxville is about 15 miles away. The Idaho National Engineering Laboratory (INEL), where reactor fuel is reprocessed to recover uranium, has a number of facilities and conducts a variety of other activities. The largest site in terms of area, INEL covers 570,000 acres in southeastern Idaho. The site boundary is about 22 miles from the City of Idaho Falls.

The Nuclear Weapons Complex is an industrial empire—a collection of enormous factories devoted to metal fabrication, chemical separation processes, and electronic assembly. Like most industrial operations, these factories have generated waste, much of it toxic. The past 50 years of nuclear weapons production have resulted in the release of vast quantities of hazardous chemicals and radionuclides to the environment. There is evidence that air, groundwater, surface water, sediments, and soil, as well as vegetation and wildlife, have been contaminated at most, if not all, of DOE nuclear weapons sites.

Although the Weapons Complex was developed in World War II as part of the Manhattan Project, a major expansion occurred in the early 1950s. Today, most of the operating facilities are more than 30 years old. Operations are in various stages of transition because of safety and environmental problems that have diverted attention from the production mission and because of uncertainty about the future of the entire enterprise.

Contamination of soil, sediments, surface water, and groundwater throughout the Nuclear Weapons Complex is extensive. At every facility the groundwater is contaminated with radionuclides or hazardous chemicals. Most sites in non-arid locations also have surface water contamination. Millions of cubic meters of radioactive and hazardous wastes have been buried throughout the complex, and there are few adequate records of burial site locations and contents. Contaminated soils and sediments of all categories are estimated to total billions of cubic meters.

Descriptions of vast quantities of old buried waste; of contaminants in pits, ponds, and lagoons; and of the migration of contamination into water supplies serve to dramatize the problem. However, so far very little quantitative characterization of each site has been accomplished.

Many factors have contributed to the current waste and contamination problems at the weapons sites: the nature of manufacturing processes, which are inherently waste producing; a long history of emphasizing the urgency of weapons production in the interest of national security, to the neglect of environmental considerations; a lack of knowledge about, or attention to, the consequences of environmental contamination; and an enterprise that has operated in secrecy for decades, without any independent oversight or meaningful public scrutiny.

Sites contaminated with radionuclides pose a unique problem because, unlike organic wastes, radionuclides cannot be destroyed by physical or chemical means; they can only decay through their natural process. Thus, alteration or remediation of the radioactive decay processes, thereby changing the fundamental hazard, is not possible.

As part of DOE's technological development program, it will be important to identify the greatest needs and the areas in which new technology can make a difference. The first step should be to identify cleanup needs and to determine those that are most urgent and serious. In this step, information about health effects should be factored in as it becomes available. For example, among the problems that DOE has already identified as particularly intractable are the following:

1. Groundwater contamination at almost all sites,
2. Plutonium in soil (e.g., at Rocky Flats and Mound Plant),
3. Silos containing uranium processing residues at Fernald,
4. Single-shell tanks containing high-level waste at Hanford, and
5. Buried transuranic waste at INEL.

Contamination of soil, sediments, surface water, and groundwater throughout the Weapons Complex is widespread. Almost every facility has confirmed groundwater contamination with radionuclides or hazardous chemicals. All sites in non-arid locations probably have surface water contamination. Almost 4,000 solid waste management units (SWMUs) have been identified throughout the Weapons Complex—many of which require some form of remedial action. Substantial quantities of radioactive and mixed waste have been buried throughout the complex, many without adequate record of their location or composition.

Presently, DOE has identified more than 1 million 55 gallon drums and boxes of waste in storage, and 3 million cubic meters of buried waste. Over the years, many of the older disposal containers have been breached resulting in contamination of the adjacent soil. Considering transuranic solid waste, approximately 190,000 cubic meters have been buried, and 60,600 cubic meters have been retrieved and stored. Mixed transuranic waste composes 58,000 cubic meters of this inventory. High-level waste stored at 4 DOE sites represent another 381,000 cubic meters of volume.

Currently, 77 million gallons of high-level waste is contained in 332 underground storage tanks as sludge/liquids. There are also small amounts, approximately 4,000 cubic meters of high-level waste, stored as granular calcined solids. Most of the high-level wastes are mixed with hazardous contaminants and are thus considered mixed wastes. The remainder of the waste in storage is low-level waste. This remainder is made up of 3,000,000 cubic meters, including 247,000 cubic meters of mixed low-level waste. Currently, no effective treatment is known to exist for 107,000 cubic meters of this mixed low-level waste.

Another of the most pressing environmental restoration needs for the DOE involve cleanup or containment of radioactive and hazardous contaminants in soils and groundwater. The DOE soils and groundwater programs were designed to identify, develop, and demonstrate innovative technology systems capable of removing or reducing potential health and environmental risks. These risks are the result of previous storage and disposal practices that left behind a legacy of radioactive and hazardous materials (including heavy metals and toxic organic compounds) in the surrounding soil and groundwater. Sources of this contamination at the DOE sites include: previous disposal of contaminated wastes in ponds, seepage pits, trenches, and shallow land burial sites; spills and leakage from waste transport, temporary storage facilities, and underground storage tanks; and unregulated discharges to the air and surface waters.

Another form of waste, representing potentially large volumes, is associated with decontamination and decommissioning of contaminated buildings and equipment. More than 500 separate facilities have been identified, and it is possible that as many as 7,000 facilities at 39 different sites could be scheduled for decontamination and decommissioning. Although materials will be recycled when possible, this activity will result in new waste generation that is immeasurable at this time. Additionally, as much as 20,000 cubic meters of mixed waste, in 100 separate waste streams, is still being generated on an annual basis from ongoing transition activities.

One of the biggest challenges facing the DOE is effective characterization of contamination. Characterization must take place before a contaminant site can be properly prioritized. To accomplish this, methods are being developed that are capable of mapping vast areas at depths up to 250 feet below ground level. Results are three dimensional images that are valuable tools for proper selection and placement of remediation technologies. Complicating remediation efforts further is the fact that techniques for accessing and removing contaminants differ in arid and non-arid environments. As a result, technologies must be demonstrated and evaluated at multiple sites.

Adequate chemical and physical characterization information is important for satisfactory management and disposal of all (both nuclear and non-nuclear) DOE wastes. For example, detailed and accurate waste characterization data are essential not only to develop appropriate and flexible pretreatment and conversion processes, but also to classify and certify wastes, both before and after pretreatment, for disposal as mandated by state and federal regulations. Similarly, characterization by physical methods of the important geologic, hydrologic, and seismic properties of candidate strata and sites is absolutely necessary to eventual disposal of certain DOE radioactive wastes in a deep geologic repository. The overriding importance of reliable waste and disposal site characterization data is a strong spur for basic research to devise new and better characterization methods and to improve and supplement existing and proven procedures.

The need for advanced and improved systems and instruments to characterize contaminated soils, sediments, etc. has been recognized in connection with DOE environmental restoration programs and activities. Some research needs such as in situ analytical techniques, portable field instruments, and advanced analytical instrumentation are common to both waste management/disposal and environmental restoration programs.

The cleanup of these nuclear waste sites is one of the most difficult tasks facing this country. Problems include legal difficulties, regulatory overlap, conflict between state and federal governments, lack of proven technical processing, enormous costs, and safety considerations.

DOE's stated goal of environmental cleanup by the year 2019 represents a formidable challenge, and currently available information does not clearly demonstrate that it can be attained. Although it may be desirable for DOE to set a completion date on which to focus its activities, 3 major barriers stand in the way of achieving this goal:

1. Decisions on cleanup levels and standards that can clarify DOE goals have not yet been established;
2. Personnel qualified to conduct characterization and remediation at DOE sites are scarce; and
3. Technologies for addressing some of DOE's more perplexing environmental problems are not currently available.

The cost estimates for nuclear waste cleanup in the United States have been rapidly rising. It has recently increased from $200 billion to $300-$400 billion. A recent (1995) estimate concluded that total cleanup costs could reach 1 trillion dollars, and take 30 years to complete. However, total costs will depend upon federal budgetary considerations.

It has been suggested by some, that lower costs could result from lowering cleanup standards, which in certain cases, are set higher than current technological capabilities. Another suggestion is to merely isolate certain sites. However, such a proposal would certainly be fought by the states involved, as well as environmentalists.

1.2 FUNDING

The Department of Energy budget request for 1996 for environment management is $6.6 billion. In 1996, Environmental Management will take over management responsibilities for a significant number of facilities formerly managed by the Department of Energy's Defense Program office. These sites include the Savannah River Site in South Carolina, the Mound Plant in Ohio, and the Pinellas Plant in Florida. Although the 1996 request of $6.6 billion is $608 million greater than the 1995 appropriation, the request includes a $843 million intra-departmental transfer of funds for managing surplus former production facilities. The base 1996 budget request of $5.7 billion is a 4 percent reduction from the 1995 appropriation of $5.9 billion.

More than 86 percent of Environmental Management's 1996 budget request goes directly to the field to remediate and manage former weapons production sites. Activities at 10 sites account for more than 80 percent of the program's budget because they include the largest sites with the most complex problems or the most urgent risks or conduct large scale technology projects. These 10 largest sites and their 1996 budget requests are as follows:

Hanford	$1,411,754,000
Savannah River Site	1,344,352,000
Rocky Flats Environmental Technology Site	639,918,000
Oak Ridge Reservation	614,510,000
Idaho National Engineering Laboratory	481,145,000
Fernald Environmental Management Project	256,330,000
Waste Isolation Pilot Plant	172,700,000
Los Alamos National Laboratory	135,995,000
West Valley Demonstration Project	122,100,000
Mound Plant	110,298,000

1.3 REGULATORY CONSIDERATIONS

Key laws and regulations governing cleanup at the Nuclear Weapons Complex are indicated below:

RCRA—The Resource Conservation and Recovery Act (RCRA) was enacted in 1976 to address the widespread contamination problem resulting from the disposal of municipal and industrial solid waste. Managed by the U.S. Environmental Protection Agency (EPA) or EPA-authorized States, the RCRA program focuses on reducing the generation of hazardous waste and conserving energy and natural resources. DOE's Nuclear Weapons Complex facilities are subject to RCRA and therefore must apply for an EPA or State permit to treat, store, or dispose of hazardous wastes or radioactive waste mixed with hazardous pollutants. Under the Hazardous and Solid Waste Amendments of 1984 (HWSA), DOE is also required to address and eliminate contaminant releases at or from its RCRA facilities within a schedule specified by EPA. This type of activity, called corrective action, is now being carried out at most weapons sites. Releases from inactive or abandoned sites or from accidental spills are not subject to RCRA, but they may be required to be remedied according to the Comprehensive Environmental Response, Compensation, and Liability Act (CERCLA).

CERCLA—The Comprehensive Environmental Response, Compensation, and Liability Act of 1980 (also known as Superfund) provides the U.S. Environmental Protection Agency (EPA) with the authority to assess contaminant releases from abandoned waste sites (such as those within the Nuclear Weapons Complex), categorize sites according to their risks, and include them in the National Priorities List if EPA considers their cleanup a national priority. Both radioactive and hazardous contaminants are included under CERCLA authority. Eight of the Nuclear Weapons Complex sites are currently listed by EPA as requiring cleanup under CERCLA. The Superfund Amendments and Reauthorization ACT (SARA) of 1986 authorize EPA to negotiate interagency agreements with other Federal agencies and States and to oversee Federal agency efforts toward developing appropriate remedies.

NEPA—The National Environmental Policy Act (NEPA) of 1970 mandates that all Federal agencies and departments take into consideration the adverse effects that their actions may have on the environment. The Council on Environmental Quality is responsible for developing the guidance for Federal agencies to comply with the Act. NEPA requires that agency actions be reviewed early in the planning process and that the process be open to public participation. This review often results in the preparation of an Environmental Assessment or an Environmental Impact Statement (EIS), usually on a specific project. An EIS prepared for an entire program of agency activities is called a Programmatic EIS (PEIS).

Most Weapons Complex sites are subject to both CERCLA and RCRA. Some sites, which have not been placed on the NPL, operate only under the regulatory jurisdiction of RCRA (i.e., Pantex, Los Alamos, Sandia, Pinellas, Kansas City). A major difference between the CERCLA and RCRA laws is that CERCLA coverage includes both hazardous and radioactive contamination, whereas RCRA and its corrective action provisions cover only hazardous waste and the hazardous portion of mixed waste. At sites subject only to RCRA authority, some radioactive materials and releases of radioactivity to the environment are regulated exclusively by DOE, subject to the Atomic Energy Act. DOE has its own set of internal directives (DOE orders) governing radioactive waste management and the limitations of radionuclide releases to the environment.

The facilities in DOE's Nuclear Weapons Complex that are contaminated with hazardous and radioactive materials are among the many public and private sites being cleaned up under RCRA and/or CERCLA. To help coordinate its activities under the two acts, DOE in 1988 developed general operating principles (called "model provisions") for its facilities to use in negotiating interagency agreements with the Environmental Protection Agency and the state agencies overseeing cleanup activities. These agreements, which vary among facilities, establish a general framework for how cleanups under the 2 acts will be coordinated.

Despite the general frameworks provided in the interagency agreements, difficulties persist in coordinating cleanup activities. For example, at facilities that became subject to CERCLA after cleanup activities under RCRA had begun, disagreements have sometimes occurred between DOE and its regulators as to how CERCLA requirements should be incorporated into ongoing cleanup activities under RCRA and how much additional paperwork is needed to document compliance with CERCLA. Furthermore, DOE and its regulators had difficulty coordinating schedules for cleanup activities under the 2 acts.

DOE has recognized these continuing difficulties and is considering actions to address them. It plans to issue additional guidance for coordinating activities under RCRA and CERCLA in 1995. It also tentatively plans to work with the Environmental Protection Agency and state regulatory agencies to improve strategies for cleanups at DOE's facilities and to address how the requirements of RCRA and CERCLA are to be met.

The requirements of both RCRA and CERCLA can apply at a DOE facility that has active and inactive hazardous waste sites. This is because a federal facility regulated under RCRA may also be listed on the National Priorities List for cleanup under CERCLA if it meets the listing criteria. EPA first included federal facilities that were subject to RCRA's corrective action requirements on the National Priorities List in 1989, and 19 DOE facilities are currently on the list. Once included on the list, facilities are subject to the cleanup actions and procedures specified under CERCLA as well as to RCRA-related requirements for corrective action established by EPA or a state regulatory agency.

Cleanups under CERCLA and corrective actions under RCRA have broadly similar objectives. Under both statutes, releases of wastes needing further investigation are identified, the nature and extent of the releases are characterized, cleanup alternatives are developed, a cleanup remedy is proposed for public comment, and the selected remedy is authorized and carried out. However, the 2 programs differ in their highly detailed sets of procedural regulations and guidelines and in the particulars of their implementation. For instance, according to environmental restoration officials at DOE, corrective actions under RCRA are generally implemented unit by unit, while cleanups under CERCLA may address contamination over a wider geographic area, such as groundwater contamination that underlies several units.

Despite efforts to coordinate activities under the 2 acts through the general frameworks of the interagency agreements, coordination problems have continued to arise. In particular, agencies have sometimes disagreed over details of implementation and have had difficulty coordinating schedules.

Some of the agencies' disagreements about particular cleanups have been resolved, but coordination problems could continue because much cleanup work remains and more DOE facilities have recently been added to the National Priorities List for CERCLA cleanups. DOE plans to provide additional guidance and to negotiate better strategies for coordinating its activities under RCRA and CERCLA. DOE's plans for developing guidance include obtaining information from facilities about approaches to specific problems that have worked well. GAO believes that DOE's efforts to apply lessons learned, to the practical difficulties involved in coordinating cleanup activities under the two acts will be beneficial.

The nature of the regulatory compliance process also affects the development of prudent estimates. Sequential regulatory requirements, such as those of CERCLA, can ultimately delay the construction of treatment, storage, and disposal facilities as well as the initiation of actual waste management and remediation operations. These regulatory difficulties may be further complicated by concurrent application of potentially overlapping environmental control statutes. The variability and uncertainty of the current regulatory process contributes to increasing costs and scheduling uncertainty.

Comprehensive standards designed specifically for the cleanup of radionuclides do not exist. The only standards designed for the cleanup of radionuclides are those for land and buildings contaminated by uranium mill tailings at inactive uranium-processing sites. In the absence of applicable cleanup standards for radionuclides, EPA and DOE identify other federal and state environmental standards that are relevant and appropriate for DOE cleanups. This process for determining cleanup requirements can be time-consuming and contentious and can result in varying levels of cleanup and public protection. Comprehensive cleanup standards are needed to allow DOE to plan and estimate costs for its cleanup program, particularly in light of upcoming decisions on cleanups.

Currently, other environmental standards for radionuclides are used for DOE's remedial actions, once the standards are identified through the CERCLA process as relevant and appropriate requirements. For example, federal standards that might be used include drinking water standards, Clean Air Act standards, and cleanup standards for soil near inactive uranium-processing sites. State standards may be used if they are more stringent than federal ARARs.

However, other existing environmental standards do not cover all radionuclides and media. For instance, apart from standards for uranium mill tailings, no federal standards exist for cleaning up radionuclides in soil. In addition, while existing standards such as those set under the Safe Drinking Water Act might be used as levels for cleaning up groundwater, drinking water standards do not exist for some radionuclides. If no federal or state standards exist for a given contaminant, CERCLA regulations state that residual contamination should generally not result in a lifetime cancer risk to an individual that exceeds a range of approximately 1 in 1 million to 1 in 10,000. EPA's guidance specifies methods for estimating cancer risk from residual contamination.

Effective May 8, 1992, all DOE mixed-waste streams fell under EPA's land disposal restrictions and, as such, can no longer be disposed of without prior treatment to destroy, separate, or immobilize the hazardous component. All mixed LLW and HLW must be treated before final disposal. In the case of mixed TRU wastes destined for deep geologic disposal, the hazardous components must not exceed established waste acceptance criteria. Most of the hazardous components of the mixed wastes have not been characterized; however, from past knowledge, they represent the entire gamut of organic and inorganic hazardous wastes. Available technology is inadequate to solve many of the problems at hand. The result is a mixed waste dilemma that poses serious legal and technical problems that need to be resolved.

Complex regulatory considerations, in conjunction with varying federal, state, and local laws are increasing costs, and delaying final cleanup.

1.4 DEPARTMENT OF ENERGY ORGANIZATION

The Environmental Management (EM) organization is responsible for DOE's waste management and cleanup efforts, and its various offices are indicated below.

The EM Office of Policy and Program Information (EM-4) serves as a central coordinate for DOE/EM public participation and program information activities. EM-4 establishes policy and guidance, and conducts and coordinates public participation activities inside and outside the agency. Its goal is to identify public concerns, needs, and objectives through two-way communications between DOE and the public before decisions are made. This interactive process improves DOE understanding of public concerns, and enhances the public's understanding of DOE decisions and subsequent technology development activities.

The EM Office of Planning and Resource Management (EM-10) supports program office financial management, procurement, and administrative activities and coordinates the annual update of the EM Strategic Plan and Five-Year Plan.

The EM Office of Oversight and Self-Assessment (EM-20) performs independent internal oversight within EM to ensure compliance with environmental and safety laws/regulations and with DOE Quality Assurance policies. EM-20 is also charged with enhancing the technical validity and cost-effectiveness of programs and projects. In addition, it is responsible for independent cost estimating functions.

The EM Office of Waste Management (EM-30) has program responsibilities for waste management at all DOE sites, including the treatment, storage, and disposal of several types of waste: transuranic, low-level radioactive, mixed, and solid sanitary. EM-30 is also responsible for the storage, treatment, and processing of defense high-level radioactive waste, waste minimization efforts, and corrective activities at waste management facilities.

The EM Office of Environmental Restoration (EM-40) has program responsibilities for remediating inactive hazardous and radioactive waste sites at all DOE installations and some non-DOE sites, including remedial actions and decontamination/decommissioning activities.

The EM Office of Technology Development (EM-50) has program responsibilities for developing better, faster, cheaper, and safer technologies for meeting DOE's 30 year goal for environmental restoration and waste management, and for managing crosscutting activities.

The EM Office of Facility Transition and Management (EM-60) plans, implements, and manages the orderly transition of facilities from their operational base to EM-30 and EM-40 and their subsequent disposition. EM-60 establishes and implements a consistent process to safely deactivate and dispose of DOE facilities.

The Office of Technology Development (EM-50) has the overall responsibility to develop technologies to meet DOE's 30 year goal for environmental restoration. Activities within EM-50 include applied research and development, and demonstration, testing, and evaluation of new and existing technologies.

In order to hold costs to an acceptable level, the Office of Environmental Restoration and Waste Management's (EM) Office of Technology Development (OTD) has been directed to initiate a national program to develop and demonstrate faster, better, cheaper, and safer means of restoring the DOE sites to conditions that will meet state and federal environment regulations. Key elements of the OTD initiative are the Integrated Programs (IPs) and Integrated Demonstrations (IDs). These components work together to identify possible solutions to major environmental problems.

When a new technology enters the OTD Research, Development, Demonstration, Testing and Evaluation Program, an Integrated Program serves as a technology filter or incubator. Through bench-scale and field-scale experiments, data are produced to evaluate a technology for full-scale demonstration. An Integrated Demonstration is an established program at an actual site, which demonstrates, tests, and evaluates related technologies individually or as a complete system to correct waste management and environmental problems from cradle to grave. IDs are funded to maintain the necessary infrastructure to ensure that tests can be performed in a rapid and acceptable manner. Each ID demonstrates the application of all aspects of a cleanup, including characterization, assessment, remediation, and monitoring to regulators, host communities, and other stakeholders to expedite public and regulatory knowledge and acceptance of a technology.

IDs are comprised of technologies with reliable bench-scale data and full-scale demonstration capability within 2 years. IPs encompass technologies that require more study and further testing. A project originating in an IP may gradually move into an ID, or a CERCLA treatability study, if new data indicate that the technology is faster, better, safer, or cheaper than currently used, baseline practices.

1.5 HOW TO GET INVOLVED: WORKING WITH THE DOE OFFICE OF ENVIRONMENTAL RESTORATION AND WASTE MANAGEMENT

The U.S. Department of Energy provides a range of programs and services to assist universities, industry, and other private organizations and individuals interested in developing or applying environmental technologies.

Working with DOE Operations Offices and management and operating contractors, the Office of Environmental Management uses conventional and innovative mechanisms to identify, integrate, develop, and adapt promising emerging technologies. These mechanisms include contracting and collaborative arrangements, procurement provisions, licensing of technology, consulting arrangements, reimbursable work for industry, and special consideration for small business.

Cooperative Research and Development Agreements (CRADAs):

EM will facilitate the development of subcontracts, R&D contracts, and cooperative agreements to work collaboratively with the private sector. EM uses CRADAs as an incentive for collaborative R&D. CRADAs are agreements between a DOE R&D laboratory and any non-Federal source to conduct cooperative R&D that is consistent with the laboratory's mission. The partner may provide funds, facilities, people, or other resources. DOE provides the CRADA partner access to facilities and expertise; however, no Federal funds are provided to external participants. Rights to inventions and other intellectual property are negotiated between the laboratory and participant, and certain data that are generated may be protected for up to 5 years.

Consortia will also be considered for situations where several companies combine their resources to address a common technical problem. Leveraging of funds to implement a consortium can offer a synergism to overall program effectiveness.

Procurement Mechanisms:

DOE/EM has developed an environmental management technology development acquisition policy and strategy that uses phased procurements to span the RDDT&E continuum, from applied R&D concept feasibility through full-scale remediation. DOE EM phased procurements make provisions for unsolicited proposals, but formal solicitations are the preferred responses. The principle contractual mechanisms used by EM for industrial and academic response include Research Opportunity Announcements (ROAs) and Program R&D Announcements (PRDAs).

EM uses the ROA to solicit advanced research and technologies for a broad range of cleanup needs. The ROA supports applied research ranging from concept feasibility through full-scale demonstration. In addition, the ROA is open continuously for a full year following the date of issue, and includes a partial procurement set aside for small businesses. Typically, ROAs are published annually in the *Federal Register* and the *Commerce Business Daily* and multiple awards are made.

PRDAs are program announcements used to solicit a broad mix of R&D and DT&E proposals. Typically, a PRDA is used to solicit proposals for a wide-range of technical solutions to specific EM problem areas. PRDAs may be used to solicit proposals for contracts, grants, or cooperative agreements. Multiple awards, which may have dissimilar approaches or concepts, are generally made. Numerous PRDAs may be issued each year.

In addition to PRDAs and ROAs, EM uses financial assistance awards when the technology is developed for public purpose. Financial assistance awards are solicited through publication in the *Federal Register*. These announcements are called Program Rules. A Program Rule can either be a one-time solicitation or an open-ended, general solicitation with annual or more frequent announcements concerning specific funding availability and desired R&D agreements. The Program Rule can also be used to award both grants and cooperative agreements.

EM awards grants and cooperative agreements if 51 percent or more of the overall value of the effort is related to a public interest goal. Such goals include possible non-DOE or other Federal agency participation and use, advancement of present and future U.S. capabilities in domestic and international environmental cleanup markets, technology transfer, advancement of scientific knowledge, and education and training of individuals and business entities to advance U.S. remediation capabilities.

Licensing of Technology:

DOE contractor-operated laboratories can license DOE/EM-developed technology and software to which they elect to take title. In other situations where DOE owns title to the resultant inventions, DOE's Office of General Counsel will do the licensing. Licensing activities are done within existing DOE intellectual property provisions.

Technical Personnel Exchange Assignments:

Personnel exchanges provide opportunities for industrial and laboratory scientists to work together at various sites on environmental restoration and waste management technical problems of mutual interest. Industry is expected to contribute substantial cost-sharing for these personnel exchanges. To encourage such collaboration, the rights to any resulting patents go to the private sector company. These exchanges, which can last from 3 to 6 months, are opportunities for the laboratories and industry to better understand the differing operating cultures, and are ideal mechanisms for transferring technical skills and knowledge.

Consulting Arrangements:

Laboratory scientists and engineers are available to consult in their areas of technical expertise. Most contractors operating laboratories have consulting provisions. Laboratory employees who wish to consult can sign non-disclosure agreements and are encouraged to do so.

Reimbursable Work for Industry:

DOE laboratories are available to perform work for industry or other Federal agencies, as long as the work pertains to the mission of a respective laboratory and does not compete with the private sector. The special technical capabilities and unique facilities at DOE laboratories are an incentive for the private sector to use DOE's facilities and contractors expertise in this reimbursable work for industry mode. An advanced class patent waiver gives ownership of any inventions resulting from the research to the participating private sector company.

EM Small Business Technology Integration Program:

The EM Small Business Technology Integration Program (SB-TIP) seeks the participation of small businesses in the EM Research, Development, Demonstration, Testing, and Evaluation programs. Through workshops and frequent communication, the EM SB-TIP provides information on opportunities for funding and collaborative efforts relative to advancing technologies for DOE environmental restoration and waste management applications.

EM SB-TIP has established a special EM procurement set aside for small firms (500 employees or less) to be used for applied research projects through its ROA. The program also serves as the EM liaison to the DOE Small Business Innovation Research (SBIR) Program Office and interfaces with other DOE small business offices.

Contact:

International Technology Exchange Division
EM-523
Environmental Restoration and Waste Management Technology Development
U.S. Department of Energy
Washington, DC 20585
(301) 903-7940

EM Central Point of Contact:

The EM Central Point of Contact can provide access to prospective research and business opportunities in waste management, environmental restoration, and decontamination and decommissioning activities, as well as information on EM-50 IPs and IDs. The EM Central Point of Contact can identify links between industry technologies and program needs, and can provide potential partners with a connection to an extensive complex-wide network of DOE Headquarters and field program contacts. It is the best single source of information for private-sector technology developers looking to collaborate with EM scientists and engineers. It provides a real-time information referral service to expedite and monitor private-sector interaction with EM.

For the EM Central Point of Contact, call 1–800–845–2096. For the EM Center for Environmental Management Information (call 1–800–736–3282).

Office of Research and Technology Applications:

The Office of Research and Technology Applications (ORTAs) serves as a technology transfer agent at the Federal laboratories, provides internal coordination in the laboratory for technology transfer, and acts as an external point of contact for industry and universities. To fulfill these purposes, ORTAs licenses patents and coordinates technology transfer activities for a laboratory's scientific departments. It also facilitates one-on-one interactions between laboratory scientific personnel and technology recipients, and provides information on laboratory technologies with potential applications in private industry for state and local governments.

For more information about these programs and services, contact:

Technology Integration Division
EM–521
Environmental Management Technology Development
U.S. Department of Energy
Washington, DC 20585
(301) 903–7928

Note: See Appendix II for additional contacts.

2

Nuclear Waste Categories

2.1 HIGH-LEVEL WASTE AND SPENT NUCLEAR FUEL

High-level waste and spent nuclear fuel originate from different sources and require distinct handling, although the Nuclear Waste Policy Act of 1982 required that they both be safely stored, transported, and finally disposed of in a mined geologic repository.

Spent nuclear fuel consists of irradiated fuel discharged from a nuclear reactor. This fuel may be reprocessed or may be considered as permanently discharged and eligible for repository disposal. Spent fuel may be generated from defense complex reactors, commercial nuclear power reactors or be classed as special fuels associated with government–sponsored research and demonstration programs, universities, and private industry.

High-level waste is the highly radioactive material resulting from the reprocessing of spent nuclear fuel or from defense production processes. This material includes liquid wastes, sludge, calcines, or other products remaining from the recovery of uranium and plutonium in a fuel reprocessing plant. Such waste contains fission products that result in the release of considerable decay energy. As a result, heavy shielding is required to control penetrating radiation and to dissipate decay heat.

The Office of Civilian Radioactive Waste Management within DOE was established in response to the Nuclear Waste Policy Act to develop, manage, and operate a safe disposal system for high-level waste. The Office of Civilian Radioactive Waste Management was given the responsibility to site, construct, and operate a deep, mined geologic repository; to site, construct, and operate a monitored retrievable storage facility; and to develop a system for transporting the waste to a repository and a monitored retrievable facility. In coordination with the Office of Civilian Radioactive Waste Management, EM is managing inventories of spent nuclear fuel and high-level waste until the repository is available.

DOE has recently announced the shutdown of reprocessing facilities at the Idaho Chemical Processing Plant and other locations in response to the changing political climate. With the declining need for reprocessing of spent nuclear fuel to recover fissile material, DOE still retains responsibility to manage this material. Plan, such as the Idaho Chemical Processing Plant Spent Fuel and Waste Management Technology Development Plan, are being prepared to address changing missions of facilities from reprocessing to developing technologies and processes for geologic disposal.

2.1.1 High-Level DOE Waste

High-level waste generated by DOE defense activities is stored in underground storage tanks at the Savannah River Site, Idaho National Engineering Laboratory (INEL), and the Hanford Site. Much of the high-level waste is alkaline liquid, sludge, salt cake, or slurry. At INEL, acidic high-level waste is dried to a calcine (a granular solid) and stored in steel bins inside concrete vaults. A small amount of high-level waste generated in commercial activities at West Valley, New York, and owned by the New York State Energy Research and Development Authority, is also stored in underground storage tanks. The total DOE high-level waste volume of about 381,000 cubic meters contains some 1.1 billion curies of radionuclides. Very little additional high-level waste is expected to be generated as a result of environmental restoration activities.

The Nation's high-level waste is to be disposed of in a deep geologic repository licensed by the U.S. Nuclear Regulatory Commission. These regulations require that, to be disposed of, high-level waste must be a durable, stable solid. To meet this requirement, many high-level waste treatment technologies were evaluated. Vitrification was selected as the immobilization technology best suited to the majority of DOE high-level waste: the process equipment performs well in remote operation and the borosilicate glass product tolerates considerable variation in waste composition. Vitrification has also been approved by EPA as the best demonstrated available technology for disposal of this waste under RCRA.

The West Valley Demonstration Project, a joint program by DOE and the New York State Energy Research and Development Authority, will vitrify the high-level waste now stored at that site. Studies to identify a suitable treatment process and waste form for calcine and liquid wastes at INEL continue. Cumulative production will be about 21,000 canisters, each containing on average approximately 2 metric tons of waste form (volume of approximately 1 cubic meter).

The vitrification process selected for the DOE high-level waste treatment facilities at the Hanford Site, Savannah River Site, and West Valley Demonstration Project incorporates high-level waste into a borosilicate glass matrix, thus reducing the mobility of radioactive and other hazardous constituents. Waste and borosilicate glass-forming materials will be fed continuously as a slurry into a glass melter and heated to temperature above 1,000°C. After becoming molten and homogeneous, the melt will be poured into stainless steel canisters. Sealed canisters will be cleaned and stored at each site pending transfer to a Federal repository for disposal.

It is not planned to vitrify all high-level waste tank contents. By pretreating the stored high-level waste, many non-radioactive substances can be separated from radioactive ones. The radioactivity will be concentrated into a small volume, and the high-activity fraction will be vitrified and disposed of in a geologic repository. The remaining large volume, low activity portion (decontaminated liquids that are now low-level waste) can be disposed of after immobilization in grout or cement.

At the Savannah River Site and West Valley Demonstration Project, waste pretreatment is relatively easy because waste at those sites is from only a few processes and is well characterized. Pretreatment is more complex at the Hanford Site because many different processes have been used over the decades to separate particular nuclides, resulting in a wide variety of waste mixtures.

2.1.2 Spent Nuclear Fuel

Spent nuclear fuel from commercial nuclear utilities will be disposed of in a mined geologic repository. DOE traditionally has chemically processed its spent nuclear fuel to recover materials for defense purposes. Recently, DOE decided to discontinue reprocessing solely to recover valuable materials. DOE spent nuclear fuel is located primarily at the Hanford Site, INEL, and Savannah River.

Although in 1990 the total amount of DOE spent nuclear fuel was less than 5,000 metric tons, there are about 100 distinct kinds of fuel. Consolidating similar fuels into the same class will reduce the number of identifiable groups that will require unique treatment to about 30 discrete classes, thus minimizing the technology development effort needed to find ways to properly prepare the spent nuclear fuel for repository disposal. Unlike commercial spent fuel, much of DOE's spent fuel will require further conditioning and packaging to meet disposal requirements.

A program has been established to investigate technologies for long-term storage and disposal of DOE spent nuclear fuel. The program will focus initially on spent nuclear fuel types now stored at INEL and later on types of spent fuel stored at other sites. The first pilot-scale facility is expected to be ready about the turn of the century and the first production facility about a decade later.

2.2 TRANSURANIC WASTE

Transuranic (TRU) waste is defined as waste contaminated with alpha-emitting radionuclides with an atomic number greater than 92 (heavier than uranium), half-lives greater than 20 years, and in concentrations greater than 100 nanocuries per gram of waste. The principal sources of TRU waste are research and development, plutonium recovery, weapons manufacturing, and decontamination and decommissioning. Currently, DOE manages approximately 251,400 cubic meters of TRU waste and 5 million curies of radioactivity. Of the 251,400 cubic meters of TRU waste, approximately 60,607 cubic meters has been generated since 1970. A long-range TRU waste management plan is being prepared.

All TRU waste generated since 1970 has been placed in long-term storage at six DOE sites. The waste is stored in retrievable form for eventual shipment and disposal at a permanent geologic repository.

TRU waste is contained in a variety of packaging, including metal drums and wooden and metal boxes, and stored in earth-mounded berms, concrete culverts, or other type facilities. It is estimated that 72 percent of the drums have been in storage for more than 10 years and 20 to 30 percent of the bermed drums contain corrosion pinholes or are badly deteriorated. Repackaging and relocating some retrieved waste will be required before shipment. Sites are planning additional storage capacity for retrieved and relocated waste pending repository availability.

All newly generated and retrieved TRU wastes must be characterized to determine their radiological and hazardous constituents. Characterization is essential to satisfy regulatory requirements and to certify that the waste intended for disposal meets the Waste Isolation Pilot Plant (WIPP) waste acceptance criteria. DOE is proceeding on the assumption that characterization requirements can be satisfied through a combination of process knowledge, statistical sampling, and physical and chemical analytical measurements. Regional and local characterization facilities must expand dramatically to support site operations and WIPP disposal phase decision and operations. To support this effort, a Transuranic Waste Characterization Plan is being developed.

Treatment of TRU mixed waste (radioactive and hazardous) might be required under RCRA to remove or reduce to acceptable levels the land disposal restriction constituents in the waste or under 40 CRF 191 (Disposal Standards) before shipment and disposal. The need to treat TRU waste is being assessed as part of the WIPP test phase.

Interim storage capacity must be sufficient to provide flexibility to respond to the site operations and activities for newly generated and stored TRU waste (e.g., retrieval, characterization, treatment, and relocation) and WIPP's availability. Each site's management responsibilities for interim storage facilities and operations are increasing dramatically in light of the delay and uncertainty in WIPP's availability.

TRU waste in interim storage at the Rocky Flats Plant (RFP) represents a special situation. Until recently, the Idaho National Engineering Laboratory served as the long-term storage location for waste from RFP and other TRU waste generators. In 1989, further plutonium processing at RFP was halted for a variety of health, safety, and environmental concerns. Then in early 1990, the Governor of Idaho announced that he would not allow receipt of TRU waste from other DOE sites for storage at the Idaho National Engineering Laboratory. Rocky Flats has ceased production of weapons components and the generation of TRU waste has decreased significantly. DOE is evaluating the establishment of a commercially owned and operated storage site through a procurement activity.

All sites are striving to maintain safe and regulatory-compliant storage of the TRU waste inventory. Some sites are increasing storage capacity for newly generated waste because of the delays in the WIPP schedule. The Hanford Site, Idaho National Engineering Laboratory, and Savannah Rive Site have major projects for characterization, retrieval, and repackaging of TRU waste. Generator sites are also participating in the development of a TRU waste management strategy, including contingency planning.

Construction of the Idaho facility began in December 1992. Hanford is proceeding with design of its facility with construction expected to begin in FY 1996. Savannah River is also in the design and construction phase.

The Waste Isolation Pilot Plant near Carlsbad, New Mexico, was constructed as a research and development facility to demonstrate safe disposal of retrievable, stored defense program waste in a geologic repository. If compliance with EPA regulations can be demonstrated, WIPP will be used for disposal of transuranic waste currently stored in Idaho, Colorado, Washington, South Carolina, Tennessee, and other States. Mined geologic disposal such as WIPP, as opposed to continued above ground storage, is expected to provide a much greater level of confidence for long-term environmental protection.

2.3 LOW-LEVEL WASTE

Low-level waste includes all radioactive waste not classified as either high-level waste, transuranic waste, spent nuclear fuel or the bulk of the by-product tailings containing uranium or thorium from processed ore. It is DOE policy (DOE Order 5820.2A) that radioactive and mixed wastes shall be managed in a manner that assures protection of the health and safety of the public, DOE, and contractor employees, as well as the environment. The policy allows small volumes of uranium/thorium by-product material to be managed as low-level waste. The same DOE policy allows waste containing naturally occurring and accelerator-produced radioactive material to be managed as low-level waste. Any low-level waste that also contains hazardous chemicals covered by RCRA requires management as a "mixed" waste.

DOE low-level waste is generated at more than 30 different sites and is disposed of at 6 sites: Savannah River, Oak Ridge, Idaho National Engineering Laboratory, Nevada Test Site, Los Alamos National Laboratory, and Hanford. DOE will continue to dispose of most of its low-level waste at DOE sites. Alternative strategies to the current disposal practices, including use of commercial disposal facilities, are being considered.

During the next several years, construction on the new low-level waste disposal facilities at the Oak Ridge Reservation will occur with a planned FY 1999 startup. The Savannah River Site will complete construction and begin full operation of all vaults in the Burial Ground Expansion and will close the current shallow land burial trench. All 16 major generators that ship low-level waste to the Nevada Test Site will resume shipments after upgrading certification processes. At that time the Nevada Test Site may become the largest burial ground in the DOE complex for defense-related, low-level waste. In the past, the Hanford Site and Nevada Test Site were called upon to dispose of limited amounts of DOE high-activity and special-case, low-level wastes in Greater Confinement Disposal. To retain this capability, these locations are in the process of evaluating the design and use of engineered disposal facilities for these wastes. An EIS for Greater Confinement Disposal is planned for the Nevada Test Site.

In regard to commercial low-level wastes, in 1980, P.L. 96.573, the Low-Level Radioactive Waste Policy Act, specifically assigned States the responsibility for providing for disposal of all low-level waste generated within their borders. This disposal will be regulated by the Nuclear Regulatory Commission and Agreement State licenses. In 1983, the Nuclear Regulatory Commission developed low-level waste disposal criteria and standards through Title 10 CFR Part 61, which categorized the waste into Classes A, B, and C based on concentration of radionuclides. Low-level waste exceeding the limits for Class C is identified as Greater-Than-Class C and is generally unacceptable for near-surface disposal.

The Low-Level Radioactive Waste Policy Amendments Act of 1985 (P.L. 99-240) further clarified States' responsibilities for the disposal of low-level waste, encouraged the formation of regional compacts, and established milestones with incentives and penalties regarding the timely development of disposal facilities. This Act requires DOE to provide technical and financial assistance to the States and compact regions in developing low-level waste disposal capacity. This Act also requires DOE to ensure the safe disposal of Greater-Than-Class C waste.

2.4 MIXED WASTE

Mixed wastes are radioactive wastes that are also contaminated with hazardous wastes regulated under RCRA. A significant portion of DOE mixed waste, including hazardous and radioactive soil in the DOE inventory, is mixed low-level waste. Compliance with the requirements for managing mixed wastes is one of the most significant issues facing DOE today.

In accordance with RCRA, EPA promulgates regulations for hazardous wastes. Implementation of hazardous waste regulations is the responsibility of EPA and authorized States. All mixed wastes are subject to RCRA hazardous waste regulations, which include requirements for treatment of waste before land disposal. Specifically, the RCRA Land Disposal Restriction regulations (40 CFR 268) require that wastes be treated to meet specific standards before they are placed in a land disposal unit that complies with the standard technical requirements for land disposal. If migration from a land disposal unit can be demonstrated not to occur for as long as the waste remains hazardous, then wastes can be disposed of without prior treatment. DOE plans to comply with RCRA standards before disposing of mixed low-level waste.

A total of approximately 247,000 cubic meters of mixed low-level waste is stored at DOE sites. DOE is working to comply with new Land Disposal Restrictions (LDR). Meanwhile, the generation of mixed waste continues at the rate of 56,000 cubic meters per year.

DOE has classified over 1,400 mixed waste streams into categories that require similar processing steps for assignment of baseline treatment. Baseline treatment schemes for each waste category have been segregated into various technical areas including: front end handling; physical/chemical treatment; waste destruction; off gas treatment; and final forms. The program requires significant funding for research, development and implementation of effective treatment and disposal technologies.

DOE's approach to mixed waste treatment and site remediation is governed by state and federal regulations for both hazardous and radioactive materials. This creates circumstances where dual regulations, which are designed to govern the management of either hazardous or radioactive wastes, apply. Inconsistencies among these regulations complicates development of methods for treating and disposing of mixed waste.

DOE is evaluating future capabilities needed for treating mixed waste. To fully meet treatment needs, technologies must be developed for some waste streams, and in other cases, the capacity of proven technologies will have to be increased. Facilities that will add to the current treatment capabilities, such as the Consolidated Incinerator Facility at the Savannah River Site, are currently in stages of design and construction.

Plant design concept studies to evaluate the mission and processing scope of future mixed waste treatment facilities will be prepared. Cost-benefit studies will be conducted to assess the options of on-site versus off-site treatment. It may be more cost-effective, for example, to establish thermal treatment facilities at only a few sites rather than at the many sites where wastes are generated. The use of commercial facilities will also be evaluated. These concept studies, cost-benefit analyses, and the outcome of the EM Programmatic Environmental Impact Statement will assist in determining the location and design of future treatment facilities.

Planning is under way for a prototype treatment facility for mixed low-level waste. Process and treatment technologies will be evaluated with respect to regulatory and waste management requirements. DOE will also begin to apply for RCRA permits for mixed low-level waste disposal facilities at Savannah River, Richland, Nevada, and Idaho.

DOE's available on-site waste storage capacity has diminished rapidly, and some of the capacity needed for mixed waste is currently being utilized to manage radioactive waste as well as RCRA-restricted hazardous waste. Fernald, Mound Plant, Oak Ridge National Laboratory, and Rocky Flats (for certain waste) are on the verge of running out of capacity for storing mixed waste. Storage capacity at 8 other Nuclear Weapons Complex facilities is expected to be reached shortly.

3

Department of Energy Locations

A brief description of Department of Energy (DOE) sites is given in this chapter, organized by states. Many of the smaller sites are not included. UMTRA (Uranium Mill Tailings Remedial Action Project), and FUSRAP (Formerly Utilized Sites Remedial Action Program) sites, are listed at the end of the chapter.

3.1 ALASKA

Amchitka Island is the only site in Alaska where Environmental Management is currently conducting work. The site was used for nuclear weapons testing and other experiments between 1961 and 1971. Contaminated soil was moved to the Nevada Test Site. Long-term hydrological monitoring is continuing on the island, pending commencement of assessment activities.

3.2 CALIFORNIA

General Atomics is a privately owned and operated nuclear facility located near San Diego, California. It maintained and operated a hot cell facility for over 30 years, primarily to conduct government-funded nuclear research and development. In 1995, General Atomics will commence decontamination and decommissioning activities that will continue through 2000.

The **General Electric Vallecitos Nuclear Center** is a privately owned research facility located in Pleasanton, California. From 1962 through 1979, the site was used for uranium and mixed oxide fuel fabrication and development. While the cell funded by the Department is no longer used, the remainder of the facility is active. General Electric plans to decontaminate the cell, which the

Department had operated to examine uranium fuel and reactor components after irradiation.

The **Stanford Linear Accelerator Center**, managed under contract between the Department and Stanford University, conducts theoretical research in high energy particle physics.

The Center continued to dispose of hazardous and radioactive waste, minimize waste generation, and commence its investigation of contaminated soil and ground water. Design for the Radioactive Management Waste Storage Facility being built on-site was approved in 1994.

The **Institute of Toxicology and Environmental Health** is located at the University of California, Davis. Research at the Laboratory originally focused on the health effects from chronic exposures to radionuclides. The present mission is confined to decontamination and decommissioning, site remediation, and related activities for eventual release of the facilities and site to the University for unrestricted use.

The **Lawrence Berkeley Laboratory**, adjacent to the University of California, Berkeley, conducts a wide variety of energy-related research activities for the Department, including energy, environment, physics, transportation, computers and communication, biology, and medicine. Past practices have left radioactive and hazardous contaminants in the soil and groundwater.

The **Lawrence Livermore National Laboratory** occupies 2 sites in northern California. The Main Site has an area of approximately 1 square mile and is located 40 miles east of San Francisco, just east of the City of Livermore. Site 300, which comprises approximately 11 square miles, is located 15 miles southeast of the Main Site. At the Main Site, U.S. Navy operations in the 1940s and subsequent laboratory activities involving the handling and storage of hazardous materials resulted in the release and off-site migration of contaminants in soil and groundwater. Operations at Site 300 involved the processing, testing, and deactivation of explosives, which contaminated both soil and groundwater at the site. Both sites have wastewater, petroleum tank systems, and transformers that must be upgraded to meet local, State, and Federal requirements.

Environmental Restoration activities continue at the **Sandia Livermore** facility adjacent to the Lawrence Livermore National Laboratory. The facility occupies an area of approximately 100 acres.

The **Atomics International** division of Rockwell began nuclear research utilizing small low-powered reactors in the northwest portion of the Santa Susana Field Laboratory in 1956. Environmental restoration and decontamination and decommissioning activities began in the early 1970s. Today, activities at 4 facilities remain to be completed.

3.3 COLORADO

The **Rocky Flats Environmental Technology Site** is located in northern Jefferson County, approximately 16 miles northwest of Denver. The site is 10.4 square miles, including a buffer zone between the site and its boundaries. With the discontinuation of nuclear weapon components production at Rocky Flats, and the consolidation of the production of non-nuclear weapons components at the Kansas City Plant in Missouri, Rocky Flats' mission now focuses on environmental remediation, nuclear material management, and deactivation and conversion of facilities for alternative uses.

Historically, the plant manufactured nuclear weapons components from plutonium, uranium, beryllium, and stainless steel. Its defense operations (nuclear manufacturing, non-nuclear manufacturing, and chemical processing for plutonium recovery) and its current environmental activities (e.g., environmental restoration, environmental monitoring, and waste management) have required a variety of support facilities and services, including analytical laboratories, waste management facilities, safety systems, safeguards and security, and the routine facilities found in any large manufacturing complex, such as a steam plant and utilities.

The main plant has 436 buildings, facilities, systems, and structures, of which 150 are permanent buildings and 90 are trailers used mainly for office space. Together they provide approximately 3 million square feet of facility space.

The facilities at Rocky Flats are divided into 2 main areas. The area on the north contains all of the facilities related to plutonium operations. Security fences and intrusion detection systems surround all buildings in which plutonium is handled or stored, and various other measures are used to provide safeguards and security. This area is referred to as the "protected area." The area to the south contains both non-plutonium manufacturing facilities, which are located in secured areas, and general support facilities, some of which are in secured areas.

In 1996, Rocky Flats will continue assessment and remedial action activities. Economic conversion of one production facility will be essentially complete, and a private company will begin operating this facility under a lease agreement in 1997.

In addition to the Rocky Flats Environmental Technology Site, Colorado has 11 sites, 2 of which were used for underground nuclear explosive experiments. The test sites, **Project Rulison**, near Parachute, and **Project Rio Blanco**, near Rifle, are each a few acres in size. The Department of Energy developed forecast baselines for each and is conducting hydrological monitoring to ensure radiological contaminants do not migrate to public water sources. Site assessments for both sites were initiated in 1994 and will continue through 1995.

Contamination from the hole drilling mud pit will be investigated. Most of the remaining sites are part of the UMTRA project.

3.4 FLORIDA

The **Pinellas Plant** of about 729,000 square feet occupies a 99.2 acre site approximately 6 miles north of St. Petersburg in Pinellas County, Florida. Pinellas County is located on a peninsula bordered on the west by the Gulf of Mexico and on the east and south by Tampa Bay. Key activities at the Pinellas Plant include the design, development, and production of special electronic and mechanical equipment for nuclear weapon applications. These products include neutron generators, specialty capacitors, thermal batteries, crystal resonators, oscillators, and clocks.

The plant had been an essential part of the Nation's nuclear weapons complex, but production of weapons-related components was discontinued on September 30, 1994. The site is now investigating which product lines could be converted to commercial and non-military applications.

The heavily populated area around the Pinellas Plant makes potential release of contaminants to the groundwater of utmost concern.

3.5 HAWAII

The **Kauai Test Facility** covers 182 acres of the Kauai Island within the U.S. Navy's Pacific Missile Range Facility. The site is a satellite for Sandia National Laboratory-New Mexico and supports the Department's research and development activities, including launching rockets carrying experimental non-nuclear payloads. Contamination of the site consists of 3 potential release sites. These include a photography laboratory, a drum storage area, and the rocket launcher pads.

During 1994, sampling was conducted at the site. Based on these samples, Sandia National Laboratory in New Mexico, which manages the site, plans to submit a request to Environmental Protection Agency Region X for approval of a No Further Action decision.

3.6 IDAHO

The **Idaho National Engineering Laboratory** is located in southeastern Idaho along the western edge of the Eastern Snake River Plain and encompasses a semi-arid area of approximately 890 square miles. The Laboratory is a multipurpose Department of Energy national laboratory providing primarily engineering expertise and operations and development support in the areas of spent nuclear fuel, waste management, environmental assessment and remediation, decommissioning and decontamination of surplus facilities, alternative energy source development, and technology transfer to government agencies and private industry.

Established in 1949 as the National Testing Station (NRTS), the INEL's initial mission was to build, test, and operate various reactors. The facility has the largest concentration of nuclear reactors in the world. From completion of the Experimental Breeder Reactor I (EBR-I) in 1951 to the present, 52 reactors have been successfully built and operated at the INEL. Seven reactors are currently operating at the INEL.

Waste Management Program activities are involved with minimization, treatment, storage, and disposal of radioactive, hazardous, mixed, and solid municipal wastes generated from current or past operations. Waste Management goals are to (1) minimize the volume of waste generated, (2) treat the waste to reduce the volume and eliminate or reduce the hazard, (3) store waste only as a short-term option, and (4) provide final disposal using proven environmentally safe methods.

To accomplish these goals, the INEL Waste Management Program instituted a waste minimization program to educate waste generators in areas of material substitution, waste handling and avoidance, and to implement a comprehensive recycling program. Treatments, such as the calcining operation conducted at the Idaho Chemical Processing Plant, minimize waste volumes and stabilize waste products. Technology development is being accelerated at the Idaho Chemical Processing Plant for calcine immobilization and preparation of spent nuclear fuel for ultimate disposal in a national repository.

Most INEL HLW is reprocessed naval reactor fuel. Acidic liquid waste is stored in underground stainless steel tanks that are housed inside concrete walls. The waste is then converted into a calcine powder and stored retrievably in stainless steel bins inside reinforced concrete vaults. There are 3,500 m^3 of HLW stored as calcine, containing 90 percent of the radioactivity, and 8,500 m^3 of liquid HLW containing ten percent of the radioactivity. The INEL waste is uniform and well characterized, but will not meet Land Disposal Requirements (LDRs).

Since the INEL HLW is in calcine form, processing the waste with aqueous solvent extraction would require dissolving the calcine in nitric acid, which would generate a large volume of aqueous LLW. As an alternative, ESPIP is exploring a glass-ceramic waste form and the possibility of pyrochemical processing. The glass-ceramic waste form would reduce the number of logs from 9,500 to 3,770. If pyrochemical processing was used with a glass waste form, the number of logs generated would be less than 900.

The Department of Energy has received spent nuclear fuel (SNF) at the Idaho Chemical Processing Plant (ICPP) for interim storage since 1951 and reprocessing since 1953. Until April 1992, the major activity of the ICPP was the reprocessing of SNF to recover fissile uranium and the management of the resulting high-level wastes (HLW). In 1992, DOE chose to discontinue reprocessing SNF for uranium recovery and shifted its focus toward the continued safe management and disposition of SNF and radioactive wastes accumulated through reprocessing activities. Currently, 1.8 million gallons of radioactive liquid wastes (1.5 million gallons of radioactive sodium-bearing liquid wastes and 0.3 million gallons of high-level liquid waste), 3,800 cubic meters of calcine waste, and 289 metric tons heavy metal of SNF are in inventory at the ICPP. Disposal of SNF and high-level waste (HLW) is planned for a repository. Preparation of SNF, HLW, and other radioactive wastes for disposal may include mechanical, physical, and/or chemical processes.

At the waste management complex, large quantities of transuranic and low-level radioactive wastes and solidified organic wastes were buried in shallow earth pits until 1970. As a consequence, plutonium may have migrated as far as 110 feet below the surface.

Many serious environmental problems had been identified at INEL. Major problem areas include (1) lack of adequate secondary containment for pipes and tanks; (2) problems related to the treatment and storage of mixed wastes, such as storing these wastes without having EPA-approved treatment technologies available; and (3) releases of radioactive and hazardous contaminants into the ground and the Snake River Plain aquifer.

Research and development efforts will continue for buried waste retrieval and waste characterization, and technical support will continue for the Waste Isolation Pilot Plant. Enhancements are planned for the Waste Experimental Reduction Facility, fuel transfers will continue at the Idaho Chemical Processing Plant, and evaporation activities are planned to be complete to reduce the volume of high-level liquid waste to be treated at the New Waste Calcine Facility. Aggressive efforts will continue to develop partnerships with private industry and universities.

In 1996, with an anticipated budget of $481 million, planned activities include the restart of the Waste Experimental Reduction Facility and the New Waste Calcine Facility. Fifty percent of the transuranic waste in the Radioactive Waste Management Complex buildings will be transferred into RCRA-approved storage modules. Construction of the remediation facility at the Pit 9 site and the retrieval enclosures facility at the Radioactive Waste Management Complex are also planned for completion. Other planned activities include assessment and disposal of surplus facilities throughout the site, fuel transfers at the Chemical Processing Plant, environmental remediation in compliance with Federal Facility Agreement and Consent Order Schedules, and start of construction of the waste characterization facility at the Radioactive Waste Management Complex.

Argonne National Laboratory-West (ANL-W) is located on the southeastern portion of the Idaho National Engineering Laboratory (INEL). The primary mission of ANL-W is research and development in support of the Nation's advanced reactor program. Reactor complexes at ANL-W include the Experimental Breeder Reactor II (EBR II), the Transient Reactor Test Facility, and the Zero Power Physics Reactor. The primary environmental management activities are managing waste streams, upgrading waste management facilities, remediating waste sites within the area, and supporting Waste Isolation Pilot Plant related activities.

3.7 ILLINOIS

Argonne National Laboratory-East (ANL-E) occupies a 1,700-acre tract located approximately 22 miles southwest of downtown Chicago in DuPage County, Illinois. ANL-E is a multi-disciplinary research and development laboratory that conducts basic and applied research to support the development of energy-related technologies. Activities at ANL-E include nuclear reactor design, synchrotron radiation accelerator design, and environmental research programs. The ANL-E primary mission is basic and applied research.

These research activities generate hazardous and radioactive waste. Major concerns are closed landfills that were used to dispose of solid and hazardous laboratory waste. Several buildings and research reactors at Argonne-East are contaminated with low levels of radiation and are undergoing or are scheduled for decontamination and decommissioning.

Fermi National Accelerator Laboratory in Batavia, Illinois, operated for the Department's Energy Research office, explores the fundamental structure of matter using high-energy particle accelerators. Remediation of spills at 22 transformer locations is required.

The remedial activities for **Palos Forest Preserve Site A/Plot M** in Cook County, the former Manhattan Engineer District site near Chicago, is continuing. Plot M was the radioactive waste disposal area located close to Site A.

3.8 IOWA

Ames Laboratory occupies several buildings on the Iowa State University campus in Ames. Ames Laboratory conducts research in material and chemical sciences, and related research in material reliability and nondestructive evaluation.

Characterization of 9 areas referred to as the "inactive waste sites" was completed. Approximately 20 cubic yards of low-level radioactive soils was removed from one of the sites, the other 8 sites required no remediation. Approximately 54,000 cubic feet of low-level contaminated soil, drums and other debris from a former chemical disposal site were removed and shipped to a commercial site in Utah in late February 1995.

3.9 KENTUCKY

The principle on-site process at **Paducah Gaseous Diffusion Plant (PGDP)**, located 10 miles west of Paducah, Kentucky, is the separation of uranium isotopes through gaseous diffusion. The process produces enriched uranium used as fuel in commercial power plants. The site encompasses 750 acres (including 74 acres of process buildings). The site is included in a 3,422-acre tract of DOE owned property.

The plant was leased in July 1993 to a newly-formed government corporation known as the United States Enrichment Corporation. The Department of Energy is still responsible for cleaning up the on-site and off-site soil and groundwater contamination, including uranium, polychlorinated-biphenyls, technetium, and trichloroethylene.

The present plans for PGDP call for completion of all investigations in FY 1996 with completion of remediation activities by FY 2015. D&D planning for PGDP has been initiated for one shutdown uranium processing facility. PGDP is expected to be completely shut down around FY 2015. Completion of the D&D at PGDP is scheduled for FY 2030. Long-term surveillance, maintenance, and institutional controls will continue indefinitely.

The **Maxey Flats Disposal Site** is a low-level radioactive waste disposal site located in northeast Kentucky, approximately 17 miles south of Flemingsburg.

The Department of Energy is one of approximately 800 identified potentially responsible parties. The Department's share of the financial liability for remedial actions and related tasks is about 40 percent.

3.10 MISSISSIPPI

Underground nuclear explosive experiments were conducted at the **Salmon Test Site** (formerly called Tatum Dome) located near Hattiesburg. An Agreement-in-Principle with the State of Mississippi regarding remediation was signed in January 1991. Remedial activities were delayed in 1993, when the landowner refused to lease the property, requesting that the Department purchase the property. The land was sold to the Federal Government at the end of 1994. Remedial activities will resume in 1995.

3.11 MISSOURI

In 1993, the Department shut down several facilities across the country and consolidated the production of non-nuclear components to the **Kansas City Plant**.

The Kansas City Plant (KCP) is part of the Bannister Federal Complex, located 12 miles south of downtown Kansas City, Missouri. Manufacturing operations are housed in 3.2 million square feet of building space. The plant mission is the manufacturing of non-nuclear weapons components involving machining, plastic fabrication, and electrical and mechanical assembly. No radioactive materials are machined or processed. Waste operations consist primarily of waste storage, off-site shipment and disposal, and on-site wastewater treatment for industrial process wastewaters. Twenty-seven sites are currently undergoing remediation.

The **Weldon Spring Site**, a 229-acre site located about 30 miles west of St. Louis, Missouri, was used by the Army as an ordnance works in the 1940s. In the 1950s and 1960s, the Atomic Energy Commission used Weldon Spring for processing uranium and thorium. The site is on the EPA National Priorities List, and DOE is conducting a comprehensive remedial action program, including long-term management of radiological wastes.

Areas to be remediated include the following:

1) Quarry - 9 acre site containing 126,630 cubic yards of radiologically contaminated soil and rubble and 3 million gallons of radiologically or chemically contaminated water.
2) Raffinate Pits - 4 waste lagoons, containing 407,930 cubic yards of raffinate sludge/soil and 57 million gallons of radiologically or chemically contaminated water.
3) Chemical Plant - 44 buildings and other structures and 347,996 cubic yards of contaminated soil and building material.
4) Vicinity Properties - Approximately 125,250 cubic yards of contaminated soil.
5) Groundwater - Nitroaromatic and radiologically contaminated groundwater at the Quarry and Chemical Plant.

3.12 NEBRASKA

The **Hallam Nuclear Power Facility** near Lincoln, Nebraska, was built and operated as a demonstration project by the Atomic Energy Commission between 1962 and 1966. The facility contains entombed components from thermal sodium-cooled graphite-moderated nuclear reactor.

Between 1967 and 1969, the facility was dismantled and placed in a safe storage condition. Surveillance and maintenance continues at Hallam to ensure that no radioactivity is being released from the facility. Semi-annual groundwater sampling and analysis and radiological survey of the entombed reactor are also conducted.

3.13 NEVADA

In the State of Nevada, there are 4 sites that require remediation. The sites are the **Nevada Test Site**, the **Project Shoal** site, the **Central Nevada Test Area**, and the **Tonopah Test Range**. From 1961 through 1973, the Project Shoal and the Central Nevada sites were off-site locations used by the

Department to conduct underground nuclear tests and experiments. Site assessments began in 1994.

The **Nevada Test Site (NTS)** is located about 65 miles northwest of the City of Las Vegas. The site encompasses 1,350 square miles of desert and mountainous terrain. The site is surrounded on 3 sides by the Nellis Air Force Range, which provides a substantial buffer between the site and the communities located in the area. The primary mission of the site has been weapons testing. Through 1992, when the President halted underground nuclear testing, the United States conducted 1,054 nuclear tests, of which, 928 occurred at the Nevada Test Site. The remaining 126 nuclear tests were conducted at other sites in and outside of Nevada.

The 1992 weapons testing moratorium resulted in a total cessation of nuclear weapons testing. The focus has now shifted to remediation of inactive sites and facilities that were contaminated during earlier testing activities. The Nevada Operations Office is working with the State of Nevada to develop a compliance agreement, under the Resource Conservation and Recovery Act and the Atomic Energy Act, addressing projected site remediation activities.

Approximately 1,100 surface and subsurface contamination-sites from nuclear test and ancillary operations have been identified. These include waste disposal sites associated with testing activities and areas where superficial soils were contaminated with plutonium as a result of destructive safety tests of nuclear devices.

The application of conventional remedial actions that are widely used at contaminated waste disposal sites may not be feasible at the Nevada Test Site because of the unique nature of the waste and releases and the depth of contamination from underground nuclear testing.

Low-level radioactive waste that originates at the Nevada Test Site, and from other Department installations, is disposed of on-site. Additionally, the site is utilized to temporarily store mixed transuranic waste from the Lawrence Livermore National Laboratory. Additional limited mixed waste disposal for waste generated off-site may be available, pending completion of the Nevada Test Site's Site-Wide Environmental Impact Statement and approval from the State of Nevada for the site's mixed waste analysis plan. An expanded mixed waste disposal facility will be constructed if the State of Nevada issues the RCRA Part B permit that is being modified to address State comments. The revised permit application for the mixed waste disposal facility is expected to be submitted to the State of Nevada by October 1995. Construction will begin within 180 days of receipt of the final permit.

The Nevada Test Site disposes of low-level radioactive waste on-site from the Site and from other Department installations. In addition, the Site stores mixed transuranic waste from Lawrence Livermore National Laboratory pending shipment to the Waste Isolation Pilot Plant. There has been considerable objection within the State of Nevada to NTS accepting wastes from out-of-state.

The **Tonopah Test Range** was established in 1957 for ballistics testing of nuclear weapons, parachute delivery systems, and other non-nuclear functions for Sandia National Laboratories. During 1994, a site assessment was conducted, as well as environmental restoration-site inventories, surface and aerial surveys, and a risk assessment. These surveys identified 14 potential release areas in need of characterization. Before sampling can proceed, a concerted effort to remove unexploded ordnance from 5 of the disposal sites must take place to protect site workers from potential hazards. This effort, initiated in February 1995, involved the remote handling and detonation of potentially live ordnance. Field sampling activities will begin in 1995 to determine the nature and extent of contamination at the identified sites.

3.14 NEW JERSEY

The **Princeton Plasma Physics Laboratory**, leased to the Department of Energy by Princeton University, conducts magnetic confinement plasma physics research and investigates the practical application of fusion power as an energy source. The principal environmental concerns are the ongoing management of hazardous and radioactive waste, underground storage tank remediation, characterization and remediation of groundwater, and planning for the disposal of low-level and mixed waste from dismantling the Tokamak Fusion Test Reactor.

The Laboratory will continue to ship and dispose of hazardous, radioactive, and mixed waste. During 1995, the Laboratory will continue routine waste management activities and will complete the Remedial Investigation report. Remedial action and alternative analysis work plans for soil and groundwater at 2 areas on-site will be prepared. Title I and II design will begin for a radioactive waste handling facility.

3.15 NEW MEXICO

Los Alamos National Laboratory (LANL) occupies about 43 square miles in Los Alamos County, approximately 60 miles north-northeast of Albuquerque and 25 miles northwest of Santa Fe. LANL is situated on the Pajarito Plateau, which is composed of finger-like mesas ranging in elevation from 6,200 to 7,800 feet. Major programs at LANL include applied research in nuclear and conventional weapons development, nuclear fission and fusion, nuclear safeguards and security, and waste management. Currently it is the only facility processing plutonium.

The University of California operates the Laboratory under a contract for the Department of Energy. Major programs include applied research in nuclear and conventional weapons development, nuclear fission and fusion, nuclear safeguards and security, and environmental and energy research. The Laboratory's core competencies are nuclear science, plasmas, and beams; modeling and high-performance computing; bioscience and biotechnology; and earth and environmental systems.

Approximately 2,250 potential release sites, aggregated into 24 operable units (OUs), are scheduled for investigation in the Environmental Restoration Program under the Hazardous and Solid Waste Amendments (HSWA) permit. Six surplus facilities are identified for decontamination and decommissioning (D&D).

During 1955, Los Alamos will continue to develop technologies and processes to address environmental issues. For example, waste treatment facilities will be designed and constructed to serve as a test bed for developing emerging technologies. The program will then use these facilities to service laboratory waste generators and demonstrate innovative treatment technologies to potential customers. The Laboratory will complete and submit Facility Investigation reports covering 350 potential release sites, complete up to 60 expedited remediation actions, and complete 2 RCRA closures in 1995. Decontamination and decommissioning work will include assessing 7 buildings housing highly explosive materials and 1 filter building.

In 1996, Los Alamos will continue its waste minimization, environmental restoration, and waste management activities at its own sites and others within the Department of Energy complex. By 1996, the Laboratory will have collaborated with Hanford and Rocky Flats to address several environmental management issues. In addition, the Laboratory will continue to collaborate with other Government agencies, academia, and industry, as well as further develop and augment methods for effective business operations.

Project Gnome-Coach and **Project Gasbuggy** are locations where underground explosive nuclear tests and experiments were conducted from 1961 through 1973. These sites are currently inactive. Site assessments are scheduled to begin in 1996. Long-term hydrological monitoring is being conducted at both sites to detect potential radiological contamination migrating to potable water sources.

From 1951 to 1967, the **South Valley Site** near Albuquerque was owned by the Atomic Energy Commission and operated by the American Car Foundry Company. Operations included electroplating, machining, painting, adhesive, and degreasing related to weapons production, reactor design, and space programs.

Albuquerque, New Mexico, is also home to the **Sandia National Laboratories**, a research and development facility with a primary mission of developing, engineering, and testing non-nuclear components of nuclear weapons.

During 1994, Sandia National Laboratories-New Mexico completed urgent environmental protection and corrective activities, including installation of a meteorological monitoring network, construction of a reactor facility liquid effluent discharge control system, and sewer system repairs. Waste management efforts focused on developing radioactive waste facilities, operations, and disposal options. Sandia successfully shipped mixed waste debris to a commercial disposal facility, marking the first time the site has shipped radioactive waste off-site for disposal.

The Sandia National Laboratories has identified 192 potential release sites that may require remedial action.

In the coming year, Sandia National Laboratories-New Mexico will focus on developing mixed-waste treatment and disposal capabilities while maintaining a strong RCRA waste management program. Efforts will continue to improve waste management facilities, and a pollution prevention strategy will be implemented.

Sandia National Laboratory-New Mexico has generated chemical, radioactive, and mixed waste at a number of off-base locations, including 1 building at the **Holloman Air Force Base** near Albuquerque.

The **Inhalation Toxicology Research Institute**, in Albuquerque, conducts studies on the health effects of inhaling potentially hazardous airborne materials that might be found in industry, the environment, or the home.

During 1994, the Institute completed and resolved all issues associated with an application for shipment of low-level waste to the Nevada Test Site. In addition, the Institute's waste management program identified off-site commercial facilities to treat its entire inventory of mixed waste. It also completed its plans to ship existing mixed waste off-site to commercial facilities for treatment and disposal.

The **Waste Isolation Pilot Plant** is located 26 miles east of Carlsbad in Eddy County, New Mexico. The site covers 10,240 acres of Federal land and is located 2,150 feet below the surface in a 2,000 foot thick salt bed with tunnels that extend over 10 linear miles. The Waste Isolation Pilot Plant provides a research and development facility to determine the suitability of the site for the safe disposal of defense-related transuranic mixed waste. The transuranic waste destined for the Waste Isolation Pilot Plant is currently in temporary storage at waste generator sites located in California, Colorado, Idaho, Illinois, Nevada, New Mexico, Ohio, South Carolina, Tennessee, and Washington.

3.16 NEW YORK

The **West Valley Demonstration Project** is carried out at the former Western New York Nuclear Service Center located in Cattaraugus County. The project is working to demonstrate safe immobilization of liquid high-level radioactive waste produced at the site using vitrification. During 1994, much of the vitrification facility building and installation of its associated equipment was completed. In 1995, the facility will be completed and inspected, and vitrification will begin in 1996. West Valley is currently reducing the volume of liquid high-level waste by treatment in the integrated radioactive waste treatment system. As a result of this treatment, over 18,000 drums of low-level waste solidified in cement have been produced and are being safely stored on-site.

Brookhaven National Laboratory is a multi-purpose research and development laboratory that directs scientific and technical efforts including physics, life sciences, and nuclear medicine research. The Laboratory is located on Long Island, about 60 miles east of New York City, and occupies about 8.3 square miles of mostly wooded area. A number of environmental activities are underway to remediate waste generated in connection with the Laboratory's work.

The **Knolls Atomic Power Laboratory** contains the Separations Process Research Unit, which was used by the Department of Energy to develop a chemical process for extracting plutonium and uranium from irradiated fuel. Once the process was developed, the operation was transferred to the Hanford Site. The buildings used for the research project will be decontaminated and decommissioned beginning in the year 2000.

3.17 OHIO

The **Fernald Environmental Management Project** is located on 1,050 acres approximately 18 miles northwest of Cincinnati, Ohio. From 1953 to 1989, the site produced uranium metals and compounds for the Nation's defense program. In 1989, all production operations were suspended. In 1991, production was permanently halted. The Fernald Environmental Management Project's main mission is remediation of the site and any off-site contamination in a timely, safe, and cost-effective manner. All intermediate removal actions have been completed to address immediate site risks. Final decisions on remedial alternatives will be made for all 5 areas by 1997.

An interim Record of Decision for decontamination and decommissioning of more than 200 structures at Fernald was signed by the Department and the Environmental Protection Agency. The tallest building on the site, a seven-story structure, was decontaminated and imploded in 1994, and field activities for the dismantlement of 5 additional structures are scheduled for 1995.

The Minimum Addition Waste Stabilization technology can process low-level and mixed waste into glass, combining several waste streams using an electrically heated melter in combination with soil washing and ion exchange wastewater treatment to minimize the use of additives. Waste loadings up to 95 percent were achieved in a 300-kilogram-per-day melter constructed and operated on-site.

Fernald is a site for the Uranium Soils Integrated Demonstration. Enhanced site characterization and precise excavation technologies will be combined with advanced uranium soil decontamination processes to produce a technology system for use at the FEMP and throughout the DOE complex for similar contamination cleanups.

The FEMP major site activities also include the cleanup at the Reactive Metals, Incorporated Extrusion Plant (RMI) in Ashtabula, Ohio. This seven-acre facility previously extruded uranium metal shapes.

The **Mound Plant**, located in Miamisburg, Ohio, was built in the 1940s to support research and development and production activities for the Department of Energy's weapons and energy programs. It occupies 306 acres of land in southwestern Ohio. Weapons production support ended in 1994, when activities across the country were consolidated at the Kansas City Plant in Missouri. The Mound Plant will continue to support the Department's nuclear energy programs.

The Mound Plant's primary missions are the manufacture and evaluation of pyrotechnic components for nuclear weapons and the surveillance testing of explosives and electrical components drawn from weapons in the stockpile. These components include detonators, times, firing sets, and actuators. Mound also recovers tritium, a crucial nuclear material, from retired weapons and ships it to DOE's Savannah River Site for recycling. In addition to its weapons complex work, Mound conducts work for other government programs. This work includes the production of power sources for space programs and of non-radioactive isotopes for commercial and medical applications.

In 1994, a CERCLA remedial investigation/feasibility study was completed of groundwater contamination by volatile organic compounds that resulted from past waste management practices. The site also began planning for removal of plutonium-contaminated sediments resulting from a previous break in a waste transfer line. Several expedited remedial actions were initiated in 1994, including a project to use bioremediation to treat contaminated soil. Decontamination and decommissioning accomplishments in 1994 include the removal of a contaminated underground waste transfer line, removal and dismantlement of most of the building where Pu-238 was formerly processed, and sampling of the building that used to store thorium.

Planned activities include developing the first CERCLA Record of Decision for remediation of groundwater contamination and commencement of design activities, site-wide remedial investigation field work in six areas, and continued planning for removal of the plutonium-contaminated sediments. Other planned activities include the dismantlement of the former plutonium processing building and several characterization activities.

Treatment, storage, and disposal of hazardous waste at the site will continue, in addition to the disposal of Mound's low-level waste. An additional 22 buildings are scheduled to be shut down this year.

The **Portsmouth Gaseous Diffusion Plant**, located on 3,700 acres in Portsmouth, separates uranium isotopes to produce enriched uranium used for fuel in commercial nuclear power plants. Portsmouth was leased in July 1993 to a newly formed government corporation known as the United States Enrichment Corporation, but the Department of Energy is responsible for remediation of pre-existing conditions.

Environmental problems, which include contamination of the aquifer beneath the site at Portsmouth, are divided into four quadrants for investigational purposes.

Environmental problems at the Portsmouth Plant involve mostly solvent contamination of the aquifer beneath the site. Solvents were used for industrial metal cleaning required to maintain the facility during operations. Plumes of groundwater contaminated with solvents extend from several locations within the plant. In addition, 2 locations were characterized to be contaminated with hexavalent chromium used as an anticorrosive in the plant cooling water systems.

The **Piqua Nuclear Power Facility** contains entombed nuclear reactor components from a reactor that was built and operated for a demonstration project by the Atomic Energy Commission between 1963 and 1966. The facility was decommissioned and dismantled between 1967 and 1969. Current site activities consist of an environmental monitoring program to ensure radioisotopes are not migrating from the entombment structure into the surrounding environment.

The **Reactive Metals, Inc. Decommissioning Project** located in Ashtabula, Ohio, originally subcontracted with Department of Energy contractors to extrude uranium for use in various nuclear applications. The environmental restoration mission at the site is to decontaminate and decommission the site for unrestricted use.

Reactive Metals, Inc. is a potentially responsible party to the **Fields Brook Superfund** site, which is adjacent to the extrusion plant. The Department of Justice will negotiate on behalf of the Department of Energy, with the Environmental Protection Agency and Reactive Metals, Inc., to reach a settlement on the Fields Brook liability.

Activities completed in 1994, included the safety evaluation for storage, transportation, and disposal of uranium oxides; completion of characterization of 3 filter buildings; and completion of characterization for several buildings. Waste shipments to the Nevada Test Site will continue to make on-site storage space available for off-site soils, and shipments of waste to off-site facilities will also occur.

Battelle Columbus Laboratories comprises two major research complexes, one in the city of Columbus and 1 in rural Madison County, Ohio. The King Avenue (Columbus) site houses corporate offices and general research laboratories. The West Jefferson (Madison County) site comprises a number of facilities formerly dedicated to nuclear research. Since mid-1943, the Battelle Memoria Institute (Battelle) has continuously performed contract research and development work at its Columbus Laboratories for DOE and its predecessor agencies.

The Battelle Columbus Laboratories are privately owned. DOE no longer needs the facilities and is obligated contractually to remove the contamination so that laboratories can be used by Battelle without radiological restriction. Fifteen buildings, or portions thereof, and associated soil areas, are radioactively contaminated as a result of work under government contract and are to be decontaminated and released to Battelle without radiological restrictions. Battelle also holds an active license from the Nuclear Regulatory Commission, which sets specific requirements for timely decontamination of these laboratory facilities.

3.18 SOUTH CAROLINA

The **Savannah River Site (SRS)** produces nuclear materials, primarily tritium and plutonium, for national defense. The SRS is located in south central South Carolina and is bordered on the southwestern side by the Savannah River. The closest major population centers are Aiken, South Carolina, and Augusta, Georgia. The site comprises 5 reactors, 2 chemical separations facilities, 1 reactor fuel manufacturing facility, and other administration and support facilities. The total area of the site is approximately 325 square miles. The production facilities occupy less than 5 percent of the site area.

The SRS EM mission is (1) to manage activities to achieve full compliance with all applicable laws, regulations, and agreements; (2) to integrate the above into all operating DOE facilities; (3) to treat, store, and dispose of the current inventory of waste; (4) to reduce the generation of new wastes; (5) to clean up inactive waste sites; (6) to remediate contaminated groundwater; and (7) to decontaminate and decommission surplus facilities. This will be accomplished over a 30-year period.

Decontamination and decommissioning (D&D) activities other than surveillance and maintenance will be started at 9 facilities with one completed during the 5 year planning period (the Savannah River Technology Center Separations Equipment Development 1 facility). This facility will be completely decontaminated and decommissioned. A total of 657 D&D candidate sites have been identified. D&D of all facilities will be complete by 2019. Determination of the types, volumes, and final disposition of waste generated from all D&D activities as well as future land-use is in progress.

It is one of the Department's bigger environmental management challenges. The Environmental Management program is expanding its role from waste storage and evaporation to waste processing, or vitrification, which is necessary for stabilization of liquid high-level radioactive waste throughout the Savannah River Site in preparation for ultimate disposal in a high-level waste repository.

All 2 year old waste tanks will be emptied and undergo D&D operations by 2019. The waste removed from the waste tanks will be processed in the Tank Farm at the In-Tank Precipitation (ITP) and Extended Sludge Processing (ESP) facilities and fed to the Defense Waste Processing Facility (DWPF) for final on-site treatment and storage. The Replacement High-Level Waste Evaporator is scheduled to start up in 1996 to handle the DWPF recycle and ESP washwater streams.

The site will continue construction of the Consolidated Incineration Facility that was designed to treat hazardous, mixed, and low-level radioactive waste. Pending the outcome of the Savannah River Site Waste Management Environmental Impact Statement, it will be determined if low-level radioactive waste will be treated. The facility is scheduled to be operational in 1996.

Four 30,000 gallon double-walled underground storage tanks with leak detection, leak collection, overfill protection, liquid waste agitators, and air monitors will be constructed. A study to re-evaluate the design of a subsurface concrete vault divided into 12 separate cells with more than 1.2 million cubic feet of storage to ensure it is a cost-effective storage option is expected to be complete in 1995. Planning is in progress to retrieve transuranic waste drums from bermed storage beginning in 1997. Work is currently underway to remove rainwater from the transuranic waste drums.

With the 1996 budget request, operation of the Consolidated Incineration Facility and New Solvent Storage Tanks will begin, and the New Waste Transfer Facility will start up. Savannah River Site will also begin closure of the Low-Level Radioactive Waste Disposal Facility.

The Site Treatment Plan, to define the site's mixed waste streams and the best treatment option for each, will be finalized and submitted to the State for approval in 1995. Savannah River Site will continue groundwater remediation and site characterization activities at contaminated sites. Interim and final remedial actions will be initiated at several waste sites.

Vitrified waste produced at the Defense Waste Processing Facility will be disposed of in a geological repository. The first canister of simulated glass at the facility was filled to test the process in 1994. Start-up testing will continue in support of beginning operation with radioactive materials in December 1995. Operation of 2 evaporator systems to reduce the volume of high-level waste resumed in 1994.

3.19 TENNESSEE

The **Oak Ridge Reservation** consists of 3 separate sites: a national laboratory, a manufacturing and developmental engineering plant, and a former gaseous diffusion plant. While each site has distinct missions for the Department, cleanup of the 3 sites and gaseous diffusion plants in Paducah, Kentucky, and Portsmouth, Ohio, are all managed as one program.

The **Oak Ridge National Laboratory** covers approximately 2,900 acres in Melton and Bethel Valleys, 10 miles southwest of the City of Oak Ridge. The Laboratory's mission is to conduct applied research and development in support of the Department's programs in energy technologies, and to perform basic research in selected areas of the physical and life sciences. Past research, development, and waste management activities at the Laboratory have produced a number of areas contaminated with low-level and/or hazardous chemical waste that will require remediation.

The **Y-12 Plant** was built in the early 1940s to produce enriched uranium by electromagnetic separation for the Nation's first nuclear weapons during World War II. A few years later, this process was rendered obsolete by the gaseous diffusion process, and the plant became the enriched uranium weapons component facility. Since then, the Y-12 Plant has become the center for the handling, processing, storage, and disassembly of all uranium materials and components. With the end of the Cold War, the Plant's mission has evolved to dismantling nuclear weapon components and serving as the primary enriched uranium repository for the United States.

The **K-25 Site** occupies a 1,700-acre area adjacent to the Clinch River, approximately 13 miles west of Oak Ridge. Originally built to enrich uranium hexafluoride for defense programs, a majority of the 125 major buildings on the site are now inactive since production ceased in 1987. The site's mission has changed primarily to environmental management. The K-25 Site is the principal waste storage facility on the Oak Ridge Reservation and houses the Toxic Substance Control Act Incinerator and the Center for Environmental Technology and Waste Management.

Also in Oak Ridge, but not on the Oak Ridge Reservation, are several sites associated with operations at the 3 facilities that require remediation. These include the Oak Ridge Institute for Science and Education, the Clinch River/Watts Bar Lake, Lower East Fork Poplar Creek, and other small privately owned sites in the area.

In 1994, Oak Ridge completed four assessments and 5 interim actions. They included assessments for 3 pilot projects. These projects are an example of the Department's contract reform initiative, and will result in a reduction of the cost and schedule of environmental restoration activities. The interim actions included the removal of material from several off-site locations. Remedial actions at 15 sites and decontamination and decommissioning actions in 4 areas were also completed, including the Electrochemical Machining Area in the 9201-4 facility at Y-12 and the decontamination and decommissioning of electrical components containing polychlorinated biphenyls at the K-25 Site.

In 1994, 140 treatment, storage, and disposal facilities operated in compliance with State and Federal regulations; 125 million gallons of wastewater were treated; a waste storage inventory of 2.1 million cubic feet was maintained; and 1.1 million cubic feet of sanitary waste was disposed. At the Laboratory, 50,000 gallons of radioactive liquid waste were solidified and prepared for shipment to the Nevada Test Site for disposal. At the Toxic Substances Control Act incinerator, 5.7 million pounds of mixed waste liquids were treated, exceeding the goals set under the site's Federal Facility Compliance Agreement. A new landfill for sanitary and industrial waste, which has enough capacity to be used into the next century, opened in 1994. Contracts were established with a private company to treat the Department's mixed waste sludge and radioactively-contaminated soils, recycle contaminated scrap metal, and reduce low-level waste volumes.

In 1995, 9 remediation and 5 decontamination and decommissioning actions will be completed, including the closure of the holding pond and the retention basin at K-25 and the remediation of seeps at a Laboratory area. Also, 10 assessments of remedial action and decontamination and decommissioning sites will be completed. The Pond Waste Management Project will complete the repackaging of 31,000 deteriorated drums containing sludge from the ponds and their subsequent placement into RCRA-compliant storage. Private sector treatment of sludge will be conducted and evaluated.

The Toxic Substance Control Act Incinerator will complete testing on combustible solid waste for renewal of the Air Permit and will meet treatment goals outlined in the Federal Facility Compliance Agreement. Another requirement of the Act is to generate a plan for treatment and/or disposal of accumulated waste on the reservation. Two contracts are scheduled to be awarded for proof-of-process treatment and disposal of mixed low-level waste sludge and soil. The final plan will be submitted in April 1995 to the regulators for approval. Shipments of solid low-level waste to the Nevada Test Site are also scheduled. Four projects will be completed to provide additional storage and treatment capacity. The demonstration phase of the Out-of-Tank Evaporator project will be initiated on waste to evaluate the potential for improved volume reduction at the Melton Valley Storage Tanks. Following the demonstration, the technology will be implemented without delay. Deactivation activities will continue on the isotope facilities at the Laboratory, and program planning will begin for additional facilities scheduled to be transferred to the Office of Facilities Stabilization in 1996.

In 1996, 6 remedial action and decontamination and decommissioning assessments, 2 remedial actions, and 3 decontamination and decommissioning actions are scheduled to be completed. These include the completion of projects to reduce the level of mercury leaving buildings at the Y-12 Plant, to reduce discharges to East Fork Poplar Creek, to demolish the Powerhouse at the K-25 site, to decontaminate and decommission the waste evaporator facility at the Laboratory. Also in 1996, Oak Ridge will initiate implementation of the requirements of the Site Treatment Plan. The first phase of construction to increase capacity at the Melton Valley Storage Tanks by 450,000 gallons of low-level waste will be completed and placed in operation. The vitrification and thermal desorption bench-scale demonstrations of mercury and volatile organic contaminants removal from mixed waste will also be completed.

3.20 TEXAS

The **Pantex Plant** is located in the panhandle of Texas, about 17 miles northeast of downtown Amarillo and 10 miles west of the town of Panhandle. Pantex includes a land area of about 16,000 acres. The plant is operated to meet DOE's responsibilities for nuclear weapons assembly, stockpile monitoring, maintenance, modifications, and retirements (disassembly). Pantex conducts research and development on high explosives in support of weapons design and development and production engineering for DOE. The facility's role within the nuclear complex is assembly, disassembly, and quality assurance of the weapons inventory.

In 1994, the site was placed on the National Priorities List, thereby requiring remediation under CERCLA authority. DOE is working closely with the regulators to integrate both the RCRA and CERCLA requirements in order to avoid costly duplication of remediation activities while meeting remediation goals.

Waste management operations at Pantex, in the near term, will add facilities to enhance capabilities to adequately handle existing waste streams. These new facilities for high explosive incineration and hazardous waste staging, treatment, and storage will be coupled with increased use of commercial off-site facilities to treat mixed waste streams. The long-range outlook for Pantex indicates increased waste generation as a result of accelerated retirement of the weapons inventory. New waste handling capacities may be required to meet this need. Environmental restoration activities at the site include the assessment of 144 solid waste management areas.

Pantex completed the Conceptual and Draft Site Treatment Plans for mixed waste and succeeded in becoming the first Department of Energy site to ship mixed waste to Envirocare of Utah for disposal. Waste management activities at the site include characterizing and tracking all waste. More than 2,300 weapon components were characterized, and 35 waste streams, identified on the State's Notice of Registration, were characterized. Pantex shipped its current inventory of low-level waste to the Nevada Test Site.

In 1995, a significant amount of characterization is planned so that Pantex can submit decisions of No Further Action on an additional 3 sites. Pantex will focus on development of mobile units for treatment of mixed waste. The site plans to ship additional mixed waste to commercial facilities for treatment and/or disposal. The Final Site Treatment Plan for mixed waste will be submitted to the State of Texas, and conceptual designs will be completed on a hazardous waste treatment and processing facility. Pantex will also continue its waste minimization efforts.

In 1996, 2 other areas will be approximately 50 percent remediated. Pantex expects to complete the design of the hazardous waste treatment and processing facility. Pantex will also request 3 additional No Further Action decisions in 1996. Remediation of 2 additional sites and Voluntary Corrective Activities on 5 sites will also continue. The weapon component characterization program to facilitate the stockpile dismantlement will proceed, and Pantex will continue minimizing waste produced on-site.

3.21 WASHINGTON

The **Hanford Site** is 560 square miles of semi-arid land in southeastern Washington, about 50 miles north of the Oregon border. The Columbia River flows through the Hanford Site. The U.S. Army Corps of Engineers selected this area to build nuclear reactors and chemical processing facilities for the production, separation, and purification of plutonium in 1943. Today, the primary mission at Hanford is the remediation of the site.

On May 15, 1989, a Tri-Party Agreement was signed between the Department of Energy, the Washington State Department of Ecology, and the Environmental Protection Agency (and amended in May 1991 and January 1994). This agreement established enforceable milestones to keep the remediation program at Hanford on schedule through the year 2024.

Activities at Hanford, which formerly focused on plutonium production, have shifted to environmental restoration, managing the wastes generated by past reactor and processing operations, and research and development for advanced reactors, energy technologies, basic sciences, and waste disposal technologies. Approximately 1,170 waste sites, grouped into 78 operable units (OUs) in 4 aggregate areas, will potentially require remediation. These aggregate areas are on the National Priorities List.

Between 1943 and 1964, 149 single-shell storage tanks were constructed and placed in service at Hanford to contain the highly radioactive, heat-producing, and chemically toxic liquid wastes resulting from the reprocessing of spent reactor fuel in connection with the nuclear weapons program. These tanks, ranging in size from 55,000 gallons to 1,000,000 gallons, last had waste added in 1980. Many of these tanks are leaking. The newest 28 tanks are double-shell tanks.

Currently, items of particular interest at the Hanford Site center on waste tank safety issues: new waste being generated by cleanup activities and its disposition; the Tank Waste Remediation System (TWRS), which includes the Hanford Waste Vitrification Project (HWVP); and expedited response actions. Major issues at the Hanford Site include:

1) Flammable gas generation, potential explosive mixtures of hydrogen gas from ferrocyanide reactions, potential organic-nitrate reactions, toxic vapors, and continued cooling required for high heat generation in tanks are issues being addressed with regard to the single and double-shell underground storage tanks at the Hanford Site.
2) The TWRS initiative will integrate efforts to characterize, retrieve, treat, and dispose of both double and single-shell tank waste with a systems engineering approach. This system includes HWVP, grout and other related Hanford waste tank activities in a coordinated system.
3) The objective of the HWVP is to convert pretreated Hanford Site defense high-level waste and transuranic waste in underground storage tanks into a solid, vitrified (glass) form suitable for final disposal in a geologic repository.
4) DOE had planned to convert the low-level waste into a cement-like product called grout and dispose of it permanently in about 240 large underground concrete vaults, however, they may be vitrified.
5) Extensive amounts of waste from cleanup activities must be characterized, packaged, and disposed of appropriately. Efficiencies in the waste characterization process are needed as are decisions on ultimate waste disposition.

Single-shell tank safety activities are aimed at reducing risks associated with waste in these tanks before transferring the waste to double-shell tanks. Pretreatment and vitrification of this waste will follow. Analytical laboratory upgrades are necessary to support this cleanup mission. These upgrades include construction of a low-level waste laboratory, construction of additional hot cells for high activity samples, and expanded use of commercial analytical services. Construction of the mixed waste storage facilities and Waste Receiving and Processing Facility (Modules I and IIa) will help support disposition of waste generated from on-site cleanup activities. Ultimate disposition remains an issue to be resolved.

Solid waste volumes of low-level mixed waste (LLMW) and transuranic/transuranic mixed (TRU/TRUM) waste and their associated container types will be generated or received at the Hanford Site for storage, treatment, and disposal at Westinghouse Hanford Company's Solid Waste Operations Complex (SWOC) during a 30-year period from FY 1994 through FY 2023.

3.22 UMTRA SITES

UMTRA (Uranium Mill Tailings Remedial Action Project) manages 24 former uranium ore processing sites that are contaminated with tailings and other by-products of uranium mining and milling operations. In addition, thousands of vicinity properties were contaminated by windblown waste or debris used in construction or landscaping. By September 1994, remediation at 13 of 24 mill tailing sites was completed. In 1995, remediation will continue at 5 sites and be initiated at 4 sites. Remedial action will be initiated at the final 2 sites in 1996. All surface contamination within the 10 States and 2 Indian Tribal lands should be remediated by 1997, except for the vicinity properties at Grand Junction.

The UMTRA sites are listed below:

Arizona
- Monument Valley*
- Tuba City*

Colorado
- Durango*
- Climax*
- Gunnison
- Maybell
- Naturita
- New Rifle
- Old North Continent
- Old Rifle
- Union Carbide

Idaho
- Lowman*

New Mexico
- Ambrosia Lake
- Shiprock*

North Dakota
- Belfield
- Bowman

Oregon
- Lakeview*

Pennsylvania
- Canonsburg*

Texas
- Falls City*

Utah
- Green River*
- Mexican Hat*
- Salt Lake City*

Wyoming
- Riverton*
- Spook*

*completed

3.23 FUSRAP SITES

Under FUSRAP (Formerly Utilized Sites Remedial Action Program), remediation actions are managed at 46 sites in 14 States where the Federal Government contracted with private firms to process or perform research in connection with the Department's atomic weapons activities. In 1994, remediation was completed at 3 sites, bringing the total completed to 16. An additional 3 sites will be completed in 1995, and another 3 sites are expected to be completed in 1996. Remediation at the balance of the FUSRAP sites is planned to be completed by the year 2016.

The FUSRAP sites are listed below:

California
- University of California*

Connecticut
- Seymour Specialty Wire*
- Combustion Engineering Site

Illinois
- Granite City Steel*
- Madison
- National Guard Armory*
- University of Chicago*

Maryland
- W.R. Grace & Company

Massachusetts
- Chapman Valve
- Shpack Landfill
- Ventron

Michigan
- General Motors

Missouri
- Latty Avenue Properties
- St. Louis Airport Site
- St. Louis Airport Site Vicinity Property
- St. Louis Downtown Site

New Jersey
- DuPont & Company
- Kellex/Peirpont*
- Maywood
- Middlesex Municipal Landfill*
- Middlesex Sampling Plant
- New Brunswick Laboratory
- Wayne

New Mexico
- Acid/Pueblo Canyons*
- Bayo Canyon*
- Chupadera Mesa*

New York
- Ashland Oil 1
- Ashland Oil 2
- Baker & Williams Warehouse*
- Bliss & Laughlin Steel
- Colonie
- Linde Air Products
- Niagara Falls Storage Site*
- Niagara Falls Storage Site Vicinity Property*
- Seaway Industrial Park

Ohio
- Alba Craft
- Associated Aircraft Toll & Manufacturing
- B&T Metals
- Baker Brothers
- HHM Safe Company
- Luckey
- Painesville

Oregon
- Albany Research Center*

Pennsylvania
- Aliquippa Forge*
- C.H. Schnoor*

Tennessee
- Elza Gate*

*completed remediation and certified in the Federal Register to be released from FUSRAP program

3.24 OPERATIONAL NUCLEAR WEAPONS COMPLEX

The current DOE nuclear weapons complex consists of 8 major facilities located in 7 states. Currently, the Complex maintains a limited capability to design and manufacture nuclear weapons; provides surveillance of and maintains nuclear weapons in the stockpile; and retires and disposes of nuclear weapons. Major facilities and their primary responsibilities within the Complex are listed below:

1) **Pantex Plant** (Amarillo, Texas) - Dismantles retired weapons; fabricates high explosives components; assembles high explosives, nuclear components, and non-nuclear components into nuclear weapons; repairs and modifies weapons; evaluates and performs non-nuclear testing of nuclear weapons.
2) **Savannah River Site** (SRS) (Aiken, South Carolina) - Tritium loading/unloading and surveillance of tritium reservoirs.
3) **Y-12 Plant** (Oak Ridge, Tennessee) - Maintains the capability to produce and assemble uranium and lithium components; recovers uranium and lithium materials from the component fabrication process and retired weapons; produces non-nuclear weapon components.
4) **Kansas City Plant** (KCP) (Kansas City, Missouri) - Manufactures non-nuclear weapons components.
5) **Lawrence Livermore National Laboratory** (LLNL) (Livermore, California) - Conducts research and development of nuclear weapons; designs and tests advanced technology concepts; maintains a weapons design program; maintains a limited capability to fabricate plutonium components; provides safety and reliability assessments of the stockpile.
6) **Los Alamos National Laboratory** (LANL) (Los Alamos, New Mexico) - Conducts research and development of nuclear weapons; designs and tests advanced technology concepts; maintains a weapons design program; maintains a limited capability to fabricate plutonium components; provides safety and reliability assessments of the stockpile.
7) **Sandia National Laboratories** (SNL) (Albuquerque, New Mexico) Conducts system engineering of nuclear weapons; designs and develops non-nuclear components; conducts field and laboratory non-nuclear testing; manufactures non-nuclear weapons components; and provides safety and reliability assessments of the stockpile.
8) **Nevada Test Site** (NTS) (Las Vegas, Nevada) - Maintains capability to conduct underground nuclear testing and non-nuclear experiments.

4

Other Locations

4.1 OTHER FEDERAL AGENCY SITES

4.1.1 Army Corps of Engineers (CoE)

The Army Corps of Engineers is responsible for various sites that utilize radioactive materials. Fourteen sites located in 12 states are under the CoE jurisdiction. Little, if any, residual radioactivity is expected at these sites due to the use of measuring instruments with sealed radioactive sources.

4.1.2 Department of Agriculture (DoA)

The DoA's Agriculture Research Center in Beltsville, Maryland is reported to have used radioactive materials in simulating the effects of atomic weapons fallout on crops. Similar tests were carried out at military bases wherein short lived radionuclides were dispersed on land, buildings, vehicles, crops and roads to assess various removal methods. Since the radionuclides used are relatively short lived, no residual radioactivity is expected at these sites. Aside from a research site in Hyattsville, Maryland, the Beltsville site appears to be the only potentially contaminated DoA site.

4.1.3 Department of Commerce (DoC)

The DoC, through the U.S. Maritime Administration, controls the Nuclear Ship Savannah which has undergone D&D and is now stationed at Charleston, South Carolina. The DoC also controls, through the National Bureau of Standards (NBS), the Center for Radiation Research at the Bureau. This Gaithersburg, Maryland site includes a reactor and several accelerators. The total number of potentially contaminated sites controlled by DoC, including various laboratories and food inspection sites, is 23. Currently, the DoC is undertaking a study to more accurately inventory sites that may be radioactively contaminated. Residual radioactivity is expected to consist primarily of fission and activation products at the reactor and accelerator sites.

4.1.4 Department of Defense (DoD)

The U.S. Department of Defense through its Departments of Army (including the Army National Guard), Navy (including the Marine Corps), and Air Force (including the Air National Guard) controls a large number of sites both in and outside the conterminous United States. Additional military sites are controlled by the Department of Transportation through the U.S. Coast Guard.

Military facilities range in size from single buildings to large forts and bases which may cover areas as large as a few million acres. These complexes cover a wide range of functions including schools, hospitals, training academies, research and development laboratories, proving grounds, bombing and gunnery practice ranges, storage depots, arsenals, air bases, naval bases, missile launch sites, forts, and manufacturing sites for weapons and ammunition. Some sites are also used for storage of strategic materials for national stockpiles.

Most of the residual radioactivity at military sites is a result of research and development testing of military munitions, testing and operation of military reactors, or accidents. Sites may be contaminated with plutonium and fission products over large areas, or may have used or stored small quantities of radioactive materials in the form of luminous dial watches, compasses, electron tubes, and lights in electric equipment. Still others have been contaminated with depleted uranium munitions, but vary widely in character.

The DoD's Defense Environmental Restoration Program (DERP) has been ongoing since 1983 to restore active (DERP) and formerly utilized defense sites (DERP/FUDS). The Defense Environmental Restoration Program has been codified into law as part of Superfund. There may be very few sites where radioactive wastes have been buried on site, but little information is available regarding deliberate on-site burials.

TABLE 4.1

Department of Defense Sites

Site	Source of Residual Radioactivity
Alaska	
Ft. Greeley	Activation/fission products
Alabama	
Redstone Arsenal	Accelerator
California	
Army Ionizing Radiation	Accelerator
Camp Parks	Sr-90 in hot cell
Camp Roberts	Depleted uranium
China Lake Naval Weapons Center	Depleted uranium
Long Beach Naval Shipyard/Base	Activation/fission products
Mare Island Naval Shipyard	Activation/fission products
Naval Electronics Lab	Accelerator
Naval Post Graduate School	2 Accelerators
Port Hueneme	Activation/fission products
San Diego Naval Base	Activation/fission products
Connecticut	
New London Submarine Base	Activation/fission products
District of Columbia	
Naval Research Lab	13 Accelerators
Naval Research Lab, Reactor	Activation/fission products
Walter Reed Research Reactor	Activation/fission products
Florida	
Eglin Air Force Base	Depleted uranium
Hawaii	
Pearl Harbor Naval Shipyard and Submarine Base	Activation/fission products

(continued)

Table 4.1: (continued)

Site	Source of Residual Radioactivity
Indiana	
Crane Naval Weapons Support Ctr	Thorium
Jeffersonville Depot	Zircon sands
Iowa	
Army Ammunitions Plant	Depleted uranium
Maine	
Portsmouth Naval Shipyard	Activation/fission products
Maryland	
Aberdeen Proving Ground	Depleted uranium
Aberdeen Pulsed Reactor	Activation/fission products
Armed Forces Radiobiology Research Institute	Activation/fission products, transuranics
Army Chemical Center	Accelerator
Diamond Ordnance Radiation	Activation/fission products
Edgewood Arsenal	Accelerator
Naval Medical Center	Accelerator
Naval Ordnance Lab	Accelerator
Massachusetts	
Army Quartermaster Depot	2 Accelerators
Michigan	
Detroit Arsenal	Accelerator
Nevada	
Nellis Air Force Base	Depleted uranium, plutonium, fission products
Fallon Naval Air Station	Shoal underground nuclear weapons test

(continued)

Table 4.1: (continued)

Site	Source of Residual Radioactivity
New Jersey	
Picatinny Arsenal	Accelerator
New Mexico	
Kirkland Air Force Base	Unknown
White Sands Missile Range (Trinity Site, Fast Burst Reactor)	Activation/fission products
New York	
Watervliet Arsenal	Accelerator
Ohio	
Wright-Patterson Air Force Base	Activation/fission products, accelerator, Am-241
Texas	
Fort Worth (Aerospace Systems Test Reactor, Ground Test Reactor)	Activation/fission products
Medina Base	Depleted uranium
Utah	
Hill Air Force Base	3 Accelerators
Virginia	
Ft. Belvoir	Activation/fission products
Newport News Naval Shipyard	Activation/fission products
Washington	
Puget Sound Naval Shipyard	Activation/fission products
Sandpoint Naval Station	Ra-226

Defense Environmental Restoration Program (Active Sites): The Defense Environmental Restoration Program (DERP) is an outgrowth of the overall Installation Restoration Program (IRP). Active sites may have segments that are inactive or which may have been decontaminated. In such cases the overall site is still considered an active site.

1) Bases - Bases can be large, sprawling complexes where many and varied activities have been carried out. Some of the military sites such as hospitals, research and development laboratories, and schools will continue in operation for the indefinite future. Others have already been taken out of service and decontaminated and decommissioned, but are still part of the active base. There are 113 military bases/camps/arsenals with expected residual radioactivity. In addition, 34 military reservations have been used to stockpile strategic materials under the management of the Government Services Administration.
2) Power Production - Most of the military nuclear reactors were designed to produce electricity and heat and, with the exception of nuclear ship reactors, have been shut down or dismantled. These power plants were typically used to service remote installations. There were 6 such sites shown in Table 4.2. Residual radioactivity at the non-operating reactors is primarily activation products. Except for the PM-3A site in Antarctica, the waste from which has been sent to the Naval Center at Port Hueneme, California, waste volumes and inventories are not available.

TABLE 4.2

Department of Defense Power Reactors for Remote Locations

Name	Location
Stationary Medium Power Plant No. 1A	Alaska
Portable Medium Power Plant No. 3A	Antarctica
STURGIS Floating Nuclear Power Plant	Canal Zone
Portable Medium Power Plant No. 2A	Greenland
Stationary Medium Power Plant No. 1	Virginia
Portable Medium Power Plant No. 1	Wyoming

3) Propulsion - The U.S. Navy has constructed approximately 150 nuclear submarines and about a dozen surface ships. To support its nuclear powered ships, the Navy has 11 shipyards, 13 tenders, and 2 submarine bases for a total of approximately 174. Residual Radioactivity consists primarily of activation and fission products. In addition, low levels of radioactivity (principally Co-60) are also usually present in harbor

sediments where ships are serviced. This is true not only for the shipyards listed in Table 4.1, but also for other nuclear ship bases such as those at Guam, Scotland, and possibly others.

4) Research Labs - The DoD has operated several small test and research reactors for simulating the effects of nuclear weapons and for other physical and medical research. Most of these have been shut down or dismantled. There are 15 such sites as indicated in Table 4.3. Residual contamination at non-operating reactors consists primarily of fission and activation products. Remediation efforts at recently shut down reactors will contend with spent fuel and fresh fission products.

TABLE 4.3

Department of Defense Test and Research Reactors

Name	Location
U.S. Navy Postgraduate School	California
Naval Research Center	District of Columbia
Walter Reed Research Reactor	District of Columbia
Radiation Effects Reactor	Georgia
Pool Type Reactor	Georgia
Army Materials Research Center	Massachusetts
Aberdeen Pulsed Reactor	Maryland
Armed Forces Radiobiology Research Institute	Maryland
Diamond Ordnance Facility	Maryland
U.S. Naval Hospital	Maryland
Fast Burst Reactor	New Mexico
Nuclear Engineering Test Reactor	Ohio
Aerospace Systems Test Reactor	Texas
Ground Test Reactor	Texas
Reactivity Test Assembly	Texas

5) Weapons Testing - There are several nuclear weapons test sites where missile, gunnery and bomb testing is performed. Tests can be both surface and underground, on-site and off-site. There are at least 2 sites where nuclear bombs were detonated, and approximately 11 sites where depleted uranium shells have been fired. In addition, there is one site where nuclear weapons have been assembled and stored, for a total of 14. Residual radioactivity from bomb testing is expected to range widely and include fission products as well as plutonium. The Nellis Air Force Base and Nellis Bombing and Gunnery Range encompass about 3 million acres, portions of which are contaminated by fallout from atmospheric nuclear weapons and weapons safety tests. Residual radioactivity can also be

present in the form of shell fragments (from projectiles that incorporated depleted uranium), storage and waste areas, and contaminated soils.

6) Accidents of Weapons Carriers - Very little information has been released by the DoE or the DoD on residual radioactivity associated with accidents involving weapons carriers in the United States. A few accidents are known to have had residual radioactivity associated with them, some on sites already contaminated with radioactivity, but essentially no unclassified information has been reported. Some accidents may also have involved other radioactive, but non-fissionable radionuclides (e.g., tritium). Estimates of the total number of weapon accidents range up to more than 50. The extent of residual radioactivity at nuclear weapons accident sites is unknown. Possible contaminants would be plutonium, enriched uranium and tritium.

Defense Environmental Restoration Program/Formerly Utilized Defense Sites (Inactive Sites): The DERP Formerly Utilized Defense Sites (FUDS) activity is managed by the U.S. Army Corps of Engineers. Included are efforts related to hazardous and toxic/radiologic wastes, ordnance and explosive waste and building demolition on lands formerly owned or used by any DoD component for which DoD is responsible.

4.1.5 Department of Health and Human Services (DoH&HS)

The DoH&HS operates the National Institutes of Health (NIH) in Bethesda, Maryland. The research sites at NIH include several accelerators, however, the combination is considered as one site at this time pending receipt of additional information. The primary contaminants consist of targets or material struck by the accelerator beam, beam stops, pipes, shielding materials, vaults, and soil surrounding the underground storage vaults. In addition to the NIH, the DoH&HS is responsible for 45 other sites located in 24 states, Puerto Rico and Washington, D.C. These sites consist mainly of research centers managed by the Food and Drug Administration and the Center for Disease Control, and branches of the Public Health Service.

4.1.6 Department of Interior (DoI)

Three inactive uranium mill sites have been identified on DoI land and are under the cognizance of the Bureau of Land Management. These sites are not included in the DoE UMTRA Program. The DoI, through the U.S. Geological Survey, also operates a Triga reactor at its site in the Federal Center in Denver, Colorado. In addition, the DoI is responsible for 84 other sites including those managed by the Bureau of Indian Affairs, Land Management and Reclamation, the National Park Service, and the Fish and Wildlife Service, most of which use sealed source radioactive devices. All told, the DoI has responsibility for 88 sites located in 35 states and Puerto Rico. There is also residual radioactivity at the DoI Albany Site, managed by DoI's Bureau of Mines, which is listed as a FUSRAP site. Several other Bureau of Mines sites are also FUSRAP sites due to early involvement in the Manhattan Project.

4.1.7 Department of Justice (DoJ)

The Department of Justice operates 4 sites, 2 in Virginia and one each in Puerto Rico and Washington, D.C. These include offices of the Drug Enforcement Agency, the Federal Bureau of Investigation, and the Institute of Forensic Sciences. Such sites include research areas and utilize measuring devices containing radioactive materials. Residual radioactivity is expected to consist of typical lab wastes (vents, gloves, coats, etc.).

4.1.8 Department of Labor (DoL)

The Department of Labor is licensed to possess nuclear materials and thus is likely to own or manage potentially contaminated sites. Through the Mine Safety and Health Administration and the Occupational Safety and Health Administration, the DoL has a total of 7 sites in 6 states under its jurisdiction. These sites utilize various measuring devices with sealed sources and typical lab equipment. Minimal residual radioactivity is expected.

4.1.9 Department of Transportation (DoT)

The Department of Transportation is responsible for 15 sites located in 10 states. Included are sites of the Maritime Administration, the Federal Highway Administration, and various U.S. Coast Guard Cutters. Such sites utilize various research labs and measuring devices with sealed sources. Minimal residual radioactivity is expected due to the nature of the materials used.

4.1.10 Department of the Treasury

The Department of the Treasury is responsible for 3 sites in 2 states and Washington, D.C. These sites include an office and laboratory of the U.S. Customs Service and a Bureau of Alcohol, Tobacco, and Firearm's site. Minimal residual radioactivity is expected since the materials in use, typical lab supplies and measuring devices, have a low likelihood of releases from their sealed sources.

4.1.11 Environmental Protection Agency (EPA)

Active Sites: The EPA manages a laboratory in Montgomery, Alabama, where surplus Ra-226 sources were stored at one time. The EPA also has a laboratory in Las Vegas, Nevada where a drum of uranium mill tailings was stored. Residual radioactivity at these EPA sites is in the form of uranium mill tailings as well as soil, equipment, piping or clothing contaminated from the leaking Ra-226 sources. The EPA also operates 26 other sites located throughout 20 states. These sites encompass various regional offices, the National Enforcement Investigations Office, research labs, and the Toxicant Analysis Center in Mississippi. Contaminants would depend upon the specific type of site in question, with typical lab waste (e.g., gloves, hoods, coats, etc.) most likely present.

Superfund (Inactive Sites): Under the Comprehensive Environmental Response, Compensation and Liability Act (CERCLA), EPA has the authority to require cleanup of most radiological releases from private and Federal sites.

4.1.12 Government Services Administration (GSA)

Prior to 1979, the GSA was responsible for managing the National Stockpile Storage Sites of strategic materials. With the formation of the Federal Emergency Management Agency (FEMA) in 1979, responsibility for the stockpiles was given to FEMA, but GSA still retains responsibility for management of the stockpiles. The stockpiles include ores of thorium, chromium, copper, cobalt, magnesium, zirconium, and other minerals. These ores are often associated with elevated concentrations of thorium and uranium. The number of sites in this category number 111. It is not known how many sites contain ores with elevated concentrations of radioactivity. The materials are contained at 29 GSA depots, 34 military depots, 14 other government depots, and 35 plants and other sites, including one office located in Washington, D.C. Residual radioactivity at the stockpiles of strategic materials are similar to that at a uranium mill. The primary concern is contamination of soil and equipment. Primary nuclides of concern are radium, uranium and thorium.

4.1.13 National Air and Space Administration (NASA)

NASA previously operated 3 reactors at its Cleveland and Sandusky, Ohio sites and an accelerator at its Cleveland site. Residual radioactivity at NASA sites is likely to be similar to that at a typical test reactor site. Waste storage rooms, hot cells, core structural and shielding components and piping, among others, are all possible sources of residual radioactivity. The primary nuclides present would be Co-60, Zn-65 and Nb-94. In addition to the sites noted above, NASA operates 9 other complexes in 8 states. These include various space and flight centers, research centers and offices. Contaminants would vary depending upon the site.

4.1.14 Postal Service (PS)

The Postal Service owns a site in Boulder, Colorado which it believes is contaminated with mill tailings. Since this is an active site, it is not included under the UMTRA Program.

4.1.15 Tennessee Valley Authority (TVA)

TVA operates several power reactors, owns an inactive uranium mill and is involved in the phosphate industry. A total of 22 sites have been identified in 3 states, the majority of which (13) are located in Tennessee. Residual radioactivity at TVA sites is varied depending upon the type of operation. Mill tailings, phosphogypsum piles and typical reactor contamination (e.g., shield, structural supports, labs, etc.) are included, as well as the radionuclides associated with those type of sites (e.g., uranium, thorium, Co-60).

4.1.16 Veterans Administration (VA)

The VA is responsible for approximately 121 sites that may require D&D, including a Triga reactor at the Omaha, Nebraska VA Hospital and an accelerator at the VA Hospital in Minneapolis, Minnesota. It is assumed that the VA operates other accelerators and radiation therapy sites through how many could not be determined from the readily available reference material. For example, the Bureau of Radiological Health is responsible for long-term storage of radium needles. VA sites would likely have residual contamination typical of reactors and accelerators. Structural supports, beam targets, shielding, and lab waste are all possible sources at these sites. Nuclides of concern include Co-60 and Fe-55.

4.2 NRC/AGREEMENT STATE SITES

Sites within this category consist of a portion of what is commonly referred to as the "nuclear industry," i.e., those civilian nuclear energy activities that require a comprehensive regulatory program to assure that they will be conducted in a manner that will protect public health and safety. Non-DoE Federal sites also require NRC or Agreement State licenses for possession of radionuclides. All sites mentioned below are, or have been, in possession of a NRC or Agreement State license in the conduct of their activities. Except in a

NRC/Agreement States; actual responsibility resides with the individual licenses. The number of sites identified as NRC/Agreement State licensees totals almost 19,000.

4.2.1 Medical Sites

Two categories of medical sites were identified and are described below: Hospitals and Medical Centers, and Nuclear Pharmacies.

1) Hospitals and Medical Centers - Typical sites may contain accelerators and use radionuclides and radionuclide devices in the diagnostics and treatment of patients. Of these, as many as 100 institutions may operate incinerators used to reduce the volume of low-level waste. There are over 5,600 sites in this category. Additionally, there is at least an equal number of medical support sites and functions (e.g., nuclear medical vans, veterinary sites, etc.) that are licensed by, or registered with, the States. Residual radioactivity at hospitals and medical centers is expected to take the form of lab bench tops, cabinets, vaults, piping, vents, etc., as well as shielding and other materials associated with the use of accelerators. Relatively minor residual radioactivity is expected.
2) Nuclear Pharmacies - Typical sites act as distributors of products between radio-pharmaceutical manufacturers and users, e.g., hospitals and medical centers. These sites utilize predominately isotopes with short half-lives, with long-lived isotopes used to a lesser extent. Those nuclear pharmacies are located within the confines of hospitals and universities. Typical residual radioactivity would take the physical form of fume hoods, filters, ductwork, and miscellaneous supplies, trash, and cleaning solutions.

4.2.2 Manufacturing Plants

Three categories of manufacturing sites indicate the diversity of ongoing activity: Radiation Device and Consumer Products, Radio-Pharmaceutical/Materials, and Radioactive Sealed Source. These categories are described below.

1) Radiation Devices and Consumer Products - Typical sites are involved in the production of products that use sealed sources, such as self-luminous products (e.g., emergency lighting signs). There are approximately 9,500 sites in this category. This figure is believed to be a slight over-estimate because it reflects the total number of industrial licensees, including State licensees. Residual contamination at these sites is associated with the release of the radioactive materials, such as tritium or krypton 85, from the sealed sources.
2) Radio-Pharmaceutical/Materials - Typical sites label compounds in batches, with each step in the process usually in a separate

enclosure. The sites may contain numerous labs for the different isotopes used and products manufactured. There are approximately 680 sites in this category. Residual contamination is expected to reside in the form of lab bench tops, hoods, vents, filters, floors where spills have occurred, and other lab equipment.

3) Radioactive Sealed Source - Typical sites manufacture sources to be used as reference standards, therapy units, and gamma irradiation sources, among others. These sites usually utilize long-lived isotopes and/or isotopes with high activities. Eleven active sites have been identified, although at least 17 are known to have been operating at one time. It is not known if the 6 outstanding facilities have been released for unrestricted use after decommissioning, are in the process of being decommissioned, or were found not to be contaminated at all. The exact number and distribution of these manufacturers could not be determined from the available reference material. Residual radioactivity takes the form of hot cells, remote handling devices, vents, and surfaces as a result of spills.

4.2.3 Non-Defense Research Laboratories

These sites can be divided into 3 categories of laboratories: those that use primarily sealed sources and/or low quantities of unsealed radioactive materials; those that use high-activity sealed sources; and those that use large curie quantities of radionuclides, some of which are long-lived in unsealed form. There are over 2,500 sites in this category. The residual radioactivity ranges from minimal, requiring disposal of small quantities of radioactive materials, to that requiring major decontamination, including removal of laboratory equipment, components, and structures.

4.2.4 Nuclear Power Reactors

This category consists of light-water reactors used in the production of commercial electrical power. Typical complexes consist of reactor, containment, cooling and power generation components. Counting the 9 Tennessee Valley Authority units, there are 119 sites of this type in the United States either shut down, in operation, or under active construction. Less than a half dozen are planned to undergo D&D in the near future. Concerns include long-lived radionuclides from activation and fission products resulting in residual radioactivity of relatively large quantities of piping, hardware and concrete.

4.2.5 Nuclear Research and Test Reactors

This category consists of non-power reactors licensed by the NRC for medical therapy and research and development. There are approximately 71 sites of this type, including several owned by Federal agencies (e.g., NASA and VA). The sites vary in size, type, and complexity. Residual radioactivity occurs in structural components (e.g., beam tubes, reactor tank walls), storage areas, and laboratories, among others.

4.3 STATE SITES

Sites within this category consist primarily of consumer product and commodity manufacturers, mines, oil and gas production sites, power plants, research sites and water treatment plants not under the authority of the NRC/Agreement States. In some cases, the States issue permits and licenses to operators. In other cases, licenses have not been issued either because the residual radioactivity levels are sufficiently low to not be a public or worker health problem, or the newness of the issues involved (e.g., NORM).

4.3.1 Manufacturing Plants

1) Radiation Devices and Consumer Products - This category includes manufactured products that incorporate by-product materials produced by the NRC/Agreement State licensees into the finished goods, such as self-luminous devices, gas and aerosol detectors, static eliminators, measuring and controlling devices, etc. The manufacturers are licensed by the State to manufacture the product, and granted a general license by the NRC to distribute the product. The manufacturer itself issues a "general license" for possession and disposal to the consumer. Residual radioactivity would be typical of that of sealed source manufacturers and would exist as a result of leaky sources.
2) Phosphate Production Plants - The phosphate cycle consists of mining, processing and product formation. Mined phosphate rock (ore) is processed by washing, flotation and drying. It is then transformed into elemental phosphorous or into phosphoric acid for fertilizers, detergents, and so on. There are approximately 24 mines and 31 processing and manufacturing sites in the United States. The process creates a slurry which subsequently is discharged onto waste piles (phosphogypsum stacks). Since the ore contains naturally occurring radioactive material (NORM), this process tends to concentrate this material and create elevated levels of contaminants. There are approximately 63 phosphogypsum stacks, making a total of 118 sites in this category.

Minerals Processing - As with uranium mining, the processing of ores rich in aluminum, copper, nickel, zinc and other minerals results in the concentration of naturally occurring radioactive materials (NORM) as a result of the techniques used. There are approximately 108 sites where concentrated NORM exists as a result of the mineral extraction process.

Uranium - These facilities extract uranium ore from above and below ground mines. Mining wastes are generally segregated into tailing waste piles and subore piles, the latter being natural materials extracted from the earth enroute to the ground depth of interest. Subore piles typically contain uranium (and uranium daughter products) in lower concentrations than is economically feasible to process. Most waste stays on-site and the amount generated depends upon the mining method used, the richness of the ore, and the economic conditions at the time. Only 4 of the approximately 3,737 uranium mines in the United States are active currently.

4.3.3 Oil and Gas Production

There are over 1.5 million oil and gas wells in the United States. Water associated with the extraction of oil and gas from the earth contains elevated levels of naturally occurring radioactive material (NORM). Over time, the insides of extraction pipes become coated with a concrete-like substance called "scale." Pipe scale can be very high in Ra-226 and Ra-228. Residual radioactivity takes the form of discarded pipes either left on-site to rust, sent to scrap yards where they may be reused, or recycled on-site where scale is removed and dumped or stored on-site in 55 gallon drums.

4.3.4 Power Plants

1) **Fossil Power** - Coal-fired units are used in the production of electrical power. The use of coal fuel results in the accumulation of naturally occurring radioactive material (NORM) in the fly ash and bottom ash which are collected and impounded on or off-site. There are approximately 1,300 utility coal-fired boilers and 51,100 industrial coal-fired boilers throughout the nation, for a total of approximately 52,400. The concern with NORM is the presence of long-lived radionuclides in potentially recyclable ash. This can be used by a variety of industries (e.g., concrete, wallboard) which may result in elevated radiation levels in structures utilizing such materials.
2) **Hydrothermal** - The pipe scale issues discussed earlier apply to the category of hydrothermal power plants as well. The number

and distribution of these plants could not be determined from readily available references.

4.3.5 Research

It is estimated that there are perhaps 1,200-2,000 atomic particle accelerators operating within the United States. Accelerators are found in every State with broad application in physics, chemistry, radiobiology, medical radiation therapy, radiation processing and sterilization, industrial radiography and ion implantation for integrated electron circuit fabrication. Of the large number of accelerators, approximately 150 have relatively high beam energy levels (>10 MeV) and are either licensed by the NRC/Agreement States or belong to the military. The larger machines create a category of waste called NARM (Naturally occurring and Accelerator produced Radioactive Materials). Those that remain are the relatively small accelerators generally exempt from NRC/Agreement State regulation. There are about 1,850 accelerators in this category. At the lower energy levels, there is insufficient energy to create significant activation products. It is expected that residual radioactivity can be readily managed by natural decay.

4.3.6 Water Treatment

There are approximately 50-60,000 water supply companies in the United States. Approximately 3,300 such companies obtain water from underground sources, about 700 of which have elevated levels of radionuclides. The process of treating these waters creates various waste forms: a sludge, ion-exchange resins, granulated activated carbon and reject water from filter backwash. If the groundwater originally had elevated levels of radioactivity, the resulting wastes would also be radioactive. Residual radioactivity takes the form of dissolved and suspended naturally occurring radioactive materials (NORM) that concentrate in the sludge. The sludge is typically dumped locally or sold to firms that produce fertilizer.

4.3.7 Other

There are certain facilities once licensed by the Atomic Energy Commission (AEC, the predecessor to NRC) or Agreement States that have since been taken over by States.

5

Storage and Disposal

Storage is defined as the emplacement of waste with the intent to retrieve it at some later time. Disposal, on the other hand, means the emplacement of waste without any intention of subsequent retrieval, or, discharge and dispersal of the waste into the environment. Recently the concept of long-term storage has been gaining support in waste management circles. From a practical aspect, long-term storage can be considered a modified form of disposal where active management of the facility is not required (as in the case of a storage facility), but the facility is monitored and appropriate remedial actions (including retrieval of the waste, if necessary) are taken if radionuclide migration is detected. Storage practices currently in use are based on matching the waste characteristics (radiation hazard and physical/chemical characteristics) to an appropriate facility design. The applicable regulatory criteria are equally important.

Some anti-nuclear advocates favor permanent storage, rather than deep disposal. Their argument is that perpetually visible waste, that requires constant monitoring and protection, is safer over the forthcoming centuries, and there will be no inadvertent intrusion as could happen with deeply disposed waste.

Most types of nuclear wastes require some type of treatment to facilitate ultimate disposal and isolation. The processes for treatment generally fall into 3 broad categories: storage to allow radioactivity to decay; volume reduction to reduce shipping, handling, and disposal costs; and immobilization to minimize the spread of radioactivity.

5.1 INTERIM STORAGE OF DOE WASTES

Typically, radioactive and mixed wastes generated at DOE sites have been or are confined in underground tanks, trenches, etc., until resources and facilities for their final disposal are available. There is concern that some natural man-made event may result in a breach of such confinement and inadvertent release of radioactive and hazardous chemicals will occur. Measures to prevent such releases or to mitigate their effects are of great interest and importance.

Transport by surface and groundwater provides the dominant mechanism for dispersion of radionuclides and other materials inadvertently released from interim confinement. Consequently, strategies to assure safe interim confinement involve isolation of confinement facilities from surface and groundwaters; stabilization of toxic species relative to water transport; and treatment of contaminated water as it exits the interim confinement location.

Defense HLW from Idaho Chemical Processing Plant (ICPP) is routinely calcined as it is generated; the dry powdered calcine is stored pending final disposal. Defense HLW from Hanford and Savannah River are stored in underground tanks. All defense HLW will eventually be vitrified in borosilicate glass in preparation for ultimate disposal.

TRU wastes are not typically incorporated into specific waste form matrices. Generally, residues remaining after volume-reduction operations such as compaction, shredding, or incineration of defense TRU wastes will be stored in 208-L (55-gallon) drums for eventual transportation to the Waste Isolation Pilot Plant, a salt geologic repository near Carlsbad, New Mexico.

Future pretreatment of DOE wastes stored at the Hanford, Savannah River, and Idaho sites may generate concentrated TRU element fractions. New forms may be required for final disposal of such concentrated TRU wastes.

Diversion of Surface and Groundwater: One approach for isolating radioactive waste confined in tanks, trenches, drums, bins, etc., from surface and groundwater is to divert water around the confinement facilities. Types of barriers that might be emplaced around confinement facilities include foam, grout sheets, electroosmotic barriers, and vitrified materials. Considerable basic research is needed to develop foam barriers with properties suitable for use in practical situations, and to identify the constraints imposed on foam formation and longevity by the chemical and hydrological characteristics of soils and sediments of the interim confinement region. Similar basic research is required in the case of microbe-produced gelatinous barriers and other types of barriers.

Physical Barriers: At least some of the radioactive and mixed wastes in interim confinement facilities could be immobilized in situ by the introduction of chemicals or cementitious grouts or polymer formers. Such treatment would greatly reduce the permeability of the waste to intruding water, but would also likely greatly complicate any subsequent removal and treatment of the waste for final disposal. Substantial basic research is needed to design new materials for polymer and grout formulations that would be sufficiently stable over extended periods of interim confinement and that would achieve the desired reduction in permeability of the wastes.

Chemical Barriers: Basic research is needed to identify and develop appropriate materials and chemical barriers that might be emplaced around interim confinement facilities containing DOE wastes. In all cases, the chemical barrier material would be emplaced to intercept any water that has come into contact with stored wastes and to stabilize the contaminants being transported by the water. Such barriers include beds of sorbents; biological materials; chemical materials that make use of chemical reactions such as precipitation to immobilize radionuclides; and ion-exchange curtains.

A particularly exciting adaptation of chemical barrier strategy involves use of columns of sorbents or ion-exchange resin as vertical drain wicks. Particular emphasis should be given to development of chemical barriers that make use of ion-exchange principles and techniques. New types of thermodynamically stable ion-exchange resins with silica matrices or new inorganic ion-exchange material should be designed and synthesized. These new ion-exchange materials will need to have high selectivities for actinides and selected fission products, e.g., Sr-90, Tc-99, and Cs-137, over Na, K, Ca, and Mg.

Colloid Formation, Interactions, and Transport: Transport of radionuclides and hazardous chemicals from breached interim confinement facilities can also occur through the formation of, or as the result of, attachment to colloid particles. There is, thus, the need for research to develop a more complete understanding of colloid formation, colloid-contaminant interactions, and of transport of colloid particles through heterogeneous porous media. Colloid-contaminant interactions should be investigated both for the purpose of enhancing sorption or incorporation as well as with a view toward eliminating motion of colloids as a mechanism for transport of contaminants.

Radiation Chemical Effects: A phenomenological parameter base should be developed to allow predictions of the production of hazardous gases or solids as a consequence of radiation fields in multi-component, multi-phase heterogeneous systems characteristic of certain DOE radioactive wastes. This parameter base should address: (1) effects of major components in wastes (at realistic concentrations) on the rate of formation and distribution of products; (2) effects of the interfaces of particulate suspensions (aluminates, silicates, and metal hydroxides), (3) effects of temperature on selected reaction rates and yields for systems that prove to be of the most importance in the production of undesirable products, and (4) delineation of the reaction mechanisms for the interaction of the major waste components, both inorganic and organic, with the radiation-produced species.

Depleted Uranium: DOE has 2 alternatives for depleted uranium—continued storage as uranium hexafluoride; and conversion to uranium metal, and fabrication to shielding for spent nuclear fuel containers.

5.2 IN SITU DISPOSAL OF DEFENSE WASTES

It presently appears that at least some of the U.S. defense waste may eventually qualify for in situ disposal in the tanks and trenches where it is presently located. Wastes in some of the 149 single-shell tanks at the Hanford site are likely candidates for such in situ disposal. Also at the Hanford site there is a large amount of TRU waste, generated in the 1944 to 1970 period, that was placed in near-surface open trenches that were later back-filled. These latter TRU wastes likely include both combustible (e.g., paper cartons, cloth, loose paper, etc.) and noncombustible (glass, metal, etc.) materials.

In situ waste disposal concepts typically involve treatment of the waste to destroy incinerables, convert nitrates and nitrites to NO_x, convert residues, insofar as possible, to a form of low water leachability, and emplacement of suitably designed and constructed engineered barriers and markers over the disposal site.

Much of the technology needed to accomplish in situ disposal of selected U.S. defense wastes is in hand. The in situ vitrification (ISV) process, a leading candidate for immobilization of certain wastes in tanks and trenches, is ready for advanced field-scale testing. Preliminary designs have been made of multi-layer barriers that might be emplaced upon waste disposed of in situ, further development of barriers, including basic research studies of barrier natural analogs, has been described. One additional area where fruitful basic research studies should be performed is to investigate reagents and technology, other than ISV, that could be used to immobilize wastes in tanks and trenches. In situ vitrification of high-level tank wastes is discussed in Chapter 13.

5.3 INTERIM STORAGE OF SPENT FUEL

Nuclear power generation creates significant amounts of radioactive waste. Highly radioactive waste, which is a small portion of the total waste produced, contains most of the radioactivity. One type of highly radioactive waste is used, or spent, fuel, which is taken from nuclear reactors after it can no longer efficiently sustain a nuclear chain reaction.

The United States banned the reprocessing of spent fuel in 1977, but lifted the ban in 1981. However, no reprocessing has been conducted since that date. Therefore, spent fuel has been stored at utilities until such time as a geologic repository becomes available. The proposed Yucca Mountain repository will not be available until 2010, or possibly 2020. Only four countries currently extract plutonium from spent fuel—France, India, Russia, and the United Kingdom. These four countries are very concerned about the plutonium diversion problem.

The DOE is also developing plans for the siting and development of a potential Monitored Retrievable Storage (MRS) facility. The MRS facility could be used to receive and store spent fuel from commercial power reactors for subsequent shipment to a repository when such a facility becomes operational. However, opposition to locating this project is quite strong.

Conventional storage technology consists of lowering fuel elements into 40 feet of water in basins with reinforced concrete walls 6 feet thick that are lined with stainless steel. The water both shield workers and cools the fuel. Some spent fuel elements have been stored as long as 25 years without harm to the elements or the environment. Each power plant has a storage basin; away-from-reactor basins are also available.

Other techniques for dry storage of spent fuel have also been investigated and may prove useful. These approaches consist of storing fuel in heavy concrete or metal vaults, casks, or caissons and cooling it with air or inert gas through forced or natural convection. Such techniques have been used successfully in other countries. Dry storage is already in place at four U.S. nuclear power plants, and certain DOE facilities.

Utilities can almost double their spent fuel storage capacity by changing the configuration of the racks that hold spent fuel in storage pools (reracking) and adding neutron-absorbing material. A utility may also "transship" fuel to another power plant in its system that has unused capacity. However, there is a limit to the amount of relief these methods can provide. Although estimates of the requirements for near-term spent fuel storage have declined because of reracking and innovative fuel cycles that reuse previously discharged fuel assemblies, significant requirements for additional storage capacity are projected over the next decade.

There is a serious problem with spent nuclear fuel at Hanford. Spent fuel has been stored for many years in 2 obsolete concrete water basins that are deteriorating, posing a considerable safety problem. It is expected that the radioactive fuel would be removed and placed in a multi-canister wet overpack, essentially placing several fuel canisters in a larger "box" filled with water. Between 600 and 1,000 of these overpacks would be needed to empty the basins. The overpacks would be transported by rail to a new vault to be constructed on the Hanford sites and held there until a processing facility is built. There, the fuel would be stabilized by drying and chemical treatment. The stabilized fuel material would be returned to this dual-purpose vault for interim storage of up to 40 years.

Government-owned spent fuel that has not been reprocessed is stored at DOE facilities awaiting final disposition. The safest, most technically sound storage of these fuel elements is borehole confinement, which is a temporary action.

Canada: Canada, which has one of the world's richest uranium deposits, does not recycle, or reprocess, its spent fuel; rather, its nuclear reactors use natural uranium in a once-through fuel cycle. Given its abundant supply of uranium, Canada has no plans to reprocess its spent fuel in the future. Canadian spent fuel is currently being stored primarily in pools at the 5 reactor sites. Some fuel is moved from pools into dry storage at the reactor sites. The Canadians plan to store waste for several decades in order to allow the heat and radioactivity to dissipate. Existing storage facilities at nuclear plants can easily be expanded to accommodate additional wet or dry storage, if necessary. From its studies and testing, Ontario Hydro has concluded that spent fuel can be stored safely in dry storage containers for at least 100 years. A study of interim storage in the 1970s concluded that on-site storage was preferable to storage at a centralized facility because, among other things, waste transportation would be avoided,and the infrastructure for operating and monitoring was in place at the reactor sites.

France: Most of France's spent fuel is stored first at reactor sites in pools for about a year and then in facilities located at the reprocessing plants until it is reprocessed. Spent fuel is transported to the reprocessing plants in specially designed casks. Transport is primarily by rail within France and continental Europe; trucks are used for short hauls, and ships transport spent fuel from countries outside the continent, such as Japan. The resultant high-level waste from reprocessing is immobilized in glass and will be stored for 30 years or more in vaults at the reprocessing facilities.

According to government officials, lack of storage is not an issue in France. Volumes of high-level waste are not large and can easily be stored in existing facilities at reprocessing plants until a repository is developed. If necessary, officials said, additional storage facilities could be easily be built at the La Hague reprocessing plant and at reactor sites.

Germany: Under German law, nuclear power plant operators must demonstrate plans for waste management 6 years into the future before nuclear plants are allowed to continue operating. The German utilities have responded by sending their spent fuel abroad for reprocessing. In the future, however, the utilities may choose to store their spent fuel for a period of time before disposing of it in a repository.

Interim storage of spent fuel and waste from reprocessing is the responsibility of the utilities. Spent fuel is currently being stored in pools at 18 reactor sites throughout Germany or in interim dry storage facilities or at reprocessing facilities in France and the United Kingdom. The highly radioactive waste remaining after reprocessing was scheduled to be returned to Germany beginning in 1994. The Germans plan to cool their waste by storing it for 30 to 40 years before disposing of it to help avoid elevating the temperature in the repository and perhaps damaging the salt formation. Spent fuel and reprocessed waste are transported between Germany and the reprocessing facilities by train, truck, and ship.

The German utilities have constructed 2 interim dry storage facilities (in Gorleben and Ahaus) for spent fuel and reprocessed waste returned from abroad. According to a recent report, the utilities plan to expand their interim storage capacity for spent fuel.

Japan: As part of their move toward energy independence, the Japanese plant to build a facility for reprocessing spent fuel from their nuclear power plants so that the recovered uranium and plutonium can be used as fresh reactor fuel. According to officials, the reprocessing facility is scheduled to begin operating by the year 2000. The Japanese have built a small reprocessing plant that will be used primarily for research and development once the larger plant is opened. Until the larger reprocessing facility is completed, most Japanese spent fuel is being shipped to France and Britain for reprocessing in these countries.

After initial storage in pools at the reactors, most Japanese spent fuel is shipped abroad for reprocessing. Japan's nuclear plants are located at coastal sites, so most spent fuel and high-level waste are transported by ship. The Japanese plan to open an interim storage facility near the future reprocessing plant, which will store reprocessed waste returned from France and the United Kingdom. The storage facility is scheduled to open in 1995 and can be expanded, if necessary. The Japanese plan to store the waste for 30 to 50 years before disposing of it in order to reduce its heat and radioactivity.

Sweden: Highly radioactive spent fuel is initially stored in pools at reactor sites for 1 to 5 years and then shipped to Sweden's central interim storage facility located adjacent to an existing nuclear plant. Opened in 1985, the underground facility will store all of Sweden's spent fuel in pools of water for 30 to 40 years—allowing the waste to cool before it is placed in a repository.

The interim storage facility will be expanded to allow the entire Swedish inventory of spent fuel to be stored in one location after nuclear power plants are shut down. At the storage facility site, Sweden also plans to build a plant for encapsulating spent fuel in the copper and steel canisters in preparation for disposal in the repository. It expects the encapsulation facility to be operational in 2006, at the earliest.

Switzerland: Spent fuel is generally stored in pools at the reactors before being sent by truck, train, or ship to the United Kingdom or France for reprocessing. To store the high-level waste returned from abroad, the Swiss utilities are planning to build a centralized interim dry storage facility. The facility can also store spent fuel, should the utilities decide against reprocessing. If necessary, this facility will also be able to store lower-level radioactive waste.

The interim storage facility is meeting considerable public opposition—over 20,000 formal objections have been raised thus far during the licensing process. Opposition groups are concerned about a variety of issues, such as the extent to which the facility is protected against disasters (e.g., an airline crash) and the large concentration of nuclear facilities in one area. This opposition may delay the development of the storage facility.

United Kingdom: Most spent fuel is reprocessed in Britain because it is considered a resource that can be recycled to use recovered uranium and plutonium. The government owns British Nuclear Fuels, a corporation that provides commercial fuel cycle services, including spent fuel reprocessing for domestic and foreign customers. However, the economics of reprocessing have recently been questioned. One of the country's nuclear utilities, Scottish Nuclear, plans to store its used fuel up to 100 years instead of immediately reprocessing it. According to officials, Scottish Nuclear finds storing its used fuel less expensive than reprocessing it. Some spent fuel may be stored longer at the reactors in dry storage.

Spent fuel is generally stored in pools at the reactors until it is sent by truck and train to the reprocessing facilities. After reprocessing, the remaining highly radioactive waste is stored at the reprocessing facility; plutonium separated during reprocessing is also stored at the reprocessing facility. To allow heat and radiation levels to decline and to allow time for evaluating disposal options, the waste will be stored for at least 50 years before being disposed of. Storage capacity at the reprocessing facility could easily hold, or be expanded to hold, the high-level waste produced from reprocessing operations during the next 50 years.

British Nuclear Fuels Ltd. is operating a reprocessing plant at Sellafield. In the process, spent fuel rods are left to cool in water ponds for up to 5 years before they are chopped into 1 to 4 inch pieces. The pieces are dissolved in concentrated nitric acid at 60-90°C, and the plutonium and uranium are recovered in a proprietary two-step process.

5.4 DISPOSAL OF SPENT NUCLEAR FUEL AND DOE HIGH-LEVEL WASTES

More than 20,000 metric tons of highly radioactive wastes are stored in 33 states at about 70 civilian nuclear plant sites and 3 Department of Energy (DOE) nuclear facilities. Because these wastes will remain dangerous for thousands of years, the Nuclear Waste Policy Act (NWPA) of 1982 charged DOE with developing an underground repository for safe, permanent disposal of the wastes. Amendments to the act in 1987 required DOE to investigate only Yucca Mountain, Nevada, as a potential repository site. This site selection has stirred considerable controversy.

DOE's interest in Yucca Mountain as a potential repository site predates NWPA. In the late 1960s, DOE began to explore the potential of several types of geologic media, including the volcanically produced rock, called tuff, in the vicinity of the Nevada Test Site and the basalt under its Hanford Reservation, to host nuclear waste repositories. Other geologic media under study included various salt formations in Louisiana, Mississippi, Texas, and Utah. In 1977, the U.S. Geological Survey (USGS), a branch of the Department of the Interior and a participant in operations at the Nevada Test Site, recommended that DOE investigate the test site as a potential host for repository. Subsequent screening of the test site led to selection in 1980 of the Yucca Mountain site. The site is located on the southwest part of the test site, on the Nellis Air Force Range, and on public land managed by Interior's Bureau of Land Management.

By 1991, DOE was estimating that its scientific investigation of a site at Yucca Mountain, Nevada, could be completed in 2001 at a cost of $6.3 billion in year-of-expenditure dollars and that, if the site proves to be suitable, a repository could be in operation in 2010. According to the General Accounting Office, DOE's investigation of Yucca Mountain will take at least 5 to 13 years longer than planned and cost more than the agency has projected.

Site characterization includes extensive field and laboratory work to collect and evaluate geologic, hydrologic, geochemical, and other information. On-site work, for example, consists of surface-based activities, such as mapping, monitoring climate, and conducting geophysical surveys and seismologic and hydrologic studies. It also includes activities conducted in boreholes and trenches that will be used for groundwater monitoring, core extraction, laboratory testing, and studies of the earth's geological structure and chemical composition and of underground water. Finally, studies will be conducted in the host rock through construction of an exploratory facility consisting of underground rooms and drifts (tunnels) excavated to and below repository depth through vertical and/or inclined shafts. In addition, DOE will design the repository and the waste package (the waste and the container in which it is packaged for disposal) and develop the information needed to support an application to NRC for a license to construct the repository.

System design highlights:

1) Use of unit trains (including piggyback cars for truck cask transporters where required) for periodic (once every ten years at each reactor) removal of old (cooled ≥ years) spent fuel from at-reactor storage facilities.
2) Buffer storage at the repository site using dual purpose transforation/storage casks of the CASTOR V/21 type.
3) Repackaging of the spent fuel from the dual purpose transportation/storage casks directly into special-alloy disposal canisters as intact fuel assemblies, without rod consolidation.
4) The filled disposal canisters are welded closed and backfilled with helium.
5) Emplacement into a repository of modular design having a maximum total capacity of 150,000 MT and an annual handling capability of 4,000 MT/yr.
6) Use of excavation techniques that minimize disturbance, both mechanical and chemical, to the geologic environment.
7) Incoloy 825 waste canisters arrayed to provide 57 kW/acre thermal loading optimized to the projected inventories.

8) Include a unit rail mounted vehicle for both the transportation and emplacement of the canister from the surface facilities to the underground repository.

The major motive force for the potential migration of radionuclides from a deep geologic repository is by water transport, e.g., by brine occlusions in salt or through the vadose zone. Much research is in progress in the United States to determine the potential for and mechanisms of transport of radionuclides from HLW emplaced in a repository located at the Nevada Yucca Mountain site. Current and past research efforts have also addressed the selection and evaluation of stable forms for deep geologic disposal of immobilized HLW; materials for construction of canisters and overpacks for waste forms, and materials for back filling repository tunnels, shafts, and other excavated areas.

It is essential that the entire deep geologic repository site remain stable for 10^4 to 10^6 years to satisfactorily contain long-lived radionuclides emplaced therein. Thus, basic research to enable confident predictions of the effects of migration of water through the disposal site must continue. Basic research of the tectonics of geological repository sites must also be performed to develop models that can be used reliably to predict geological changes that will occur over very long times and to predict the effect such changes will have on water flow through the disposal site.

There is a need for more basic studies on the repository-site-specific behavior of spent fuel as a waste form. Such studies should address:

1) The reaction kinetics of UO_2, U_4O_9, U_3O_7, and UO_3 with aqueous solutions under both oxidizing and reducing conditions as a function of temperature and solution composition.
2) The effects of solid solution of the rare earth elements and other actinides on aqueous reaction kinetics of uranium oxides.
3) The oxidation kinetics of both single crystal and polycrystalline (ceramic) uranium oxides at low ($<150°C$) temperatures as a function of grain size, grain boundary characteristics, and atmospheric composition.
4) The in-reactor processes leading to the segregation of fission products into separate solid and gas phases and identification of these phases in spent fuel.
5) Other in-reactor processes that affect the microstructure of the fuel and the distribution of radionuclides within the fuel rod.
6) Thermodynamic data on solid and aqueous species produced by the aqueous corrosion of spent fuel.

Because of the long times periods of concern in deep geological disposal of HLW, a detailed understanding must be obtained of reactive transport processes involving the radioactive waste and the geological strata. These processes include both advective plus diffusive transport along fractures and other imperfections in the geological medium as well as transport in unfractured rock matrices and transfer of material between fractures plus matrices. Fundamental reactions involving interactions between the radioactive materials and water also need to be elucidated to assess the consequences associated with both the expected low water flux anticipated at a good repository site as well as the extremely improbable events such as the catastrophic ingress of water into a repository site.

Basic research is needed to determine thermodynamic properties for all materials involved in possible interactions in the waste form-waste package-back fill-geologic strata system. Such fundamental knowledge is needed to satisfactorily develop systems which will ensure radionuclide containment for long periods of time.

High-level nuclear waste (HLW) contains 2 groups of radioactive nuclides with very different characteristics: 1) the "heat generators" which produce heat during a few hundred years, and 2) the long-living radioactive nuclides, especially the "transuranes," which remain radioactive for a very long time with practically no heat production. The presence of each group leads to special demands for waste disposal. Separation of the HLW into 2 waste categories, a heat generators category, and a transuranes category, may lead to a simpler, safer and more efficient disposal of the waste.

Canada: AECL has developed a generic design for a repository that would be altered to suit a specific site and has been supported by data obtained from its underground research facility. The repository would be 500 to 1,000 meters deep in the granite of the Canadian Shield. Used fuel would be incased in titanium or copper canisters with a minimum life expectancy of 500 years; the canister material has not yet been selected. Clay would be used to surround the canisters, and a mixture of clay and other geological material would fill the repository openings. Officials said that as the reference design is altered to fit the characteristics of a specific site, it may include a canister with a greater life expectancy, in part to help alleviate public concern. Once sealed, the repository is planned to be a passive system; it would not require monitoring, maintenance, or control. Waste retrieval would be possible—although difficult and expensive—for at least several hundred years while the containers remained substantially intact. Because the province of Ontario has most of Canada's nuclear reactors and nuclear waste, the repository is expected to be sited in Ontario.

According to Canadian official, building a repository is not urgent because Canada's spent fuel can easily be stored in existing facilities at the reactor sites. Also, the officials said they would spend the time necessary to gain as much public acceptance as possible. Under the current tentative schedule, the government plans to decide around 1996 whether and how to proceed with developing a repository. A repository would then be available no earlier than 2025.

France: France has not yet chosen a repository design for its highly radioactive wastes. Although firm specifications and criteria are yet to be established, a complementary multiple-barrier system is generally envisioned for the repository. This system would include a canister containing the waste, a backfill material surrounding the canister and filling the repository tunnels, and the geology, which would serve as the final barrier to the release of radiation. The system's actual design will depend on the chosen geology and site-specific information. ANDRA officials believe that a long-lived waste canister might help gain public acceptance; however, such a canister might be more than is technically necessary to demonstrate the repository's safety. Researchers are also considering placing a surrounding wrap, known as an overpack, on the waste canisters to help make them retrievable.

Originally, the French were considering four types of geology—clay, granite, schist, and salt—for siting a repository. Now they are primarily considering granite and clay. As required under the 1991 legislation, ANDRA plans to conduct studies at underground laboratories. The French plan to select 2 sites with the assistance of a negotiator, who will work with the local populations to try to ensure public support.

Germany: The Germans are studying the suitability of a salt formation near the town of Gorleben as a deep geologic waste repository. If the site proves acceptable, they plan to construct a repository and begin accepting waste at the facility in 2008. The Germans have a long history of experience with salt mining and have also conducted various studies of salt in an underground laboratory since the 1960s. Given the nation's experience and the abundance of salt deposits in Germany, the German government decided in the 1970s to move forward with the development of a salt repository, according to officials. Two other repositories—one under construction and one already built—will be used for storing lower levels of waste. Once these facilities are full, all types of nuclear waste will be stored at Gorleben. Officials said that the German government favors moving waste into deep geologic repositories as soon as technically possible because it considers repositories the safest place for waste disposal. There has been considerable public opposition to the Gorleben site, and the project could be abandoned.

Japan: The Japanese plan to build a deep geologic repository for high-level waste and are in the second of 4 stages begin in 1976 and designed to accomplish this goal. During this stage, the Japanese plan to select potential candidate sites for the repository. However, an organization has not yet been named as responsible for the repository's construction and operation. Officials said that because they plan to store their waste for 30 to 50 years to allow it to cool before disposal, they sense no immediate urgency to dispose of it and do not anticipate the need for a repository until 2030 or later.

The Japanese have not decided on the specific details of their repository's design, but they envision using a multi-barrier approach. Partly because of difficult geologic conditions, the Japanese are studying the use of long-lived engineered barriers in their repository design, such as a thick canister overpack (surrounding wrap) and a clay backfill. The Japanese are also studying partitioning and transmutation, which would change long-lived, highly radioactive waste into shorter-lived elements. However, these remaining elements would still be dangerous and would require disposal.

Sweden: Sweden plans to dispose of its spent fuel in a deep geologic repository and is searching for a suitable repository site. Swedish nuclear waste policy is based on the premise that the utilities that create the waste are responsible for managing and disposing of it safely. In 1977, the government required nuclear utilities to demonstrate a safe method for disposing of spent fuel before it would license new nuclear plants for operation. To satisfy this law, the utilities developed a concept for disposing of spent fuel that involves burying the waste in long-lived containers deep in the Swedish crystalline rock.

Switzerland: Switzerland is studying crystalline rock and clay formations to determine the feasibility of using them as a geologic repository for highly radioactive wastes. If possible, Switzerland would prefer to dispose of its relatively low volume of waste abroad in an international repository, primarily because this alternative would be more economical than building a domestic facility. However, Switzerland recognizes that an international repository is highly unlikely under the current political environment, so it is planning to build its own repository. If Switzerland moves ahead with a repository, the waste management organization plans to propose a site by the year 2000, construct an underground laboratory on the site, and open a repository sometime after 2020.

United Kingdom: The United Kingdom is deferring decision on the final disposal of highly radioactive waste for at least 50 years. Government officials believe that the United Kingdom will eventually dispose of the waste in a geologic repository, but the government will make its decision to do so at a later time. In the interim, the British will reprocess most spent fuel and store the resultant high-level waste. The United Kingdom has a relatively low volume of highly radioactive waste, which can easily be stored. The British believe storage offers the technical advantages of allowing the radioactivity to decay and the waste to cool.

Considered more pressing than the need for a high-level waste disposal facility is the need for a repository for lower-level radioactive waste. The United Kingdom reprocesses its spent fuel domestically, and while this process reduces the volume of high-level waste, it creates a significant amount of lower-level radioactive waste. A potential site for a lower-level waste repository is under investigation. Current plans are to commission this repository by about 2007.

5.5 STORAGE AND DISPOSAL OF TRANSURANIC (TRU) DEFENSE WASTE

Prior to 1970, transuranic waste was disposed of in the same manner as low-level waste—by shallow land burial; since 1970, however, it has been retrievably stored (mostly in 55-gallon metal drums placed on concrete or asphalt pads) at several sites including Idaho (61 percent, the largest volume), Oak Ridge (which has most of the TRU waste that must be remotely handled because of its high radioactivity), Hanford, Rocky Flats, Los Alamos, and Savannah River. A portion of the stored TRU mixed waste is in containers that are reaching their design lifetime of 20 years.

Transuranic waste (including mixed TRU waste) from Rocky Flats, Mound Plant, and other weapons sites was shipped to Idaho until September 1989, when the Governor closed State borders to additional TRU waste.

The Waste Receiving and Processing (WRAP) Facility at Hanford is being designed for construction in the north end of the Central Waste Complex. The WRAP Facility will receive, store, and process radioactive solid waste of both transuranic (TRU) and mixed waste (mixed radioactive-chemical waste) categories. Most of the waste is in 55-gallon steel drums. Other containers, such as wood and steel boxes, and various sized drums will also be processed in the facility. The largest volume of waste is TRU in 55-gallon drums that is scheduled to be processed in the Waste Receiving and Processing Facility Module 1 (WRAP 1). Half of the TRU waste processed by WRAP 1 is expected to be retrieved stored waste and the other half newly generated waste. Both the stored and new waste will be processed to certify it for permanent storage in the Waste Isolation Pilot Plant (WIPP) or disposal.

In December, 1979, Congress authorized construction and operation of the Waste Isolation Pilot Plant (WIPP) near Carlsbad, New Mexico, for disposal of defense-related TRU wastes. Construction of WIPP began in 1981. The Waste Isolation Pilot Plant is located 26 miles east of Carlsbad in Eddy County, New Mexico. The site covers 10,240 acres of Federal land. The repository is located 2,150 feet below the surface in a 2,000-foot-thick salt bed with tunnels that extend over 10 linear miles.

DOE's plans call for transportation of transuranic waste to WIPP when it opens, with waste from Rocky Flats and INEL among the earliest shipments. The waste will have to be transported over long distance in a fleet of trucks, each carrying 3 shipping containers (which were granted a Certificate of Compliance by the NRC in August 1989); each container would, in turn, hold 14 waste drums. It will take 20 to 30 years for weapons site or yet-to-be-generated waste to be disposed of at WIPP. Waste would remain on-site until its turn to be sent to WIPP.

DOE plans to dispose of all transuranic waste (including TRU mixed waste) now retrievably stored in the Weapons Complex at WIPP, a geologic repository excavated from salt formations. Construction of a substantial portion of WIPP was completed in 1989. According to DOE, WIPP has the capacity to handle newly generated as well as presently stored transuranic waste. DOE's current program for managing stored transuranic waste contemplates the construction of 6 new facilities at various sites for processing, treating, and certifying transuranic waste prior to shipment to WIPP.

In 1986 and 1987, scientists' discovery that brine (water saturated with salt) was seeping onto the walls of WIPP's underground area—when the facility was expected to be dry—raised questions about the facility's suitability for the disposal of transuranic wastes. A panel of New Mexico scientists advanced the theory that EPA's disposal standards might be violated at WIPP because (1) the repository would become saturated with brine soon after closure, (2) the interaction of the waste and the brine would stimulate the production of gases within the disposal rooms, (3) the combination of gas build-up and salt "creep" (i.e., the inward movement of the surrounding rock to fill in open spaces) would eventually pressurize the gases in the repository, and (4) the pressurized gases would drive contaminated wastes out of the repository and into the general environment. Possible ways that such wastes could escape were through fracturing in the salt and adjacent rock formations or through inadvertent human intrusion, such as oil exploration sometime in the future.

The test phase at WIPP was scheduled to begin in 1993, however it was abandoned in favor of laboratory tests, and further Federal review. Therefore, first shipments of waste to WIPP will probably not occur until the end of this decade. The new plan will use non-radioactive simulated wastes in carefully controlled laboratory experiments, as well as actual waste. The simulated wastes will be easier for experimenters to measure in detail, while the actual wastes will help fine-tune the results and detect any unanticipated phenomenon not disclosed by the simulations.

The WIPP is surrounded by reserves of potash, crude oil, and natural gas. These are attractive targets for exploratory drilling in the distant future, which could disrupt the integrity of the transuranic waste repository.

5.6 STORAGE AND DISPOSAL OF LOW-LEVEL WASTES

Low-level radioactive waste (LLW) is radioactively contaminated industrial or research waste such as paper, rags, plastic bags, protective clothing, cardboard, packaging material, organic fluids, and water treatment residues. It is waste that does not fall into any of the 3 categories previously discussed. Its classification does not directly depend on the level of radioactivity it contains: LLW itself is divided into three classes: A, B, and C.

LLW is generated by government facilities, utilities, industries, and institutional facilities. In addition to 35 major DOE facilities, over 20,000 commercial users of radioactive materials generate some amount of LLW. LLW generators include approximately 100 operating nuclear power reactors, associated fuel fabrication facilities, and uranium fuel conversion plants, which together are known as nuclear fuel-cycle facilities. Hospitals, medical schools, universities, radiochemical and radiopharmaceutical manufacturers and research laboratories are other users of radioactive materials which produce LLW. The clean up of contaminated buildings and sites will generate more LLW in the future.

Throughout the 1980s and the early 1990s, commercially generated low-level waste was routinely disposed of in three facilities in Nevada, South Carolina, and Washington. However, Nevada closed its facility on January 1, 1993. The facility in Washington was closed to generators in all but 11 states in two compacts on January 1, 1993, and on July 1, 1994, South Carolina closed its facility to all but 8 states, but then on July 1, 1995 added more states. As of January 1995, 11 states had plans to develop disposal facilities for commercially generated low-level waste, and the state of Washington planned to continue operating its existing disposal facility. Altogether, these 12 facilities would serve waste generators in 47 states. The states that are developing these new facilities estimate that they will complete the facilities between 1997 and 2002; however, only four candidate sites have been selected, and no facility is being constructed. Moreover, the remaining states do not have plans to develop disposal facilities.

The Low-Level Radioactive Waste Policy Act of 1980 and subsequent amendments direct states to take care of their own LLW either individually or through regional groupings, referred to as compacts. The states are now in the process of selecting new LLW disposal sites to take care of their own waste. The selection process for these new sites is complex and varies because of many factors including the regulations for site selection. This selection process will be affected by EPA's new LLW standard.

As the United States embarks upon a major effort to clean up its nuclear defense facilities, a large quantity of low-level waste (LLW) will be generated. This LLW must be managed and ultimately placed into final disposal. Much of this waste is expected to exceed certain limits defined in U.S. regulations (Title 10, U.S. Code of Federal Regulation, part 61) called Class C. The waste which exceeds Class C, called Greater-than-Class C (GTCC), poses a major challenge to waste managers. Each GTCC waste form must be placed into costly geologic disposal unless separate approval is obtained from the United States regulator to place it into less costly land burial.

The possibility exists for fabricating packaging from contaminated scrap metal (which would otherwise be part of the waste inventory) and for using these packaging for storage, transport and disposal of GTCC in near-surface burial facilities without reopening or repacking. This approach is appealing and should lead to major safety and cost benefits.

Management of GTCC will also require, to some extent, storage and transport prior to its final disposal. A further LLW stream exists in the United

States also stemming from the prior operations of United States defense facilities, viz., radioactively contaminated and irradiated scrap metal which has been accumulating over the past 40 years. Similarly, as clean up, decontamination, and decommissioning proceeds, this contaminated scrap metal inventory is expected to grow rapidly.

For waste contaminated with only shorter-lived radionuclides, interim storage can be used to contain the waste until the radioactivity has decayed to background levels and the waste can be disposed of as non-radioactive waste. For most LLW, however, a certain portion of the radioactive content will outlive the design life of currently used storage structures (typically about 50 years). Thus, the waste will have to be retrieved and sent to a disposal facility. Storage facilities also supplant disposal facilities to accommodate surge situations resulting from mismatch of production, transportation, and disposal schedules.

In the United States, a large portion of solid LLW generated by utilities operating nuclear power plants and by other industrial sources is shipped to commercial LLW disposal sites, either directly or through broker firms (which run the collection and transportation service). A variety of storage techniques are also used at various generation sources.

For most solid waste with radiation fields of less than 1 rem/h, above ground concrete storage buildings provide a cost-effective storage facility. The waste is generally prepackaged, e.g., in drums or in stackable metal, concrete, or wood containers.

For higher-activity waste, concrete trenches, concrete monoliths/radblocks, and concrete tile holes have been used. Ontario Hydro has also developed quadricells, inground storage containers (ISCs), and large dry storage modules (DSMs).

Concrete trenches in shallow ground are used for storage of large quantities of relatively higher-activity solid LLW. The floor of the trenches is sloped to a sump to allow monitoring of the waste with respect to any potential water ingression into the trench. The filled trenches are covered with precast concrete lids. For example, inground cylindrical bunkers built with reinforced concrete, typically 6 m in diameter and about 4 m deep for the storage of higher-activity LLW originating from medical/industrial use of radioisotopes.

Radblocks can be manufactured off-site on a modular basis. They consist of portable concrete modules, each with 4 or 5 cylindrical cavities where waste components can be stored.

Concrete tile holes can be used for high-activity LLW, such as spent cartridge filters and ion-exchange resins from nuclear reactors. The contact radiation field for waste accommodated in tile holes is less than 100 rem/h. Concrete tile holes are vertical inground facilities typically about 0.7 m in internal diameter and about 3.5 m deep, and they are built in arrays. Each tile hole can accommodate two ion-exchange columns or cartridge filters that can be directly bottom unloaded from the shielding flask into the tile hole. Loaded tile holes are backfilled with high slump concrete to form a monolithic cylindrical structure. Retrieval requires the removal of the one-piece tile hole monolith.

Whereas the principle of dilute and disperse is applied in many cases for managing gaseous radioactive waste, and in some cases liquid radioactive waste, it is not applicable to the management of solid waste. Disposal of solid LLW is based on the principle of confinement. The high-activity LLW, similar to HLW, must be isolated from the human environment for a long time. Most solid LLW is, however, low-activity waste and does not require a highly engineered facility intended to achieve long-term isolation of hazardous waste.

For well-segregated solid LLW, disposal in shallow-ground trenches can be a cost-effective means of disposal so long as human health and safety and the environment are protected. The success of shallow-ground disposal depends on the nature of the geologic medium, the site setting, the nature and form of the waste, and the site closure design.

In the United States, 3 commercial radioactive waste disposal sites—Barnwell, South Carolina; Beatty, Nevada; and Hanford (Richland), Washington—had received the bulk of the commercially generated solid LLW.

The experience at these sites has not identified any problems related to site characteristics or disposal practices. The Beatty and Richland sites are located in arid regions whereas the Barnwell site is located in a humid region. At Barnwell, the unlined trenches dug in clayey soil (about 33% clay content) are typically 300 m long, 30 m wide, and 7 m deep. The base of the trenches is sloped and kept at least 1.5 m above the highest water table. A drain and collection system is used for sampling and monitoring; no migration of radionuclides has been observed. Packaged wastes are emplaced in the trench, and sandy soil is used to fill the void spaces between the packages. Filled trenches are covered with a minimum of 0.6 m of compacted clay, which in turn is covered by about 1 m of soil overburden. The topsoil cover is graded to promote runoff and then seeded with grass.

A variety of designs similar to the Barnwell site are being used or studied in the shallow-land disposal (SLD) concept, with variations in trench cover design, trench floor design, waste packaging, waste emplacement strategies, and sampling and monitoring equipment. One of the most important factors for the performance of shallow-land disposal facilities is the selection of a proper site. The problems experienced at several sites now closed (for example, Maxey Flats, Kentucky; West Valley, New York; and Sheffield, Illinois) can be traced to inadequate site characterization.

Newer designs that have evolved as an alternative to simple shallow-land burial rely heavily on the engineered features of the facility rather than the natural characteristics of the site. A brief overview of such facilities is provided below.

Above ground vaults (AGVs) are engineered concrete structures at grade level and are mostly site-independent. They are designed to be intrusion-resistant and to withstand long-term weathering. The facility design generally consists of multiple disposal cells that are individually monitored for water releases. The concept is similar to the long-term storage concept and offers the advantages of comprehensive monitoring of any releases and allows remedial actions to be taken, if necessary. The design relies almost entirely on the integrity and longevity of the concrete structure, which leads to some technical uncertainty as to the long-term performance of the facility. Because of the potential for migration of radionuclides, through the surface water pathway and via direct dispersion in the environment, (in case of a breach of the facility), the potential impacts on human health and safety and the environment could be higher than if the vault is located below ground.

Below ground vaults (BGVs) are engineered concrete structures built below the ground surface. These vaults are designed to be compatible with local soil characteristics. The structures are intrusion-resistant and provide engineered barriers to potential radionuclide migration. Being located below ground, the soil provides an additional natural barrier to radionuclide migration, thus minimizing any potential impacts from release of radionuclides. The shell of the structure (walls and roof) is constructed of reinforced concrete, and the floor is either concrete or made of natural materials. For example, *CRNL*'s intrusion-resistant underground structure (IRUS) will use a specially engineered floor of high-sorption capacity buffer materials.

Quadricell is an above ground storage structure designed to contain bulk quantities of ion-exchange resins that are collected at nuclear stations in large storage tanks. Quadricells can also accommodate highly radioactive reactor core components. Being totally above ground, they have the advantage of being site-independent. Each quadricell module consists of 2 independent reinforced concrete barriers: (1) an approximately cubic structure, 6 m x 6 m x 5.5 m, that is internally separated into 4 cells: and (2) 4 inner cylindrical concrete vessels that are placed within the cells.

A cumulus is an above ground disposal facility designed to provide multiple engineered barriers to isolate radionuclides and allow for complete monitoring of leachates. The engineered barriers include a concrete pad on which the waste is placed, a synthetic underpad liner, the primary waste containers, concrete disposal vaults to house the primary waste containers, and a multilayered cap to be constructed after the pad is filled. An example of a tumulus can be found at Oak Ridge.

Intermediate-depth disposal (IDD) is suitable for higher-activity LLW. It has been generally practiced (e.g., at Oak Ridge National Laboratory, Oak Ridge, Tennessee) using shafts drilled about 100 m into the ground. Disposal at depth reduces the radiation levels at the ground surface and virtually eliminates the potential for inadvertent human intrusion. Shaft disposal is sometimes also referred to as greater confinement disposal (GCD). The Canadian concept for intermediate-depth disposal involves a facility constructed in rock at a depth of about 200 m to accommodate LLW for which the potential radiation hazard will remain even after 500 years.

Both solid and liquid forms of low-level waste are produced at many DOE sites. Solid LLW is often packaged and buried in near-surface facilities without being incorporated into a waste form matrix. Liquid LLW typically is mixed with cement to produce a concrete waste form for disposal. A major disadvantage of concrete as a form for disposal of at least some DOE liquid waste is its relatively high susceptibility to water leaching of certain radionuclides (e.g., Tc-99, Se-79) and hazardous chemicals (e.g., NO_3^-). In some cases, concrete has proven to be too soft. Basic research studies to develop improved cementitious waste forms include:

1) Fundamental research to identify and develop tailored concrete forms for disposal of large volumes of DOE defense liquid LLW solutions needs to be performed. Such studies should address formulations that significantly (factor of 10 to 100) lower the water leachability of waste components such as Tc-99, Se-79, and NO_3^-; the properties of sulfur polymer cements for such use are especially singled out for study.
2) Solid LLW that been reduced to a homogeneous form such as LLW incinerator ash may also be amenable to incorporation into a concrete matrix for final disposal. Concrete is potentially very useful in this application because, unlike the case with TRU waste in concrete, gas generation is not a serious problem. Research to develop improved cementitious waste forms for immobilization of LLW incinerator ash is needed.

The Hanford Site Permanent Isolation Barrier Development Program is developing an in-place disposal capability for low-level nuclear waste at the Hanford Site. Layered earthen and engineered barriers are being developed that will function in what is currently a semi-arid environment for at least 1,000 years by limiting the infiltration of water through the waste.

Chem-Nuclear Systems, Inc. (CNSI) has been selected by several state waste compacts to design, construct and operate new LLW disposal sites. They will receive waste generated at commercial sites (power utilities, commercial processors, hospitals, etc.), with volumes ranging from 200,000 to 550,000 cubic feet per year. As currently planned, these facilities will be operational for from 20 to 50 years.

The new LLW site operational procedures will draw upon Chem-Nuclear's 20 years of waste disposal experience at the Barnwell, South Carolina disposal facility. During this time, from 1-2 million cubic feet of waste were received each year at Barnwell, with waste from throughout the entire U.S. The basis of the new designs is multiple engineered barriers which augments the natural features of the site and the solid form of the waste as shipped by the generator.

The design concept is referred to as the "Triple Safe" concept, since it is composed of 3 distinct engineered barriers (or safety features). This design has been adapted from disposal technology developed in France. Experience at l'Abube will be factored into the new U.S. site designs, leading to greater confidence of the site operational safety and waste containment.

The functions of the "Triple Safe" engineered safety features are as follows:

1) **Concrete Overpacks:** The waste containers are placed within thick walled, reinforced concrete containers. A fluid concrete grout is used to encapsulate the waste container within the overpack. Each overpack contains 100-400 cubic feet of waste. Two types of overpacks are used which are cylindrical and rectangular in shape. The intent is to maximize the waste packaging efficiency within the modules.
2) **Concrete Modules:** Individual concrete overpacks are closely packed within thick walled (typically 2 feet thick), reinforced concrete modules. Each module is fully sealed after loading with overpacks, and typically contains 50,000-100,000 cubic feet of waste. The modules are isolated from ambient conditions during the loading operations.
3) **Earth Cap:** Typically, 40-50 modules (in 2 rows) are grouped and covered with a multi-layered engineered earth cap. The earth cap soil layers passively divert surface water from the concrete modules. The earth cap is composed primarily of natural materials (clay, sand, vegetative soil) obtained from the disposal site or nearby. The modules are completely covered with the earth cap with a cap height extending above the module roof of 7 to 10 feet.

Tanks: Throughout the years of operations, low-level radioactive wastes have been accumulated and stored in underground tanks. These wastes were neutralized to minimize corrosion and concentrated to conserve tank volume. Many of these tanks have leaked in the past, usually to an unknown extent. Liquids have been removed from many inactive older tanks and either treated for disposal by other means or concentrated into other and newer tanks. Radioactivity, however, usually remains in the tanks due to the pressure of residual sediments and sludge that could not be removed completely by the methods used to get the liquids from the tanks.

Pits and Trenches: Low-level liquid-radioactive waste seepage pits and trenches have been used for the disposal of liquid-radioactive waste. The liquids and sludge were placed into shallow seepage pits that allowed excess water and some radionuclides to disperse into groundwater and surface water. Within the suspended sludge and proximate soils, large inventories of fission products were retained along with a variety of TRU isotopes, and activation products.

Use of waste seepage pits and trenches was discontinued at Oak Ridge in 1966, after the hydrofracture method of disposal of low-level waste went into operation. For long-term containment of pits and trenches, radionuclide source-term identification and long-term performance testing of materials and construction methods are needed. A principal science and technology need for in situ trouting is the development of delivery, mixing, and dispersion systems. Grout-waste compatibility, as well as long-term performance modeling under field weathering, would also need to be determined. In situ extraction requires the development of fluid extractants that could mobilize the radionuclides of concern. Technology for injection, disposal, and recovery of extractants in the pits and trenches would also be needed. In situ vitrification is the most promising technology for the pits and trenches. The most significant technology need is modification of off-gas handling techniques to deal with volatilization of cesium during the vitrification.

Remedial alternatives considered for the pits and trenches include containment, in situ treatment, retrieval, ex situ treatment, and disposal. Information is available for the following technologies: capping/barrier overview—containment; ISV, grouting, and extraction—in situ treatment; remote excavation—retrieval.

Ex situ treatment and disposal technologies are available in other problem areas. Additional containment technologies (such as cryogenic barriers and slurry walls) and in situ technologies (such as freezing and heating) are available.

Hydrofracture: As part of waste disposal operations conducted at Oak Ridge, wastes were injected into low-permeability shale about 300 m below ground. The process, known as hydrofracture injection, was conducted during the 1960s through the 1980s. Radioactive wastes were mixed with grout and other additives to form a slurry then pumped down injection wells and into the Pumpkin Valley Shale formation.

The hydrofracture problem area includes only the underground component at the 4 sites. It consists of the wells, grout sheets and contaminated subsurface media, and about 150 observational wells. The latter were used to measure changes to the geology around the injection site, and most of these wells do not penetrate the grout sheets.

The problem area was divided into 2 subproblems: the well bores and the grout sheets. The well bores present a potential pathway for vertical migration of any mobile waste from the grout sheets to aquifers above or below the grout sheets. The grout represents a potential waste problem because the potential exists that some of the waste may not have been immobilized with the grout or that a phase separation of the grout mixture may have left a liquid slurry capable of leaching or migrating to an aquifer.

Impoundments: Unlined impoundments have been used at Oak Ridge to collect, treat, and, in some cases, dispose of process wastewater and storm water. These impoundments, also called waste basins or ponds, represent the earliest form of waste management at the laboratory. Many of the impoundments are located at low elevations and are therefore in direct contact with the groundwater.

Pipelines: Leaking pipelines have also been a problem. Remediation of contaminated soil, without destruction of the buildings and to other pipelines, calls for the use of innovative methods. The methods used at remediation sites have consisted exclusively of either capping the pipes and/or filling them with grout and leaving the pipe and contaminated soil in place, or excavating the soil and pipe and disposing of the soil and pipe material.

5.7 OCEAN DISPOSAL

The concept of disposing of high-level waste or spent fuel by burial in suitable geologic media beneath the deep ocean floor is a potential alternative to geological disposal on land. It is based on the same general principle: the main objective is to isolate waste from the biosphere in suitable geological strata for a period of time and in conditions such that any possible subsequent release of radionuclides in the environment, will not result in unacceptable radiological risks, even in the long-term. Seabed disposal is different from sea dumping which does not involve isolation of low-level radioactive waste within a geological strata.

In the seabed concept a multi-barrier system would be involved, including a suitable waste form such as glass and the use of corrosion-resistant packages. Deep seabed sediment formations would be chosen in order to contain radionuclides after the waste package fails through corrosion and the radionuclides are released from the waste form by leaching. Such sediments would be made up of very fine-grained particles with the ability to absorb and impede the movement of most waste radionuclides. Sites in the ocean would have to be chosen on the basis of the characteristics of the seabed sediments. They would need to be free from erosion and located away from the edges of tectonic plates where seismic or volcanic movements could disrupt a repository and expose the waste packages. Sites would also be located away from continental margins to avoid areas containing potential mineral and biological resources and away from areas of active pore water movement.

In practice the waste packages would have to be transported to a site or sites with 4,000 to 6,000 meters of water depth. Methods of land and sea transportation would be similar to those in use today. Burial in the sediments could be made by 2 different techniques: penetrators or drilling emplacements. In the case of penetrators, packages could be implanted about 50 meters into the sediments. Penetrators weighing a few tons would fall through the water, gaining enough momentum to embed themselves in the sediments. Penetrators would be placed about 200 meters from each other. At that spacing, a repository capable of accepting wastes for about 10 years at the current world production capacity would require an area of about 500 square kilometers.

Wastes could also be implanted using drilling equipment based on that in use in the deep sea for about 20 years. By this method, stacks of waste-filled packages would be placed in 800 m deep holes with the uppermost package about 300 meters below the ocean floor.

In both option, the waste package would protect and contain the wastes during transportation and emplacement operations, and for 500 to 1,000 years after emplacement. Long-term containment, for thousands of years, would be provided by the barrier properties of the sediment.

The Seabed Working Group was formed under the auspices of the Nuclear Energy Agency. Members are countries conducting research in seabed disposal.

Repositories built by tunnelling crystalline rock under the sea is another alternative concept. This concept is being developed in Sweden for the Swedish final repository for reactor waste at Forsmark.

Although the general feasibility of seabed disposal has been established, further research in the following areas would be needed before actual schemes could be implemented:

1) The magnitude and possible causes of pore water migration in the sediments; hole closure behind a penetrator; role of existing or subsequent faults; role of layered sediments.
2) Adsorption properties of the sediments, especially for long-lived fission products, but also for actinides; chemical speciation of radionuclides in the sediments near the waste package.
3) Ocean mixing in continental slope and coastal areas to study transportation accidents in these zones.
4) Deep sea biological activity and its role in redistribution of materials in the ocean; deep sea fish pathways from sediments to surface waters at the disposal area.
5) Engineering aspects of transportation; emplacement and recovery actions in case of accidents.
6) Field or laboratory validation of the models used in the assessment.

The concept of ocean disposal of low-level radioactive wastes is not new. A sizable amount of these wastes was disposed at sea between 1946 and 1970.

The radioactive wastes that have been disposed at sea were usually in concrete-filled drums or containers. Three sites were used in the Atlantic Ocean. One was 12-15 miles from the coast in 300 feet of water near Massachusetts Bay. The other 2 were in water deeper than 6,000 feet, one 150 miles off Sandy Hook, NJ, and the other 105 miles off Cape Henry, VA. Two sites were used in the Pacific about 48 miles west of San Francisco.

Ocean disposal could be considered for tailings and other radiologically contaminated soils that are free of other hazardous wastes. This alternative should not be considered for enhanced radioactive materials or concentrated residuals. Stabilization techniques could be applied to the waste before emplacement to provide for more security against leaks. For those materials contaminated with hazardous chemicals, the potential danger to marine biota must be evaluated.

Ocean disposal offers the opportunity for extreme isolation of low-level radioactive waste. Transportation of the contaminated materials will involve transfer between land and sea. If the radioactive contaminants should be released, the potential for dispersal and dilution is immense. With international consensus emerging against ocean dumping and the current moratorium, this disposal practice may not be a viable option for the future.

5.8 UNDERGROUND MINE DISPOSAL

Underground disposal in existing mines may provide secure and remote containment. The radioactively contaminated wastes could be excavated and transported without treatment to the mine site, pretreated for volume reduction, or solidified to facilitate transport and placement, thus reducing associated costs. Movement of radionuclides into groundwater must be investigated and prevented.

Mine containment of hazardous waste in Europe has been successful. Multipurpose use of a mine for hazardous waste and for low-level radioactive waste might be considered and would likely reduce the per unit costs of waste disposal.

Underground mine disposal would not be appropriate for radiologically contaminated bulk liquids or non-containerized waste.

Mine disposal might be considered for use for a variety of radionuclide and matrix types. It could be used to dispose of wastes with or without prior treatment, although volume reduction and/or solidification or vitrification might facilitate the process. Wastes that have been concentrated by extraction or separation techniques may be particularly appropriate for mine disposal, since they are likely to be more radioactive, requiring disposal that is more remote and more secure.

The mine disposal of hazardous radioactive waste may be among the more costly disposal alternatives, particularly if a mine must be excavated for only that purpose. Wastes must be excavated and transported with the associated permit and safety concerns. The use of an abandoned mine would involve the cost of reconstruction and may pose safety hazards. Also, the groundwater must be protected.

5.9 DISPOSAL OF NATURALLY OCCURRING (NORM) AND ACCELERATOR-PRODUCED RADIOACTIVE MATERIALS (NARM)

Naturally Occurring Radioactive Materials (NORM): Naturally occurring radioactive materials (NORM) generally contain radionuclides found in nature. Once NORM becomes concentrated through human activity, such as mineral extraction, it can become a radioactive waste. There are 2 types of naturally occurring radioactive waste: discrete and diffuse. The first, discrete NORM, has a relatively high radioactivity concentration in a very small volume, such as a radium source used in medical procedures. Estimates of the volumes of discrete NORM waste are imprecise, and the EPA is conducting studies to provide a more accurate assessment of how much of this waste requires attention. Because of its relatively high concentration of radioactivity, this type of waste poses a direct radiation exposure hazard.

The second type, diffuse NORM, has a much lower concentration of radioactivity, but a high volume of waste. This type of waste poses a different type of disposal problem because of its high volume. The following are 6 sources of such naturally occurring radioactive materials:

1) Coal Ash
2) Phosphate Waste
3) Uranium Mining Overburden
4) Oil and Gas Production Wastes
5) Water Treatment Residues

Accelerator-Produced Materials: Accelerator-produced radioactive waste is produced during the operation of atomic particle accelerators for medical, research, or industrial purposes. The accelerators use magnetic fields to move atomic particles at higher and higher speeds before crashing into a preselected target. This reaction produces desired radioactive materials in metallic targets or kills cancer cells where a cancer tumor is the target. The radioactivity contained in the waste from accelerators is generally short-lived, less than one year. The waste may be stored at laboratories or production facilities until it is no longer radioactive. An extremely small fraction of the waste may retain some longer-lived radioactivity with half lives greater than one year. There are no firm estimates of the amount of this type of radioactive waste; however, it is generally accepted that the volume is extremely small compared to the other wastes discussed.

There are currently no federal regulations covering disposal of NARM with high radioactivity concentrations. Few states have regulations, and those regulations are inconsistent. The EPA has initiated studies to more accurately characterize the radiological hazards posed by NARM.

5.10 PLUTONIUM

Storing plutonium poses problems for DOE. Plutonium metal reacts with oxygen, hydrogen, and water vapor. As a result of these reactions, plutonium fines and plutonium hydrides are sometimes formed, they are pyrophoric, having the potential to spontaneously ignite. For any prolonged period, therefore, plutonium should be stored in an oxygen and moisture-controlled environment. Plutonium also should not be stored in direct contact with organic materials, such as plastic. The radiation from plutonium can cause the organic material to decompose, producing hydrogen and other substances.

The hydrogen can react with the plutonium to produce plutonium hydrides. Finally, although the plutonium metal used in weapons consists mostly of the isotope plutonium-239, small amounts of other plutonium isotopes are also present. One of them, plutonium-241, will decay to americium-241, an isotope that emits a type of radiation that is more difficult to shield against. This process can increase the risk of radiation exposure to workers.

As of December 1994, about 12.8 metric tons of plutonium was being stored at Rocky Flats in 4 basic forms—plutonium metal, plutonium oxides, plutonium contained in liquids, and plutonium residues.

The plutonium metal is nearly pure plutonium or is alloyed with other metals. It was used to fabricate various parts of nuclear weapons. About 6.6 metric tons of plutonium metal, consisting of over 3,000 items, is currently stored in several different buildings within the Rocky Flats complex.

When operations ceased at the plant, DOE stored the plutonium metal in containers (cans made of stainless steel or tin-plated steel) that (1) were not airtight or (2) had seals that were not designed for long-term storage. According to DOE officials at Rocky Flats, the containers could permit oxygen and/or moisture to enter and react with the plutonium, possibly creating pyrophoric material. Also, according to DOE headquarters officials, oxygen could enter the containers and create plutonium oxides that could expand and rupture the containers. The plutonium has been stored in this manner for over 5 years.

In addition, an undetermined amount of the stored plutonium metals may be in direct contact with plastic. The radiation from the plutonium could react with the plastic and cause hydrogen to form. The hydrogen could then react with the plutonium to form plutonium hydrides, which are pyrophoric. The plutonium metal was packaged in plastic because plant official anticipated restarting operations within a few months, and they considered plastic to be safe for this period of time.

The Safety Board cited storage of plutonium metal in contact with plastic, stating that it is "well known that plutonium in contact with plastic can cause formation of hydrogen gas and pyrophoric plutonium compounds leading to a high probability of plutonium fires."

Plutonium oxide is formed when plutonium metal oxidizes. This material was formed during the past production operations or as a result of plutonium metal reacting with air. About 3.2 metric tons of plutonium oxide are stored at Rocky Flats in more than 3,000 containers. At the time operations were shut down in 1989, approximately 97 percent of the plutonium oxides had been thermally treated to remove pyrophoric components. Once plutonium oxide is thermally treated and properly sealed, it is better suited for longer-term storage.

The remaining 3 percent of the oxide, which has not been thermally treated, is stored in stainless steel cans in glove boxes with inert atmospheres.

Heat detectors and alarms were placed in the glove boxes as required to detect spontaneous ignitions. According to DOE officials at the plant, some of these cans probably contain oxides mixed with small plutonium fines. The potential risk in storing these cans involves breach of containment and dispersal of the plutonium oxide within the glove box. Contractor officials at Rocky Flats state that without some external stimulus, the danger of spontaneous ignition of the plutonium oxide that has not been thermally treated is minimal. The containers, along with the reduced oxygen atmosphere in the glove boxes where the containers are stored, greatly reduce the potential for fire.

About 30,000 liters of liquid solutions, containing about 0.1 metric tons of plutonium, are stored at Rocky Flats. When Rocky Flats was operating, the liquids were routinely processed to recover the plutonium and were not generally stored for long periods of time. When Rocky Flats was shut down, the liquids were contained in plastic containers, tanks, and pipelines in several buildings, where they currently remain.

In April 1993, the Los Alamos Technology Office at Rocky Flats, a contractor to DOE, conducted a study to determine the hazards of continuing to store plutonium-bearing liquids at Rocky Flats. The Los Alamos Technology Office concluded that the plutonium stored in tanks and bottles in 7 of the Rocky Flats buildings presents a safety hazard from leaks and/or spills and the associated increased risk of workers being exposed to radiation. The study states that as the containers age, the incidence of spills and leaks will increase and could result in increased exposure of workers. The study concluded that continued storage of the plutonium solutions was inadvisable and recommended converting them into solid form promptly. In a June 1994 DOE headquarters review, officials found the plutonium solutions in plastic bottles to be particularly hazardous because the plastic was becoming brittle from reacting with plutonium.

According to staff of the Defense Nuclear Facilities Safety Board, another important risk arising from continued storage of plutonium solutions in deteriorating equipment is accidental criticality—that is, an accidental nuclear chain reaction. The staff points out that as the equipment and infrastructure deteriorate, it will become more difficult to take representative samples, control the chemistry of the solutions, and move the solutions from one place to another. The staff believes that this awkwardness and uncertainty will tend to make accidental criticality more likely.

The fourth category of plutonium stored at Rocky Flats is 2.9 metric tons of plutonium contained in about 100 metric tons of residues. These residues, the by-products of past production operations, consist of ash, salts, slags, graphites, and other materials. They are contained in over 20,000 packages in 5,000 metal drums located in various buildings at Rocky Flats. If operations had been restarted at Rocky Flats, much of the residues could have been reprocessed to recover the plutonium. A June 1994 DOE headquarters review found a number of potential problems with the storage of residues. These included fire hazards, radiation exposure, and gas buildup in the drums.

The national criteria for storage have been incorporated in a DOE technical standard that should be published in 1995. The criteria establish specific guidelines for DOE facilities to follow. The criteria specify that the stored plutonium be either solid metal or oxide (powder or solid) and that it be retrievable for future use. The criteria also specify the maximum permitted quantities of plutonium per container. Furthermore, the criteria require that

1) Plutonium metal be stored in a size and configuration that makes it less prone to being pyrophoric,
2) No plastic or organic materials be in direct contact with stored plutonium,
3) Plutonium be encased in 2 protective barriers meeting stringent storage and/or transportation criteria, and
4) Plutonium oxides be thermally treated to remove pyrophoric materials and minimize moisture content, and then be packaged so that the oxides do not reabsorb any moisture.

DOE officials believe that if these criteria are used, plutonium metals and oxides can be safely stored for 50 years.

Plutonium storage practices at Department of Energy (DOE) facilities evolved over decades during which the objectives of Department programs were to support nuclear weapons development and production. These storage practices reflected a desire to primarily maintain plutonium in metal form for prompt recycling into weapon components. Weapon-grade plutonium generally was considered to be either "in-process" or "in-use." Prevailing procedures and safety requirements addressed only short-term storage. The end of the Cold War and the new arms control agreements are leading to the retirement of large numbers of nuclear weapons resulting in an excess of plutonium that will require management. A new standard establishes safety criteria for safe storage of plutonium metals and plutonium oxides at DOE facilities. Plutonium materials packaged to meet these criteria should not need subsequent repackaging to ensure safe storage for at least 50 years or until final disposition.

An important part of the Department of Energy Program is the development of facilities for long-term storage of plutonium. The design goals are to provide storage for metals, oxides, pits, and fuel-grade plutonium, including material being held as part of the Strategic Reserve and excess material. Major activities associated with plutonium storage are sorting the plutonium inventory, material handling and storage support, shipping and receiving, and surveillance of material in storage for both safety evaluations and safeguards and security. A variety of methods for plutonium storage have been used, both within the DOE weapons complex and by external organizations.

Storage concepts include floor wells, vertical and horizontal sleeves, warehouse storage on vertical racks, and modular storage units. Issues/factors considered in determining a preferred design include operational efficiency, maintenance and repair, environmental impact, radiation and criticality safety, safeguards and security, heat removal, waste minimization, international inspection requirements, and construction and operational costs.

Since plutonium can be diverted to weapons use, security is an important consideration in any disposition of excess plutonium. A black market may emerge overseas.

The Institute for Energy & Environmental Research (IEER), in Maryland suggests the plutonium should be vitrified in glass logs. Before vitrification, it should be mixed with another material such as depleted uranium to make it difficult for subnational groups to extract and use the plutonium in weapons. However, IEER points out, nations taking the vitrification route could still extract the plutonium from the logs if, in the future, it becomes economical for use as an energy source.

In another proposal by the Natural Resources Defense Council (NRDC), the U.S. would convert the plutonium into mixed oxide (mox) fuel, a mixture of plutonium and uranium oxides. The fuel could then be sold to utilities in Europe and Japan that are now paying companies in the U.K and France to reprocess their fuel. In exchange, the U.S. would store the utilities' spent fuel.

This path has 2 advantages, says NRDC. It would put U.S. plutonium into a form that is not easily reused in weapons. At the same time, it would stop the worldwide buildup of civilian plutonium now taking place as a result of reprocessing.

Plutonium storage practices at Department of Energy (DOE) facilities evolved over decades during which the objectives of Department programs were to support nuclear weapons development and production. These storage practices reflected a desire to primarily maintain plutonium in metal form for prompt recycling into weapon components. Weapon-grade plutonium generally was considered to be either "in-process" or "in-use." Prevailing procedures and safety requirements addressed only short-term storage. The end of the Cold War and the new arms control agreements are leading to the retirement of large numbers of nuclear weapons resulting in an excess of plutonium that will require management.

Both the near-term and long-term disposal of plutonium pose unresolved national and international policy issues. Of special concern is weapons-grade plutonium being separated from dismantled nuclear weapons. Monitored storage, material reutilization, isotopic dilution or transformation, and space expulsion are the primary disposal alternatives. Plutonium and uranium can be effectively demilitarized by mixing with lower isotopic grades. Vitrification and storage of all plutonium have the support of those who oppose burnup in nuclear reactors. Reutilization and demilitarization could be encompassed by once-through burnup as mixed-oxide fuel in existing nuclear-power reactors, or by closed-cycle annihilation in special government reactors. Because reactor-grade plutonium has been habitually avoided in making weapons, burnup would be a very effective measure in denying expeditious military use of plutonium by nuclear-weapons states.

Reactor-based plutonium burnup could be accomplished by the Advanced Light Water Reactor with plutonium-based ternary fuel, the Advanced Liquid Metal Reactor with plutonium-based fuel, the Advanced Liquid Metal Reactor with uranium-plutonium-based fuel, and the Modular High Temperature Gas-Cooled Reactor with plutonium-based fuel.

INEL investigated the feasibility of using plutonium fuels (without uranium) for disposal in existing light water reactors and provided a preconceptual analysis for a reactor specifically designed for destruction of weapons-grade plutonium

Brookhaven has developed a new concept termed ADAPT for the rapid and virtually complete burning of plutonium. ADAPT employs a high current CW linear accelerator (linac) to generate neutrons in a lead/D.0 target. The neutrons are then absorbed in a surrounding subcritical blanket assembly, that hold small graphite beads containing the plutonium to be burned. The graphite beads are coated and sealed to contain all fission products, including the noble gases. After destruction of virtually all of the original plutonium loading, the fuel beads are discharged and sent to a geologic repository for ultimate disposal.

5.11 DEPLETED URANIUM DISPOSAL OPTIONS

The Department of Energy (DOE), Office of Environmental Restoration and Waste Management, has chartered a study to evaluate alternative management strategies for depleted uranium (DU) currently stored throughout the DOE complex. Historically, DU has been maintained as a strategic resource because of uses for DU metal and potential uses for further enrichment or for uranium oxide as breeder reactor blanket fuel. This study has focused on evaluating the disposal options for DU if it were considered a waste. This report does not declare these DU reserves a "waste," but is intended to provide baseline data for comparison with other management options for use of DU.

5.12 URANIUM BLEND DOWN

Westinghouse Savannah River Company was asked to assess the use of existing Savannah River Site (SRS) facilities for the conversion of highly enriched uranium (HEU) to low enriched uranium (LEU). The purpose was to eliminate the weapons potential for such material. Blending HEU with existing supplies of depleted uranium (DU) would produce material with less than 5% U–235 content for use in commercial nuclear reactors. The request indicated that as much as 500 to 1,000 MT of HEU would be available for conversion over a 20–year period.

The low estimated cost per kilogram of blending HEU to LEU in SRS facilities indicates that even with fees for any additional conversion to UO_2 or UF_6, blend down would still provide a product significantly below the spot market price for LEU from traditional enrichment services.

5.13 MIXED WASTES

Mixed waste is waste that contains both hazardous waste and radioactive material (source, special nuclear, or by-product material as regulated by the Atomic Energy Act of 1954). Mixed waste is classified by DOE according to the type of radioactive waste that it contains as either mixed low-level waste (MLLW), or mixed transuranic waste (MTRU). DOE's high-level waste (HLW) is assumed to be mixed waste because it contains hazardous components or exhibits the characteristic of corrosivity.

Currently, 72% of DOE's mixed waste is high-level waste (HLW), 20% is mixed low-level waste (MLLW), and 8% is mixed transuranic (MTRU).

Established processes are being implemented by DOE for studying, designing, constructing, and ultimately operating disposal facilities for HLW and MTRU wastes (specifically the HLW repository in Nevada, and the Waste Isolation Pilot Plant in New Mexico).

HLW is managed at four sites (the Hanford site in Washington, the Savannah River site in South Carolina, the West Valley Demonstration Project in New York, and the Idaho National Engineering Laboratory in Idaho). HLW will only be transported from these sites as a stable solid waste form ready for disposal.

DOE's current policy is that defense related MTRU waste will be disposed at the Waste Isolation Pilot Plant (WIPP) using the No Migration Variance and will not require treatment to meet the land disposal restriction standards.

Approximately 32% of DOE's current inventory and mixed waste projected to be generated over the next five years is MLLW. Wastes classified as either "remote handled" or "contact handled," and "alpha" wastes that contain between 10 and 100 nCi/g of alpha-emitting transuranic isotopes are included as MLLW. These "alpha" wastes require additional precautions to protect against worker exposure and environmental release.

The largest portion of MLLW is aqueous liquids and aqueous slurries, such as wastewater. Because transporting this waste for treatment would be difficult and costly, on-site treatment is planned. Less than 5% of the MLLW is proposed for off-site treatment. For the remaining 12%, treatment options have not yet been assigned. Mobile treatment units will also be used.

Currently there are no active permitted mixed waste disposal facilities operated by DOE for disposal of residuals from the treatment of MLLW. Through the Site Treatment Plan development process, DOE and State and

Federal regulators have formed working groups to evaluate issues related to disposal of treated MLLW. These work groups have defined criteria to evaluate the sites subject to the FFC Act in order to identify sites that may be suitable for disposal of these residuals. Evaluation of these facilities and determination of potential disposal locations is continuing.

6

Soils and Sediments

The selection of a remediation technology at a specific site depends on the physical and chemical properties of the contaminant, the properties of the environment at the site, and the degree to which contamination is to be reduced or confined. Perhaps the most difficult aspect to evaluate is the characterization of the site. In general, both saturated and unsaturated subsurface zones exhibit heterogeneous properties. Costs associated with drilling wells, analyzing large numbers of chemical samples, and performing other activities required to characterize a site are expensive and time consuming. It should be noted that the cost of implementing an inadequate remediation technology can be even more expensive and time consuming.

Use of mathematical models can reduce data requirements, but models have limited capabilities to reflect the heterogeneity of subsurface environments. Furthermore, application of techniques to predict changes in chemical properties resulting from changes in environmental conditions from point to point in a site have been limited. The only way to compensate for these limitations is to overdesign the adopted treatment technologies. Most chemical and biological processes are overdesigned to account for operations under less than optimal conditions. Because the environment being dealt with is so uncertain, the margins by which overdesign is included must be increased greatly over traditional process design.

Techniques for remediating arid sites are somewhat different than those used for humid sites.

The methods of remediation will vary, but given the hazards of the waste, robotics and remote handling techniques will be necessary to reduce the risk to the worker.

A recent Pacific Northwest Laboratory (PNL) study identified 59 waste sites at 14 DOE facilities across the nation that exhibit radionuclide contamination in excess of established limits. The rapid and efficient characterization of these sites, and the potentially contaminated regions that surround them represents a technological challenge with no existing solution. In particular, the past operations of uranium production and support facilities at several DOE sites have occasionally resulted in the local contamination of surface and subsurface soils. Such contamination commonly occurs within waste burial sites, cribs, pond bottom sediments and soils surrounding waste tanks or uranium scrap, ore, tailings, and slag heaps.

The objective of the Uranium In Soils Integrated Demonstration (of the DOE) at Fernald is to develop optimal remediation methods for soils contaminated with radionuclides, principally uranium (U), at DOE sites. It is examining all phases involved in an actual clean up, including all regulatory and permitting requirements, to expedite selection and implementation of the best technologies that show immediate and long-term effectiveness specific to the Fernald Environmental Management Project (FEMP) and applicable to other radionuclide contaminated DOE sites. The demonstration provides for technical performance evaluations and comparisons of different developmental technologies at FEMP sites, based on cost-effectiveness, risk-reduction effectiveness, technology effectiveness, and regulatory and public acceptability. Technology groups being evaluated include physical and chemical contaminant separations, in situ remediation, real-time characterization and monitoring, precise excavation, site restoration, secondary waste treatment, and soil waste stabilization.

One of the most important science and technology needs is an accurate and complete characterization of contaminated soils and sediments. It is a gamble selecting the most appropriate remediation technology if properties of the contaminant and the soils and sediments are unknown. In many instances, the distribution and chemical form of the contaminant within particle-size fractions of the soils and sediments is important in determining the success or failure of a specific technology.

Technologies considered for the management of contaminated soils and sediments are broadly divided into containment and isolation, and treatment processes. Because the contamination of soils and sediments often includes a variety of contaminants (organic compounds, toxic metals, and an almost unlimited list of radionuclides), the application of both types of processes to many of the soils may be required. For example, the prevailing management strategy for contaminated soils is containment in low-level radioactive waste burial grounds. One alternative would be treatment to achieve volume reduction, which would lower containment costs. The selection of a technology for the management of contaminated soils and sediments is highly dependent on the type of contaminant(s), its distribution in the soils and sediments (both in terms of depth within the profile and distribution among particle-size fractions), and its chemical form. Knowledge of these properties requires extensive characterization. Because of the lack of characterization data and the potential presence of a wide variety of contaminants in many of the soils and sediments, the descriptions of treatment technologies are also varied.

Because near-term needs are related primarily to contaminated soils resulting from remediation of pipelines and waste tanks, emphasis has been placed on technologies involving ex situ treatment over containment. In the case of long-term needs, the emphasis on in situ treatment and containment technologies generally prevails.

Many of the DOE radioactive and mixed wastes are characterized by their very large volume and relatively low concentrations of radionuclides. There are strong economic incentives to partition these wastes into a large fraction, which can be properly disposed of in inexpensive near-surface facilities, and a much smaller fraction, which must be immobilized for expensive final disposal in a deep geologic repository. Such partitioning is commonly referred to as "waste pretreatment."

Depending upon the particular type of waste, pretreatment operations typically involve one or more of the following operations: (1) dissolution of solids; (2) processes for removal of various radionuclides; (3) processes for destruction of hazardous and other organic constituents; (4) processes for removal or destruction of toxic anions (e.g., NO_3^-, Fe, etc.); and (5) processes for removal of toxic metals (e.g., Hg, Cd, etc.). Both aqueous-based technology and pyrochemical technology may be used to solubilize radioactive solid wastes and to remove actinides and, possibly, other radionuclides (e.g., Sr-90, Tc-99, I-129, and Cs-137) from liquid wastes and dissolved solid wastes. For mixed wastes, dissolution and radionuclide removal operations can, in some instances, be conveniently performed either before or after destruction of organic materials.

In addition to radionuclides, the soils are also, in most cases, contaminated with hazardous organics and metals. Many DOE sites have soils contaminated by large plumes of volatile organic compounds (VOCs). Since the book is devoted to radionuclide clean up, the subject of VOC remediation is not discussed.

At sites having old surface impoundments that accepted waste, or where surface water contamination is known to exist, sediment contamination is either suspected or confirmed. The extent of contamination is not fully known, but some off-site migration may have occurred, and DOE is beginning to examine the extent of both on-site and off-site sediment contamination. This includes site-specific and waste-specific information concerning the environmental fate and transport of constituents in contaminated sediments. DOE is removing or stabilizing in situ contaminated sediments from some units in an attempt to clean and close those units.

In Situ Techniques: In situ treatment processes can remediate subsurface contaminants without excavation of the contaminated soils or extraction of the groundwater. The contaminants of interest can either be treated in place or transferred to the surface via a secondary carrier phase for subsequent treatment. In situ chemical/physical treatment processes can either be applied as an alternative to in situ bioremediation or as a pre- or post-treatment in conjunction with biological treatment. In situ chemical/physical treatment can be used in environments where microorganisms fail to thrive, can treat recalcitrant organic compounds and inorganics, and can accomplish treatment more rapidly and extensively than in situ bioremediation. The overall goal of the DOE In Situ Chemical/Physical Treatment Subprogram is to develop a portfolio of in situ remediation technologies that employ chemical/physical processes for treatment of contaminants in situations common across the DOE complex.

The predominant sources of contamination at DOE facilities appear to be the various liquid and solid waste management units. These include liquid waste disposal facilities (e.g., land treatment units, surface impoundments, retention ponds, burning pits, french drains) and buried waste deposits (e.g., pits, trenches, and landfills). Other sources of contamination include leaking waste pipelines, high-use areas (areas surrounding waste treatment facilities, test firing sites), and leaking underground storage tanks. Leachates from the land treatment units and burial sites have in many cases contaminated subsurface soil and groundwater. In addition, contamination of surface water sediments has occurred due to off-site releases. The most prevalent contaminants are: (1) radionuclides, (2) chlorinated hydrocarbons, and (3) anions (specifically nitrates). Mixtures of contaminants are also common.

In situ chemical/physical treatment involves additions to or alterations of the subsurface that change the chemical and/or physical properties of the subsurface environment. In situ remediation technologies are increasingly being sought for environmental restoration, due to the potential advantages that in situ technologies can offer as opposed to more traditional ex situ technologies. These advantages include limited site disruption, lower cost, reduced worker exposure, and treatment under obstructed structures and at depth. While in situ remediation technologies can offer great advantages, many technology gaps exist in the application of in situ chemical/physical treatment. The technology gaps include inadequate information, particularly at the field-scale, in areas of performance potential, implementation constraints, limitations to applicability, and verification of performance.

Few in situ technologies are available to remediate contamination located in the area between a landfill and the groundwater. This vadose zone, unsaturated soil zone, is an important area because it provides a barrier between the landfill and groundwater. While the vadose zone can effectively isolate and contain some contaminants, other contaminants may move quickly through this zone. When the vadose zone becomes contaminated with fast-moving pollutants, such as volatile organics, scientists worry that pollutants may reach groundwater before intervention can take place.

Successful in situ remediation methods can achieve substantial cost reductions. For example, the recent ROD for Pit 9 at the Idaho National Engineering Laboratory containing 14,000 m^3 of buried TRU waste estimated the cost of excavation/disposal (the baseline technology) at $24,000 per cubic meter. The cost of using in situ techniques would be about $2,000 per cubic meter.

The **DOE Subsurface Manipulation** subprogram evaluates physical control systems for avoiding dispersal of contaminants, or assisting dispersal of treatment agents during in situ remediation. Technologies may include electrokinetic migration of contaminants, hydraulic isolation, auger or jet mixing, hydro- or cryofracturing, pneumatic fracturing, and vacuum-vaporizer well systems. Examples of technologies currently under development in these areas include: Frozen Soil Barriers, Reactive Barriers, In Situ Magnetic Separation, Uranium Biosorption, In Situ Redox Manipulation, Electrokinetics, and In Situ Corona Discharge.

The **DOE Containment** subprogram seeks to develop in situ technologies for pollution containment. It supports investigations of new barrier materials, containment absorbers/neutralizers, and emplacement methods for barrier formation without soil excavation. These technologies could provide short-term containment while the source plume is being remediated, or long-term containment for sites that present no immediate health/environmental risk or that require development of new remediation methods.

The **DOE Treatment** subprogram develops and evaluates in situ technologies for destruction, enhanced removal, extraction, and immobilization of groundwater/soil contaminants. Supplementary process monitoring and control technologies are also being addressed in this subprogram. R&D in this area will lead to the demonstration of biological, chemical, and physical treatment technologies to destroy or immobilize contaminants without harming the environment.

Barriers: The use of barriers is one of the most common and cost-effective methods for the disposal of waste materials. Wastes disposed under the Resource Conservation and Recovery Act (RCRA), for example, are disposed in excavations protected with various liner materials to prevent or minimize the escape of waste leachates to the groundwater. Waste disposal sites under RCRA are closed using a barrier cap constructed of low-permeability materials. A cap is intended to provide a barrier to the ingress of water, thereby limiting the quantity of leachate that may form. Wastes disposed under the Atomic Energy Act (AEA) also employ barrier materials intended to ensure safe containment while the radioactive constituents decay to safe levels. Barrier materials employed for disposal of AEA wastes include special packaging materials and thick covers of native soil. Barriers are also frequently used as part of remedial actions conducted at existing waste sites regulated under the Comprehensive Environmental Response, Conservation, and Liability Act

(CERCLA). Barrier types that have been employed under CERCLA include caps installed over the waste sites and low-permeability structures around contaminant plumes.

The purpose of a barrier is to contain or limit the spread of contamination to within the confines of the waste disposal zone. Several naturally occurring mechanisms that induce migration of contaminants must be addressed in designing an effective barrier system. These include 1) penetration by burrowing animals and plant roots, 2) erosion by the wind and surface water runoff, 3) infiltration, advection, and dispersion of subsurface water, and 4) diffusion. Infiltration, advection, and dispersion mobilize contaminants as a consequence of the gravitationally induced movement of both unsaturated and saturated zone groundwater. The moving groundwater carries contaminants at rates that depend on the properties of the individual contaminants and the geochemical and hydrogeological properties of the local soil. Diffusion involves the molecular or ionic migration of contaminants through materials. Diffusion in the soil air space can be a significant mechanism for dispersal of contaminants, but only for gaseous species. Diffusion in the water-filled pores or capillaries of the soil is relatively slow, but can be a significant factor when surface water infiltration is essentially precluded.

Various barrier technologies have been developed or are under development that show promise for counteracting the four general waste spreading mechanisms. These technologies may employ physical, thermal, chemical, and/or biological methods. A barrier system with several barrier technologies may be necessary to provide adequate protection, especially in cases of highly variable waste and soil conditions.

Barriers can be designed to be impermeable to water flow (hydraulic barriers), or can be semi-permeable, allowing water to pass, but retaining the pollutant. Both types are being studied in a DOE subprogram.

Impermeable barriers made with clays or cement/clay mixtures are widely used in construction. These barriers are effective in slowing water flow, but their use at contaminated sites can be limited by the need to excavate (and dispose of) contaminated soil from the placement trench. Clay may also be chemically attacked by leachates from the waste material, leading to degradation of the plugging effect of the clay and diffusion of contamination. Proper moisture content must be maintained to prevent shrinkage cracks in the clay. These deficiencies may be overcome through development of new barrier concepts, materials, and construction techniques. New synthetic binders and polymers are being evaluated for long-term stability and effectiveness as sealants. Inorganic grouts are also being studied for use with or without clays.

Developing semi-permeable barriers that control contaminant mobility without affecting groundwater flow is a major goal of the DOE Containment Subprogram. By placing a substance in the barrier zone that absorbs or reacts with the target contaminant(s), pollutants can be physically trapped or chemically converted to a harmless form. Capacity and long-term effectiveness of such barriers are principal concerns of this research.

Forming barriers in situ by injection from the surface can decrease construction and waste disposal costs and can be useful for replenishing barriers that have lost their effectiveness over time. Development of barrier emplacement methods that do not involve soil excavation would be an important advancement of this technology. Frozen barriers in both non-arid soils are being evaluated as part of the R&D effort.

A need exists to develop and field test a means of mechanically mixing or emplacing barrier-forming liquids and solids into the soil beneath waste sources, to create a permeable or impermeable continuous barrier. Soil properties range from clays typically found on the east coast, to the sandy/gravelly soil found at Hanford. Some waste sites feature the presence of cobbles, boulders, and agglomerated soils. The required depth of the barrier typically ranges from 10 to 60 feet. Special cases exist where a barrier may be required at a depth greater than 200 feet. The required area of a barrier beneath a waste site ranges from a fraction of an acre to more than 100 acres.

The primary challenge in this need area is achieving uniform mixing or emplacing of the soil and barrier-forming materials necessary to create a continuous barrier. The barrier emplacement technology must be consistent with constraints posed by the barrier-forming materials (e.g., Portland cement sets over time and cannot be remixed). A horizontal barrier must also be capable of being joined to a vertical barrier to create a leak-tight basin for the case when an impermeable barrier is required. A technology capable of creating V-shaped barriers will be considered by DOE. The technology must be consistent with the requirement that (1) the integrity of the barrier be verifiable during or immediately after its installation, and (2) the performance of the barrier can be monitored over its design lifetime.

There is a need to develop and adapt technologies for verifying that emplaced subsurface barriers meet design criteria and then perform as planned over the design lifetime. The verification of integrity of impermeable barriers is extremely challenging when the level of leakage permitted is very low. It may be difficult to measure the integrity of cold joints in certain barrier applications, such as grout barriers. Monitoring the long-term performance of a subsurface barrier is made difficult by the generally large size (up to 5,000,000 ft^2) of subsurface barriers.

Techniques and methods for evaluating emplaced subsurface barriers require development and demonstration. The development of subsurface barriers has barely advanced to the field testing phase; thus, there is little information on the quality of joints created between individual segments of subsurface barriers. Moreover, little information exists on the homogeneity of the barrier within segments installed in field tests conducted to date.

Barrier verification technologies must be able to detect the location and measure the magnitude of barrier discontinuities. These barrier discontinuities should be measured to check for conformity with design requirements, e.g., hydraulic conductivity and reaction with contaminants. Researchers must recognize that important discontinuities may exist on a micro-scale (fractions of an inch). Thus, surface gases and liquids for verification testing may be a promising basic approach that requires development and refinement. Barrier monitoring technologies must enable observations or predictions of loss of barrier performance.

Non-destructive monitoring techniques include:

1) Direct Current Resistivity
2) Electromagnetic Conductivity
3) Spectral Gamma-Ray Logging
4) Ground-Penetrating Radar
5) In Situ Subsurface Parameter Monitoring
6) Cross-Hole Seismic Tomography
7) Neutron Probe
8) Down-Hole Temperature Logs

Landfills typically contain waste materials that include solid waste materials, such as construction debris and off-spec manufacturing intermediates or final products. Residual radioactivity and chemical contamination on these items do not warrant excavation and ex situ treatment, nor in situ stabilization treatments. Development of secondary subsurface contaminant features that modify existing subsurface conditions may provide additional contaminant mobility reduction, in combination with surface covers, to further minimize the risk of residual contamination. Technologies are sought that will provide long-term residual contaminant migration control with passive methods that may include inducing a negative subsurface water balance, or ventilation methods to reduce the build up of contaminant vapors.

6.1 MECHANICAL SEPARATION TECHNIQUES

Physical liberation and separation methods are used widely in processing ore and coal. These processes are well characterized, and considerable information is available on their operation. These methods are excellent candidates for use in volume reduction of soils contaminated with low levels of radioactivity and have been demonstrated to be effective in tests with soil from the Montclair site. Physical separation can significantly lower the cost of remediating sites with radioactive soils by reducing the volume of soils that must be disposed of. For this reason, soil separation technologies should be considered during the feasibility studies for Superfund and other sites. Soil characterization will provide preliminary information on the feasibility of volume reduction, liberation, separation, and collection of clean and contaminated fractions. Bench-scale test results effectively lead to a preliminary design that will correlate well with field equipment. The equipment, commonly used in the coal and ore industries, is commercially available or relatively easy to manufacture and operate.

Many DOE sites have metal-contaminated surface soils that present significant environmental problems. An estimated 20-25 million cubic feet (600,000-750,000 m^3) of plutonium-contaminated soil exists at the Nevada Test Site (NTS), and the adjacent Tonopah Test Range. One area at the Idaho National Engineering Laboratory (INEL) contains 12 million cubic feet (360,000 m^3) of soil contaminated with plutonium and other heavy metals. Other sites with high plutonium or uranium contaminated soil volumes include Operable Unit 2 at the Rocky Flats Environmental Technology Site (RFETS), Operable Unit 5 at the Fernald Environmental Management Project (FEMP), Los Alamos National Laboratory (LANL) and the DOE facility at Hanford, Washington.

The cost of disposing large volumes of contaminated soil in land disposal facilities is high. At the NTS, the current disposal fee for bulk wastes is $10 per cubic foot, excluding excavation, handling, and transportation expenses. Therefore, the minimum cost for disposing of plutonium-contaminated NTS soil could be on the order of $200 million. Conventional technologies often produce waste volumes several times larger than in-place contaminated volumes. Without new technologies, projected costs could increase ten-fold. New cost-effective technologies are needed for heavy metal contaminated soils that address soil and vegetation removal, volume reduction, and waste disposal.

The Heavy Metals Contaminated Soils Project (HMCSP) will evaluate 7 off-the-shelf technologies, routinely used by the mining industry, to conduct treatability studies on soils from DOE sites. These technologies include:

1) Knelson Centrifugal Concentrator
2) Carrier-Assisted Flotation
3) Air-Sparged Hydrocyclone
4) Campbell Centrifugal Jig
5) U.S. Naval Academy Air Classification
6) High Gradient Magnetic Separation System
7) Dissolved Air Flotation (Denver Cell)

The studies use test soils contaminated with plutonium, uranium, or other heavy metals to determine technical and economic feasibility of various physical soil decontamination processes. These tests will provide scientists with data on the physical and contaminant characteristics of the test soils to design better treatment processes for specific contaminated soils. If successful, the volume of radiation-contaminated soils needing further treatment and disposal will be reduced by 80%.

In a recent test at Hanford, a plant was specifically designed for use as a physical separations unit and consisted of a feed hopper, wet screens, hydrocyclones, as well as settling and dewatering equipment. The plant was supported in the field with prescreening equipment, mobile generators, air compressors, and water storage tanks.

Two soil types were treated during the testing: a natural soil contaminated with low levels of uranium, cesium, cobalt, and heavy metals, and a natural soil contaminated with uranium carbonate material that was visually recognizable by the presence of a green sludge material in the soil matrix. The "green" material contained significantly higher levels of the same contaminants. Both source materials were treated by the plant in a manner that fed the material, produced clean gravel and sand fractions, and concentrated the contaminants in a sludge cake. Process water was recycled during the operations. The testing was extremely successful in that for both source waste streams, it was demonstrated that volume reductions of greater than 90% could be achieved while also meeting the test performance criteria. The volume reduction for the natural soils averaged a 93.8%, while the "green" soils showed a 91.4% volume reduction.

6.1.1 Campbell Centrifugal Jig

The Campbell Centrifugal Jig (CCJ) is a patented new technology developed by TransMar, Inc., of Spokane, Washington, to separate fine, heavy mineral particles from gangue material (i.e., waste).

The experimental approach for this study will focus on combining 2 widely used methods of heavy particle separation (jigging and centrifuging) to cause separation in a liquid medium. The result is the centrifugal jig, which combines the continuous flow and pulsating bed of the standard jig with the high acceleration forces of a centrifuge to segregate and concentrate particles from 150 microns to as small as 1 micron.

In this study, separation will be accomplished by feeding appropriately sized slurried material (~50 mesh) into the centrifugal jig through a hollow shaft at the top. This material impinges on a diffuse plate, which has vanes to distribute the material radially to the screen under the influence of gravity. The material will also be centrifuged by the rotating screen. Pulsing will cause the heavier particles to migrate through the jig bed and screen to become concentrates. These concentrates can be recycled or treated for disposal, while particles with lower specific gravity are flushed downward across the jig bed and become the tailings.

Montana College of Mineral/Science & Technology has equipped a pilot plant to evaluate the Series 123 CCJ (capacity 1 to 3 tons per hour). Tests are currently underway to determine the removal efficiency for radionuclides from soil, and for purite from mine tailings waste.

6.1.2 Air Sparged Hydroclone

The Air Sparged Hydroclone (ASH) flotation is a new particle separation technology that has been under development at the University of Utah and Advanced Processing Technologies, Inc. This technology combines froth flotation principles with the flow characteristics of a hydroclone, such that the ASH system can perform flotation separations in less than a second. This feature provides the ASH with a high processing capacity, 100-600 times greater than the capacity of conventional flotation or columns.

The ASH consists of 2 concentric right-vertical tubes, a conventional cyclone header at the top, and a froth pedestal at the bottom. The inner tube has a porous wall (plastic, ceramic, or stainless steel) through which air is injected. The outer non-porous tube simply serves as an air jacket to provide for even distribution of air through the porous inner tube. The slurry is fed tangentially through the conventional cyclone header to develop a swirl flow of a certain thickness in the radial direction (called the swirl layer thickness) adjacent to the porous wall, leaving an empty air core centered on the axis of the ASH. This swirl flow shears the injected air to produce a high concentration of small bubbles. Hydrophobic particles in the slurry collide with these bubbles, and after attachment, lose some of their tangential velocity and centrifugal momentum, and are transported radially into a froth phase that forms at the surface of the air core on the cyclone axis. The froth phase is stabilized and constrained by a froth pedestal at the underflow and thus moves towards the vortex finder of the cyclone header and is discharged as an underflow product through the opening between the inner porous wall and the froth pedestal.

6.1.3 Centrifugal Gravity Concentrator

The Centrifugal Gravity Concentrator utilizes the principle of hindered settling combined with centrifugal action. This is made possible with a proprietary mechanism of a water-jacketed perforated cone fed through a hollow shaft-hydraulic device. Gravity concentration devices, like the Centrifugal Gravity Concentrator, depend upon differences in particle size, particle specific gravity, or both (i.e., particle mass) for their effectiveness.

The UNR unit is a centrifugal bowl concentrator with a water jacket around the bowl, essentially a modified centrifuge. Feed slurry enters the rotating ribbed bowl where heavier particles are trapped between the ribs. Compaction of the material between the ribs is prevented by injecting water through holes in the bowl. The water fluidizes the bed and allows heavier particles to continuously displace lighter particles. The wager addition is the key to the performance of the Centrifugal Gravity Concentrator. The degree of fluidization controls the effectiveness of separation.

6.1.4 Tall Column Flotation

The effectiveness of mechanical flotation devices decreases in ultra-fine particle size largely because of the large bubblesize (as large as 1 mm) and turbulent conditions present in the cell. The column flotation technology has been very popular and effective recently in the flotation of ultra-fine particles. Considering the fact that a large fraction of radionuclides is present in the 38 micron size soil and its concentration is in parts per billion (ppb) range, it is conceivable that a combination of conditions such as fine bubble size (30-60 microns), quiescent conditions, and froth drainage mechanism prevalent in the column will result in a selective separation of discrete ultra-fine radionuclides from contaminated soil.

The column flotation is a tall device, having at least a length-to-diameter ratio of 10:1. The reagentized slurry is fed at the upper portion of the column and travels downward.

The hydrophobic particles attach to the rising stream of fine bubbles generated at the bottom of the column. The swarm of air bubble-laden particles are further washed at the top of the column to minimize the entrainment of unwanted material (in this case, clean soil). The radionuclide-enriched soil fraction overflows at the top of the column and hydrophilic clean soil unattached to bubbles is collected at the bottom.

6.1.5 Automated Mechanical Flotation (Denver Unit)

Flotation is a physic-chemical process in which one mineral constituent can selectively be separated from another on the basis of surface properties. This is achieved by adding controlled additions of chemical reagents at a predetermined pH, thereby selectively altering the surface characteristics of radionuclide enriched particulates. This treatment renders soil particles contaminated with radionuclides as hydrophobic (water repellent). Phase separation is then followed by passing air through reagentized slurry. Air bubbles selectively attach to radionuclide-enriched soil particles and are levitated to the surface in the form of froth. The separation of soil particles contaminated with radionuclides thus renders the remaining soil clean.

The Automated Mechanical Cell, developed by UNR, is a modification of the Denver D-12 VAC adjustable automated froth removal system and a controller to maintain constant pulp-froth interface. The modification to the Denver unit is in the mounting of the main shaft.

6.1.6 High Gradient Magnetic Separation

Los Alamos National Laboratory, in conjunction with its industrial partner Lockheed Environmental Systems and Technology Co., is exploring a promising new technique that could be used to remove radioactive contaminants from soils. The technique, high-gradient magnetic separation (HGMS), takes advantage of the fact that all actinide compounds are slightly magnetic. Much of the contaminated soil contains plutonium and uranium oxide particles; these slightly magnetic particles are attracted by very strong magnetic fields and thus can be separated from the mostly non-magnetic soil. The availability of reliable superconducting magnets, which create very strong magnetic fields, makes HGMS an attractive method for extracting actinide contaminants. Preliminary experiments with magnetic surrogates and modeling of the process have yielded encouraging results. Contaminated soil samples from DOE sites are now being tested, and the partners are working to develop the process for full-scale site remediation.

6.1.7 The Sepor System

The U.S. Naval Academy (USNA) Sepor system is a commercially available, bench-scale air separation technology being evaluated by the USNA to remove heavy metals from soils.

The system produces 2 effluent streams, 1 containing predominately smaller-sized particles and referred to as the fine discharge, and the other containing predominately larger particles, referred to as the coarse discharge. It is the goal of the evaluation to concentrate most of the heavy metal in one of the effluent streams, so as to reduce the volume of contaminated soil requiring site removal.

6.1.8 Acoustics

A DOE project is to develop an acoustic-based technology to improve remediation of contaminated soil and groundwater. Applying an acoustics excitation field (AEF) to contaminated soil will increase soil permeability and contaminant mobility, which will enhance removal or treatment of subsurface contamination and significantly decrease the time needed for remediation.

6.1.9 Overburden Removal

There is a need to reduce the total amount of soil that requires treatment by removing precise incremental layers of either contaminated or clean soil. The objective of the DOE Overburden Removal project was to demonstrate that discrete thicknesses of overburden soil can be removed with precision and accuracy and that fugitive dust can be controlled during excavation. The overburden removal system is a Caterpillar EL300B excavator, fitted with an innovative end-effector. The end-effector is specially designed to remove incremental layers of soil from the area of excavation.

6.1.10 Other Physical Techniques

The radioactive contaminants in soils and tailings in many cases are associated with the finer fractions. Thus, size separation may be used to reduce the volume of concentrated material for disposal, leaving a cleaner fraction. Physical separation may be used with extraction to further reduce contaminant volume. Four physical separation technologies are screening—both wet and dry, classification, flotation, and gravity concentration.

Screening: Screening separates soil (or soil-like material) on the basis of size. It normally is applied to particles greater than 250 microns. The process can be done dry or by washing water through the screen. Screening is not efficient with damp materials, which quickly blind the screen. It may be particularly effective as a first operation to remove the largest particles, followed by other methods.

Application: Appropriate for all soils, can separate fractions as low as 50 micron in size. Advantages: Simple and inexpensive method. Disadvantages: Noisy. Dry screening requires dust control. Wet screening will require separation of contaminants from water.

Classification: Classification separates particles according to their settling rate in a fluid. Several hydraulic, mechanical, and non-mechanical configurations are available. Generally, heavier and coarser particles go to the bottom, and lighter, smaller particles (slimes) are removed from the top. Theoretically, classifiers could be used to separate the smaller particle fractions, which may contain much of the radioactive contamination in waste sites. Classifiers could be used with chemical extraction in a volume reduction process.

Application: Effective for sandy soil with low clay and humus content. Advantages: Low cost, reliable, high continuous processing capabilities. Disadvantages: Humus and clay soil are hard to separate by classification.

Flotation: Flotation is a liquid-froth separation process often applied to separate specific minerals (particularly sulfides) from ores. The process depends more on physical and chemical attraction phenomena between the ore and the frothing agents, and on particle size, than on material density. Ordinarily, flotation is applied to fine materials; the process often is preceded by grinding to reduce particle size.

Application: Effective for extraction of radium from uranium mill tailings. Advantages: If the particle fraction containing the contaminants can be collected by the froth, then flotation is a very effective tool. Disadvantages: New additives may have to be developed to permit successful flotation separation for radioactively contaminated materials.

Gravity Concentration: Gravity concentration is an old and proven technology that takes advantage of the difference in material densities to separate the materials into layers of dense and light minerals. Separation is influenced by particle size, density, shape, and weight. Shaking and other motions are employed to keep the particles apart and in motion. Gravity separation can be used in conjunction with chemical extraction.

Application: Limited to those soils in which the contaminants are relatively coarse and capable of resisting breakage and sliming. Advantages: Highly efficient and proven process for a wide range of applications. It gives a high-grade concentrate over a wide range of particle sizes and functions well with most soil types. Disadvantages: Low capacity. High throughput requires multiple decks, clean water. Must ensure there is no slime buildup in recycle water.

6.1.11 Dust Suppression

Dust suppression materials have been evaluated for soil removal that will not adversely impact subsequent soil treatment processes such as soil washing. Natural polysaccharides, beet starch and potato starch, were tested for their ability to fix surface soil and suppress dust generation. The test showed that the natural polysaccharides are generally economically favorable to synthetic products, have the potential to fix soil contaminants in a soluble matrix, and can be easily broken down during soil treatment processes.

The Electrostatic Curtain addresses the problem of containing airborne dust contaminated with Pu-239 and Am-241.

The Electrostatic Curtain uses grounded conducting plates to form the walls of an inner containment structure to capture charged contaminated dust particles. The grounded conducting plates are also used in a ventilation system upstream from a HEPA filter to neutralize charged dust particles entrained in an air stream drawn from within an enclosure. A double enclosure with a ventilation system was used for the experiments.

Electrostatic curtains can provide in-depth contamination control during TRU waste handling operations. Removal efficiencies as high as 99% have been obtained in ventilation systems.

The electrostatic curtain technology minimizes dispersal of contaminated dust during excavation and retrieval. This technology maintains a safer work environment in contaminated environments.

The Contamination Control Unit (CCU) is a field deployable self-contained trailer mounted system to control contamination spread at the site of transuranic (TRU) handling operations. This is accomplished primarily by controlling dust spread. This demonstration was sponsored by the U.S. Department of Energy's Office of Waste Technology Development Buried Waste Integrated Demonstration. The CCU, housed in a mobile trailer for easy transport, support 4 different contamination control systems: water misting, dust suppression application, soil fixative application,and vacuuming operations. Assessment of the CCU involved laboratory operational performance testing, operational testing and contamination control at a decommissioned Idaho National Engineering Laboratory reactor, and field testing in conjunction with a simulated TRU buried waste retrieval effort at the Cold Test Pit.

6.1.12 Volume Reduction Case Study

There is a volume reduction system being operated at Johnston Atoll, a site with large volumes of plutonium-contaminated soil. The system combines wet and dry volume reduction. The latter method is very successful because contamination at Johnston Atoll is not uniformly distributed—a condition common for most contaminated soils. Contaminated and uncontaminated soils are interspersed as a result of non-uniform initial disposition, weather, vegetation, traffic, or previous clean up efforts. Excavating only the contaminated soils from a site is difficult because excavation equipment, such as bulldozers, is not able to remove just the contaminated spots, and operators of the equipment have little experience in soil clean up. Site managers also are inclined to excavate large soil quantities to ensure that all contaminants have been captured. As a result, large volumes of clean soil typically are excavated along with contaminated soil. Volume reduction procedures, which separate or sort clean soils and contaminated soils to different paths, reduce the volume of soil requiring wet corrective action.

Two methods typically are used to analyze soils at sites contaminated with radionuclides: (1) the removal method, in which samples are drawn at various locations across the site and analyzed in a laboratory; and (2) the in situ method, in which a radiation detector is used to estimate an average contaminant concentration for an area much larger than the size of removal samples. The Johnston Atoll clean up plant employs a third method, which combines the best features of the other 2 methods. This method, known as the conveyor method, conveys all suspect soil beneath detectors under well-defined conditions and automatically sorts clean soil from contaminated soil.

First, excavated soil is screened to remove large rocks. These rocks, which have a relatively large volume with respect to their contaminated surface area, typically are cleaner than the sand and soil fines. As a result, their presence lowers the average radioactivity concentration of the soil. Removal of oversize rocks by screening is an effective volume reduction technique. The rocks must be crushed, however, to ensure that they are clean. Once separated out, large rocks pass through a crusher, which reduces their size and allows radionuclides on their surfaces to be detected more easily.

After the screening process, several devices are used to sort soils based on their levels of radioactivity. These sorters have an array of radiation detectors on 3-ft wide conveyors that analyze batches of soil. Each batch is approximately 4 inches wide, 1 foot long,and 3/4 inches deep,and is counted for

2 seconds. The detectors trigger gates that direct each batch of soil either to a contaminated path or to a clean path.

After soils are separated into clean and contaminated paths, soils on the contaminated path are subdivided further to separate uniformly contaminated soil fines from contaminated particles. Contaminated particles are defined as those having more than 5,000 becquerels (Bq) of radioactivity, which is equivalent to a pure plutonium oxide particle about 70 microns in diameter. As soon as a contaminated particle is identified, it is diverted to a drum. Contaminated fines continue on to a washing system, which includes a spiral classifier and a settling pond. This system separates the very finest, highly contaminated, soils from the larger, less-contaminated, fines.

6.2 SOIL WASHING

Soil washing uses a combination of physical separation and chemical extraction technologies. Contaminated soil or tailings are mixed with water and/or extraction reagents. The clean coarse particle sizes are separated from the liquid containing the fines and radioactive material by a combination of physical separation methods. The radioactive material would then be extracted from the liquid by standard water treatment processes such as filtration, carbon treatment, ion exchange, chemical treatment, and membrane separation.

The main advantages of using water are that it is very inexpensive, completely non-toxic, uses ambient temperatures, and utilizes simple extraction vessels. The technique can be used to dissolve some radionuclide salts. It can be used as a pretreatment technique to reduce interference at subsequent extractions.

Soil washing/extraction is a broad term that involves size segregation of soil fractions using water as solvent to selectively extract specific contaminants (organic compounds, metals, and/or radionuclides) and using a variety of chemical leaching media. The major objective is volume reduction, when water is used, and the transfer of contaminants from the soil phase to a liquid phase, when specifically designed extraction media are used.

Each site must be characterized: type of soil, type and combination of contaminants, treatment possibilities, types of additives, side effects of the additives on soil washing equipment and personnel. All site debris has to be sorted, separated, and prepared prior to any soil washing activity.

Always recycle wash water, but not the additives. There are problems with precipitating radionuclei out of the washing solution.

Water should be used as the baseline for any soil washing methodology:

1) Do not use ionized water.
2) Do not use water that has gone through a filter.
3) Tap water may not be preferable.
4) Must use local (indigenous) water.

Representative soil washing additives:

1) Natural citric acids (biodegradable).
2) EDTA (13% solution).
3) Sodium sulfide (precipitates lead from soil washing solution).

The higher the concentration of gravel and sand, as compared to clay and silt, the more amenable the soil is to washing.

The information listed below must be collected and considered before implementing soil scrubbing and physical separation procedures.

1) Nature of the soil: sandy, clay, humus.
2) Nature of the particle: size, shape, specific gravity, mineralogical and chemical properties.
3) Radionuclide distribution with particle size.
4) Nature of the contaminant-chemical and physical properties.

Removal of uranium from heavy textured soils by conventional soil washing processes is ineffective because of the sorption of uranium on the high silt and clay content of these soils. A chemical extraction technique, one that selectively extracts uranium without causing serious physicochemical damage to the soils, is required.

One soil washing process was tested at the Montclair Superfund site in New Jersey.

This technology was designed to reduce the volume of soils contaminated with low concentrations of radionuclides. The process is used with soils in which radioactivity is concentrated in the fine soil particles and in friable coatings around the larger particles.

The soil washer used attrition mills to liberate the contaminated coatings and then uses hydroclassifiers to separate the contaminated fines and coatings. Next, a filter press dewaters the contaminated portion in preparation for off-site disposal. The clean portion remains on site, reducing the high costs of transporting and burying large volumes of low-level radioactive soil.

The result was a 56 % volume reduction of 40 picoCuries/gram soil, with the clean portion at 11 picoCuries/gram. The soil washer also achieved steady-state operations for 8 hours, with little operator assistance, at the rate of approximately 1 ton/hr.

As an example of a soil washing system, the BioTrol® Soil Washing System (U.S. Patent No. 4,923,125) is a water-based, volume reduction process for treating excavated soil. Soil washing is based on the premise that: (1) contaminants tend to be concentrated in the fine size fraction of soil (silt, clay, and soil organic matter), and (2) contaminants associated with the coarse soil fraction (sand and gravel) are primarily surficial. The objective of the process is to reduce the volume of soil that requires treatment by concentrating the contaminants in a smaller volume of material while producing a washed soil product which meets appropriate clean up criteria.

Following debris removal, soil is mixed with water and subjected to various unit operations common to the mineral processing industry. Process steps can include mixing trommels, pug mills, vibrating screens, froth flotation cells, attrition scrubbing machines, hydrocyclones, screw classifiers, and various dewatering operations.

Intensive scrubbing is the technology at the core of the process. For the gravel fraction, scrubbing is accomplished with a mixing trammel, pug mill, or ball mill. For the sand fraction, a multi-stage, counter-current, attrition scrubbing circuit with inter-stage classification is used. This scrubbing action disintegrates soil aggregates, freeing contaminated fine particles from the sand and gravel fraction. In addition, surficial contamination is removed from the coarse fraction by the abrasive scouring action of the particles themselves. Contaminants may also be dissolved as dictated by solubility characteristics or partition co-efficients.

These three mechanisms: (1) dispersion and separation of contaminated fine particles, (2) scouring of coarse particle surfaces, and (3) dissolution of contaminants each operate to varying degrees, depending upon the characteristics of the soil and contaminant(s).

To improve the efficiency of soil washing, the process may include the use of surfactants, detergents, chelating agents, pH adjustment, or heat. In many cases however, water alone is sufficient to achieve the desired level of contaminant removal while minimizing the cost.

A significant reduction in the volume of material which requires additional treatment or disposal is accomplished by separating the washed, coarse soil components from the process water and contaminated fine particles.

The contaminated residual products can be treated by other methods. Process water is normally recycled after biological or physical treatment. Options for the contaminated fines can include off-site disposal, incineration, stabilization, or biological treatment.

As another example, in the pilot-plant testing at Rocky Flats, the plutonium-contaminated soil was washed in a rotating drum washer using a pH 11 NaOH solution as a washing agent. A trammel screen was used to separate the coarse particles (+5 mesh), and a vibrating screen was used for further particle separation (+35 mesh). This was followed by use of a hydrocyclone and classification to separate +10 micron particles. Centrifugation and ultrafiltration were employed to separate the fine contaminants. The water was sent back for recycle without any purification.

The EPA Soil Washing System, developed by the EPA Risk Reduction Engineering Laboratory at Edison, NJ, uses a scrubber extraction process to clean soil. Pilot studies were performed to select the equipment for the EPA soil washer. Three unit operations were developed and proved by testing:

1) *Water Knife Concept* - A thin, flat, high-speed water jet breaks up clumps of soil and scrub contaminants from larger soil particles like stone and gravel. Testing showed that this concept is very effective.

2) *Rotary Drum Screener* - A rotary drum was employed as a pretreatment to mix the soil with the extractant and separate the soil into 2 particle size categories (+2mm and -2mm).

3) *Extraction and Separation Concept* - A 4 stage counterflow extraction train was designed and built to treat the -2mm soil fraction separated by the drum screener. Each stage consists of a tank, stirrer, hydrocyclone, and circulating pump. The pump moves the soil from one stage to the adjacent stage. The hydrocyclone discharges the soil slurry in the next stage and returns the extractant. The extractant flows by gravity as a tank overflows in a stream from one tank to another, counter to the direction of the soil. Fresh extractant is added to the fourth stage, and spent extractant is removed from the first stage.

DOE, EPA, and DoD sites need to develop products that can easily be applied during retrieval operations to suppress dust generation and contamination spread, specifically in cases where there is a need to reduce generation dust on large, disturbed areas where synthetic chemicals are not permitted.

DOE conducted a demonstration project to evaluate inexpensive dust suppression materials that will not adversely impact subsequent soil treatment processes, such as soil washing. Natural polysaccharides, beet starch and potato starch, were tested for their ability to fix surface soil and suppress dust generation. The test showed that the natural polysaccharides are generally economically favorable to synthetic products, have the potential to fix soil contaminants in a soluble matrix, and can be easily broken down during soil treatment processes.

6.3 CHEMICAL EXTRACTION

Chemical extraction generates 2 soil fractions. One fraction contains the concentrated radioactive contaminants and may require disposal; the remaining material is analyzed for residual contamination and evaluated for replacement at the point of origin or at suitable alternative sites. The various applicable chemical extraction techniques include agents, as well as other techniques discussed here.

Selective extraction and leaching of contaminants has the potential to clean soil to acceptable levels and significantly reduce the volume of contaminates to be disposed of.

Physical separation methods are quite effective for contaminated soils in which a large fraction of the contamination is concentrated in a small volume of soil that can be separated by density or size. However, at some sites such as Fernald, the uranium contamination is associated with all size and density fractions of the soil. Consequently, it appears that traditional physical separation methods will not be applicable, and that any soil washing strategy will depend upon a chemical extraction process.

Past and current soil remediation technologies have been directed towards either ex situ or in situ processing for soil treatment. Contaminated soils may contain a variety of radionuclides and heavy metals which require different remediation techniques. Current technologies are very expensive and complex, and frequently generate secondary wastes. An integrated system approach utilizing a physico-chemical treatment train might provide a cost-effective alternative to stand-alone, individual soil treatment technologies for the removal of uranium and related non-radioactive heavy metals. Such a process should remove uranium from soil to an acceptable residual level of less than 35 pCi/g, and non-radioactive heavy metals to target levels promulgated by the EPA.

Removal of uranium from heavy textured soils by conventional soil washing (defined here as a physical separation process that relies on size fractionation and concentration of contamination in the fines) is ineffective because of the sorption of uranium on the high silt and clay content of these soils. True restoration of soils with a high fraction of fines is highly dependent on the application of a chemical extraction technique that will selectively extract uranium from soils without impairing them with serious physico-chemical damage. The effectiveness of uranium extraction will depend upon how well the particular chemical reagent can contact and dissolve the specific uranium form. Extractants that alter the oxidation-reduction potential of the extraction environment may be useful in that such changes weaken crystalline structures enhancing the dissolution of uranium from mineral phases.

Analyses have shown that the uranium exists primarily in particulate form. It is associated with the sand and silt fractions of the soil, but some samples also have uranium in the clay fraction. More than 80 percent of the uranium is in the hexavalent oxidation state. In general, hexavalent uranium has greater solubility than uranium in other oxidation states. Thus, strong oxidizing agents may not be necessary as part of a chemical remediation scheme.

Much of the particulate uranium exists in discrete, crystalline mineralogical phases. Uranium bearing phases identified include: (1) uranium absorbed onto iron oxides; (2) uranium silicates; (3) uranium phosphates; (4) uranium oxides; (5) calcium uranium oxide; and (6) uranium contained within a calcium fluorite phase. Particles of uranium (IV) phases have also been identified: (1) uranium silicide and (2) uranium oxides.

Characterization of residual uranium waste forms in treated soils show a slightly higher ratio of tetravalent to hexavalent uranium. This suggests the extracts being applied are less effective at removing tetravalent uranium. There is a decrease in size of particulate hexavalent uranium. In particular, a phase identified as meta-autunite was never seen in treated soils. All treatment technologies seem to lead to a more dispersed, finer-grained contamination.

Treatability tests are being conducted on the following processes: (1) carbonate, citric acid and CBD (sodium citrate/sodium carbonate/sodium dithionite) leaching (ORNL); (2) aqueous biphasic separation (ABS) (ANL); extraction utilizing organic chelators (LANL); (3) the Westinghouse/SEG soil washing process; bioextraction (INEL); and (4) column leach tests to allow comparisons between batch extraction and heap leaching (LANL). The Environmental Restoration program at Fernald has been investigating sulfuric acid. Lab-scale tests (using soil initially at approximately 500 ppm uranium) resulted in the following approximate extractions: (1) 80% with carbonate (2) 65-79% with citrate (3) 95% with CBD (4) 95% with sulfuric acid (5) 80% with ABS involves the selective partitioning of either solutes or colloid-size particles between 2 immiscible aqueous phases (6) polyethylene glycol being one of the phases.

ABS can successfully separate particles ranging from 50 μm to 20 nm. Effectiveness is dependent upon the degree to which uranium is present as discrete particulates, 65-85% with TIRON (TIRON combined with the reducing agent dithionite was able to remove 85%). Heap leaching using potassium bicarbonate/carbonate or sodium bicarbonate/carbonate was able to remove approximately 80-85% of the uranium after a 24 hour leach. The SEG process combines physical separation equipment with ammonia carbonate extraction. SEG tests indicate uranium was soluble in sodium hypochlorite/ammonia carbonate solutions. Microorganisms indigenous to Fernald soils were investigated; however, no beneficial effects to uranium extraction were detected. Tests of the microorganism *Thiobacillus ferrooxidans*, a bacteria, show enhanced (slightly faster) sulfuric acid extractions.

6.3.1 Extraction with Inorganic Salts

Radioactive contaminants can be extracted by thoroughly mixing soil and mill tailings in an inorganic salt solution. The slurry is filtered, separating the extractant from the solids. The radioactive contaminant is separated from the extractant by ion exchange, co-precipitation, or membrane filtration.

An increasing ratio of salt solution to solid, as with water, plays a positive role in the effectiveness of the salt solution in removing radionuclides from ore tailings and soils. Multistage extraction increases the effectiveness of the extraction essentially by increasing the ratio of solution to solid.

A review of the literature indicates a broad range of results with the use of salt solutions to remove radium and thorium from mill tailings and soils. In many cases the effectiveness of a given salt appears to be related to several obvious variables, such as the nature of the tailings or soil (geochemistry, radionuclide concentration, method of extraction, particle size distribution, and chemical composition), the concentration of the salt solution, temperature, pH, solid to liquid ratio, time, and temperature.

Application: Should not be used as a pretreatment to an acid extraction process. The presence of sulfate and hydroxide in soils and tailings will negatively impact the efficiency of radium and thorium removals. Advantages: A high percentage of radium and thorium removal is possible. Simple extraction vessels are needed. Recycling of salts is possible. Disadvantages: Large amounts of salts may be required. Some salts, such as chloride, may be environmentally undesirable.

6.3.2 Extraction with Mineral Acid

In these processes, the ore is ground to 28 mesh and mixed with water to form a slurry. The slurry is pumped into an acid leach circuit, maintaining a pulp consistency of 50 percent solids. The solids are separated from the leach liquid by physical methods. The radioactive material is removed from the leach solution by ion exchange, solvent extraction, or precipitation.

Sulfuric acid, rather than hydrochloric or nitric acid, is commonly utilized for leaching in uranium extraction due to its less corrosive nature and lower costs.

Application: Removes most of the metals, both radioactive and non-radioactive. Advantages: High percentage of radium removal is possible. Uranium and other metals would also be removed. Disadvantages: Increased operating and capital costs due to expensive reagents, higher operating temperatures, and the corrosion resistant material required. The resulting chemically leached material may create a harmful waste stream.

In a nitric acid process developed at INEL, soils are screened, classified, and placed into a leaching unit with hot nitric acid. Contaminants—cessium-137, cobalt-60, and chromium—are removed from the leachate using a system of ion exchange, reverse osmosis, precipitation, or evaporation. In a similar process, contaminants are sequentially exposed to milder leachates such as oxalic acid and hydrogen peroxide. This process is designed to remove successive layers of weathering deposits from surfaces of the soil particles.

The process produces sludge from leaching and precipitation, large-grained material from the screening plant, and residuals from the other processes. Ultimate disposal options include solidification, calcining leachate, and storage of residuals.

A pilot-scale test of the process was completed late in 1992 at the DOE's Idaho National Engineering Laboratory (INEL) Superfund site. Testing results indicated excellent removal efficiencies for cobalt-60 and chromium, utilizing either the sequential extraction or the hot nitric acid. Cesium-137 could be removed only with successive dissolution steps in nitric acid. Approximately 30 percent of the soil matrix was co-dissolved in order to achieve release of most of the cesium-137.

Another acid digestion process is used to carbonize organic waste using hot (250° C) concentrated sulfuric acid and to then oxidize the waste, forming carbon dioxide, using nitric acid. In addition to the CO_2, final products are H_2O, HCl, and sulfate residues. This process could be adapted for treatment of common types of LLW, but at present it is used for plutonium recovery from PCM (probably GTCC-LLW) resulting from HLW processes. The off-gas is treated before release.

Advantages: Useful pretreatment process in recovery of plutonium from plutonium-contaminated waste; process forms plutonium sulfate, acids are recyclable, no fly-ash formation, no afterburner is needed. Disadvantages: Highly corrosive nature of the acids, off-gas treatment system is complex, possible nuclear criticality risk when processing PCM.

6.3.3 Extraction with Complexing Agents

This process differs from acid extraction in that complexing agents like EDTA (ethylenediaminetetraacetic acid) are used instead of mineral acids.

Application: Radium from soils with low concentrations of thorium. Advantages: High percentage of radium removal. Low reagent concentrations required, reagent can be recycled, reducing operating costs. The process works at ambient temperatures, and many of the reagents are innocuous. Disadvantages: Reagents are expensive, process would not remove thorium.

6.3.4 Citric Acid Process

Citric acid and citrate/dithionite have been used as extractants.

Citric/dithionite extraction procedures have removed >90% of the uranium from both storage pad and incinerator soils at Fernald. Citric acid extraction (at pH <5) has removed 90% and 50% of the uranium from the Fernald storage pad and incinerator soils respectively.

Citric acid extraction followed by 2 carbonate extractions containing $KMnO_4$ removed > 80% of the uranium from the incinerator soil.

Use of citric acid for soil decontamination is very expensive compared to the other competing extractants. The ability to recycle citric acid could reduce the overall cost of chemicals and may have a substantial impact on the overall process cost. A technology is needed that can break the uranium/citrate complex down without destroying the citric acid ability to complex uranium, allowing citric acid to be further utilized for extraction purposes.

The CBD process leaching process involves the use of sodium citrate, sodium carbonate, and sodium dithionite.

6.3.5 Carbonate/Bicarbonate Process

Carbonate/bicarbonate is highly selective for uranium. The latest results indicate only negligible quantities of other compounds (non-uranium bearing) were removed from the soil during leaching.

It is not necessary to separate the soil into its size fractions (i.e., gravel, sand, clay, and silt) before it can be leached.

The liquid-to-solid ration can be kept quite low (1 to 2 liters per kg soil per 24 hour day) to minimize (eliminate) the need for dewatering of the soil after treatment.

Uranium removal is a function of carbonate/bicarbonate concentration. Carbonate extractions removed 70 to 90% of the uranium from Fernald storage pad soil.

Aqueous solutions of carbonate/bicarbonate of either potassium or sodium are being tested because they are the reagent of choice in the uranium mining and processing industry. Future reagents, however, may depart from carbonate/bicarbonate.

Pilot-scale heap leaching tests are envisioned where contaminated soil will be excavated and placed (heaped) on an impermeable pad on the surface of the ground. The pad will be sloped toward a sump at the bottom edge of the heap. Selected leaching reagent(s) will be pumped to and distributed on top of the heap with a drip irrigation system or aerial sprayers. Reagent will travel down through the soil, solubilizing and mobilizing the contaminant(s). The leachate will then be collected from the sump and pumped to a leachate treatment and regeneration system. This system will remove the contaminant(s) from the leachate and regenerate the leaching reagent for return to the top of the heap. The process will be continued nonstop until the contaminant(s) in the soil have been reduced to EPA standards.

Heap leaching for soil clean up is an adaptation of a proven mining method for removing precious and semi-precious metals from low-grade ore. In the mining industry, thousands of tons of ore are processed daily.

6.3.6 Chelation

This approach is focused on the use of siderophores (microbial iron chelators) and biomimetic analogs as mobilizing agents for uranium. Based on the use of these chelators, together with redox chemistry, chemical extraction of the uranium can be performed under mild conditions. This process will produce an aqueous leach solution that will be treated to remove and concentrate the solubilized uranium. If required, the treated leach solution will be recycled.

Inputs for the process are contaminated soil and a neutral aqueous solution containing chelators and mild redox agents. In some cases, it may be beneficial to pretreat the soil with a physical separation process or another chemical leaching process to either pre-concentrate the uranium or to remove a fraction of the uranium not amenable to this process.

The outputs of the process are soil from which the majority (90-99%) of the uranium contamination will have been removed, and an aqueous leach solution containing the solubilized uranium.

Another process involves the selective chelation and extraction of lanthanides and actinides with lariat crown ethers in supercritical fluids technology. Proton ionizable crown ethers were shown to be highly selective for complexation with lanthanides and actinides.

6.3.7 Aqueous Biphasic Separation

ABS can successfully separate particles ranging from 50 μm to 20 nm. Effectiveness is dependent upon the degree to which uranium is present as discrete particulates, 65-85% with TIRON (TIRON combined with the reducing agent dithionite was able to remove 85%).

Polyethylene glycol is used in the Aqueous Biphasic Extraction process. Immiscible aqueous solutions containing PEG are mixed with uranium contaminated soil. Following mixing, particulate uranium selectively partitions to one phase, while soil partitions to the PEG phase. Removal of soil from the PEG is difficult. A technique is needed to efficiently remove clean soil from PEG for recycle of the PEG to reduce chemical costs, to reduce secondary waste treatment/disposal burden for spent PEG solutions, and to facilitate production of clean soil suitable for landscaping purposes.

TIRON is very expensive, compared to the other extractants being evaluated for use in treating uranium contaminated soils. The ability to recycle TIRON will reduce the overall cost of chemicals, which may have a substantial impact on the overall process cost. A technology is needed that is capable of breaking the uranium-TIRON bond without destroying the ability of TIRON to bond with uranium, therefore allowing it to be further utilized for extraction purposes.

6.3.8 Separating Radionuclides from Extractants

There are 2 chemical techniques for removing radionuclides from the pregnant liquor.

Precipitation and Coprecipitation - By addition of chemicals the radionuclides can be precipitated. Several stages of precipitation at controlled pH are used. The pH is readjusted in the precipitation tank near the end of the circuit. The slurry from the precipitation tank is dewatered in thickeners and followed by filtration. The filter cake, containing the concentrated radionuclide, is then ready for disposal. Precipitation, however, produces products with impurities. This may not be a problem on cleaning soils and tailings. However, in extraction of uranium from ore, solvent extraction or ion exchange is used before precipitation to obtain a purer product.

Solvent Extraction - Solvent extraction is an efficient method for separating uranium on a commercial scale. There are no commercial solvent extraction processes to extract radium or thorium. The solvent extraction, as applied to uranium extraction plants, consists of a two-step process. In the first step, termed "extraction," the dissolved uranium is transferred from the feed solution (or aqueous phase) into the organic or solvent phase. The second step, called "stripping," recovers the purified and concentrated uranium product into a second aqueous phase after which the barren organic is recycled back to the extraction step. The aqueous and organic solutions flow continuously and countercurrently to each other through the required number of contacting stages in the extraction and stripping portions of the circuit. The uranium is recovered from the second aqueous solution by precipitation.

The extraction of metal from the aqueous solution and its eventual transfer to another aqueous solution (the strip liquid) involves the use of various reagents (extractants, diluents, and modifiers) and requires a suitable vessel to bring about intimate contacts between the different liquids. The extractants are the reagents in the solvent that extract the metal ions. Extractants that are used in recovery of uranium from acid leach solutions are alkylphosphoric acid, amines, tri-n-butyl phosphate (TBP) and trioctyl phosphine oxide (TOPO).

6.4 COVERS AND LINERS

Covers can be applied to any site to reduce atmospheric emissions and to reduce disturbance of and percolation through the waste as well as reduce runoff and soil erosion. Liners and liner replacements are associated with landfills. Both engineered surface covers and subsurface barriers must be compatible for particular soil and environmental conditions.

The use of covers and liners as engineered barriers to provide containment can be relatively effective, especially over the short term. However, they are generally much less effective in providing long-term containment, especially for corrosive wastes. A successful liner replacement technology would be very useful to help assure long-term containment.

In situ underground installation of liners (for example, to replace failed liners) is difficult to achieve and to verify.

With the exception of radioactivity, there are no safety issues regarding covers/liners other than that normally associated with such construction projects. In situ replacement of failed liners could be very hazardous, in terms of waste as well as opening stability (unless done totally by remote control).

Containment technologies, including surface caps, are essential to reduce the potential for contaminant migration from the landfill by an alteration of the surface and/or subsurface soils. The process of selecting containment cover technologies for landfills requires consideration of many complex and interrelated technical, regulatory, and economic issues. A decision support system is needed to integrate the knowledge of experts from scientific, engineering, and management disciplines to help in selecting the "best" capping practice.

Hydrologic analysis might identify a particular barrier design as "better" in controlling runoff (and erosion) from the site, but at the expense of increasing water infiltration into the landfill. A method to decide whether the increased infiltration will significantly enhance the potential of deep percolation and concomitant migration of solutes toward groundwater, and whether this enhanced migration has relevance in light of other factors, such as thickness of the unsaturated zone, potential use of the water, climate, etc., can be very useful.

Most landfills require both above-ground and below-ground barriers. In arid environments, capillary barriers are often used in containment systems. Incorporating dry barriers into the containment systems would allow inexpensive isolation in many circumstances, and extend the probable life of the capillary barrier. The dry barrier concept addresses a number of issues associated with landfills. If a low-maintenance dry barrier can be incorporated into the design, the cover design can be improved, and perhaps its longevity can be extended. Dry barriers used as liner can serve as both a redundant barrier to liquid flow and as a means of stripping gas-phase contaminants. For existing landfills on alluvial deposits, it may be possible to use an existing coarse layer as a dry barrier.

Field-tested migration barrier cover designs, tailored to the climate, can serve as the sole containment technology or as a component of an integrated barrier system that incorporates other barrier concepts, along with cover, to contain wastes. In addition, the hydrologic control exerted by the cover can be used to establish optimum moisture conditions in the waste backfill to improve performance of other treatment technologies such as in situ vitrification (ISV), vapor extraction, and other in situ treatment technologies.

LANL has conducted the basic research and begun to field test various landfill cover designs, and has had some success in reducing erosion and percolation of water into underlying waste under local climatic conditions. However, tests for some of these barrier concepts in other climatic conditions (i.e., at Hill Air Force Base in Utah) and for wastes other than radionuclides have just begin. Factors such as climate, soils, vegetation, and waste composition are important site attributes that affect both the design and the performance of migration barrier cover systems. Field testing will evaluate the performance levels of each cap in preventing water percolation into the waste and in preventing soil erosion.

The toxicity of nuclear materials creates problems both in their isolation from groundwater and in the clean up of existing pollution sources. In addition, the strong adsorption of many radionuclides to soil particles makes physical extraction slow and expensive. Innovative technologies are needed that protect human workers from exposure to the materials while controlling the migration of contaminants. Remote barrier installation and mining systems might be able to fill that need.

A project out of Morgantown Energy Technology Center is an adaptation of shaft sinking and long wall mining methods used by the coal industry. Modifications will be made to commercial mining equipment to perform the tunneling operation to block out a panel which would then be mined by the longwall mining method. During the following extraction, the barrier consists of a synthetic membrane and, if needed, a leachate recovery system. The mat will be welded together to form a continuous barrier. Once completed, the containment system will prevent ingress of groundwater via the sides and bottom of the barrier while meteoric water will be controlled by a surface barrier. Total water flux through the system will be much lower than under conventional containment scenarios. In addition, if needed, a well system can be installed to periodically relieve pressures and treat the resulting leachate.

Another innovation is the Hanford Protective Barrier. Multi-layer surface barriers that use natural earthen materials overlying bio-intrusion layers of rock, asphalt, fine soil, sand, gravel, and riprap are expected to reduce infiltration or deep percolation to less than 0.5 mm (0.02 inches) of water. The layering also reduces root penetration, preventing uptake and translocation of hazardous materials to the surface vegetation.

Surface covers are one of the most widespread remediation and waste management options in all climates. Barrier layers to limit percolation through cover systems are principal features of engineered, multi-component cover designs. Conventional barrier layer components developed for humid climates have limitations in dry climates. One alternative barrier layer is a capillary barrier, which consists of a fine-over-coarse soil arrangement. The capacity of capillary barrier to laterally divert downward moving water is the key to their success. Another alternative is a dry barrier, in which atmospheric air is circulated through a coarse layer within the cover to remove water vapor. Incorporating a coarse layer which stores water for subsequent removal by air flow reduces the requirements for the air flow velocity and increases the applicability of the dry barrier.

6.5 EX SITU VITRIFICATION

A pilot test program using the Vortec 20 tons per day CMS vitrifier, handled dry granulated material with average particle size up to 600 microns. The resulting glass passed TELP and ANS standards for zirconium (a uranium surrogate).

This project consists of 3 phases. Phase I includes the identification of the physical and chemical properties of soils likely to require remediation at DOE sites, performance stemming on an existing 15-20 tons/day combustion and melting system (CMS) test facility, and development of a conceptual design of a sub-scale integrated system. The second phase includes the design, fabrication, and construction of an approximately 25 tons/day integrated facility. The third phase includes the design, fabrication, construction, and integrated testing of the 100 ton/day CMS facility at a DOE site.

Another task will demonstrate the potential value of a remediation approach in which multiple technologies (soil washing vitrification) are integrated and multiple waste streams are blended.

In this study, soil and sludges from the Weldon Spring, Missouri site and storm sewer sediments from the Oak Ridge, Tennessee site will be used for the testing of the system concept. Weldon Spring raffinate sludge will be included since this material, due to the low silica content of the sludge and the insolubility of its contaminants which are distributed through the bulk of each particle rather than on the surface, is not amenable to either soil washing or in situ vitrification techniques. Oak Ridge Y-12 storm sewer sediments will be included since they are characterized by both a high silica fraction and a high contaminant content. The particular batch selected for this study averaged 4,000 ppm mercury, 630 ppm uranium, 1.2 ppm thorium, and 24 ppm PCBs.

Vitrification tests demonstrated that glasses having a high waste loading could be produced from the wastes at Weldon Spring.

Raffinate sludges, due to their low silicate but high calcium, magnesium, and fluoride content could play the role of fluxes if appropriate blending schemes and glass composition formulations can be developed.

Testing on the Oak Ridge material showed that both mercury and uranium could be removed from the sediment by soil washing techniques using chemical extraction, reducing the volume of the waste stream by 80 percent.

For the contaminant-enriched minority fraction, thermal desorption was used to separate the mercury from the uranium, so as to produce a waste stream suitable for stabilization by vitrification.

A system that combined soil washing, thermal desorption, and vitrification on Oak Ridge wastes produced clean soil (about 90% of the input waste stream), non-radioactive mercury, and a glass waste form.

The estimated processing cost for such a system was in the range of $260 to $420 per ton of waste processed.

6.6 IN SITU TECHNOLOGIES OVERVIEW

In situ technologies are becoming an attractive remedial alternative for eliminating environmental problems. In situ treatments typically reduce risks and costs associated with retrieving, packaging, and storing or disposing waste and are generally preferred over ex situ treatments. Each in situ technology has specific applications, and, in order to provide the most economical and practical solution to a waste problem, these applications must be understood.

In situ treatment processes can remediate subsurface contaminants without excavation of the contaminated soils or extraction of the groundwater. The contaminants of interest can either be treated in place or transferred to the surface via a secondary carrier phase for subsequent treatment. In situ chemical/physical treatment involves additions to, or alterations of, the subsurface that change the chemical and/or physical properties of the subsurface environment. In situ remediation technologies are increasingly being sought for environmental restoration, due to the potential advantages that in situ technologies can offer as opposed to more traditional ex situ technologies. These advantages include limited site disruption, lower cost, reduced worker exposure, and treatment under obstructed structures and at depth.

This section presents an overview of 31 different in situ remedial technologies for buried wastes or contaminated soil areas. Some of those devoted to radionuclides are discussed in more detail further on.

6.6.1 Biological Treatments

Biological treatments utilize the natural activity of micro-organisms (primarily bacteria, actinomycetes, or fungi) to remediate polluted soils and groundwater. While biological treatments often require a longer period for remediation than other treatment alternatives, they have the potential to completely destroy organic contaminants and are relatively inexpensive. Following is a summary of several biological treatments.

Bioremediation: The natural activity of micro-organisms is used in the bioremediation process to decontaminate soils and groundwater polluted with organics. Effective micro-organisms are often found in small quantities at a contaminated site and, through nutrient enrichment, can be multiplied and encouraged to accelerate the natural degradation process. If the proper organisms are not already present, often they may be introduced. Bioremediation can be applied to chlorinated solvents and non-chlorinated organic contaminated water, soil, sludge, sediment, and other types of materials.

Bioaccumulation: Biological techniques can also result in the precipitation and immobilization of metals. Metals such as Fe, Cu, Zn, and Pb can react with hydrogen sulfide produced by anaerobic microbial activity and form insoluble metal sulfides. Although the toxicity and volume of the metals will not be changed, insoluble metal sulfides will not dissolve and therefore the possibility of their migration will be significantly reduced. Bioaccumulation has been applied to metal-contaminated soils, groundwater, and surface water.

Dual Auger System: This technology uses a dual auger system to drill into contaminated soils and inject micro-organism mixtures, water, and nutrients. This process is applicable to soils contaminated with organics. Soils at depths greater than 100 feet (30.48 m) can be treated.

Radionuclides: Biological processes are being investigated for decontamination of uranium contaminated soils; fungal metabolism is also being considered. Micro-organisms can bring about dissolution or immobilization of radionuclides under the proper conditions.

6.6.2 Containment Technologies

Containment technologies are used to reduce the mobility of contaminants. Containment may be used in conjunction with other in situ technologies to assist in the remediation of the site, or they may be used to control the migration of contaminants until an appropriate remediation technology is selected. However, containment does not treat the contaminants, and contained sites still require monitoring.

Containment barriers usually include walls, floors, and caps composed of various types of materials. Barriers may be formed from numerous materials such as concrete, polymers, vitrified soil, and frozen soil.

Bottom Sealing: Using a horizontal or directional drilling method, bottom sealing involves grout injection techniques to place horizontal or curved barriers beneath a hazardous waste site to prevent downward migration of contaminants. Once in place, the barrier acts as a floor and seals the bottom of the waste site.

This technology has possible applications in all soils, including silts, clays, and weak rocks. It can be used with most contaminants including inorganics, organics, metals, mixed, high-level, low-level, and TRU waste. It is used in soils that are contaminated with liquid waste that have the potential of migrating downward.

Capping: The capping process is used to cover buried waste materials to prevent their contact with the surface environment and groundwater. Generally, capping is performed when extensive subsurface contamination at a site prevents excavation and removal of the wastes due to potential hazards and/or unrealistic costs. Capping may be used for water, liquids other than water, gas, and/or soil contaminated with organics, metals, and/or radionuclides.

Polymer Concrete Barrier: This containment technology uses high strength, impervious polymer concrete to create an in situ barrier. Sealant materials are used that consolidate an earth/sand/gravel matrix into a high strength, impervious polymer concrete useful for the formation of barriers in the earth. These materials have very good chemical resistance and are typically 2 or 3 times stronger than structural concrete. This technology is effective for the containment of most contaminated waste. Residual risk from the untreated waste is greatly reduced once contained within a perimeter barrier with a sealant cap over the top (may also be composed of polymer concrete). This containment barrier could be used in conjunction with other in situ technologies.

Cryogenic Barrier: This type of barrier is formed by installing freezing pipes around the circumference of a contaminated site. A refrigerant fluid is pumped down the outside pipe and returned through the inner pipe. The double wall design allows the entire volume between walls to freeze, thus containing the site. If necessary, another in situ treatment could then be applied with little risk of contaminant migration.

This technology can be used to isolate or contain all types of contamination and can be used on all media states in which freeze pipes can be installed. It appears to be more cost effective to use this technology for temporary rather than permanent containment because of the high operational costs. Under certain circumstances, containment for a relatively short period of time is sufficient in itself. Cryogenic barriers are compatible with most other in situ technologies.

Fluidized-Bed Zeolite System: This system utilizes zeolite and particulate/solution polymer based grouts for in situ stabilization and isolation of radioactive and hazardous chemical waste materials that have been disposed in or near underground waste disposal and containment structures. The fluidized-bed will provide chemical fixation by mechanically homogenizing and incorporating waste tank residuals (tank bottoms and sludges) with granular zeolite (or equivalent) materials. Then particulate and solution polymer-based materials are incorporated into the interstitial void volume of the granular zeolite and surrounding geologic media to provide chemical isolation and physical stabilization.

This system could be used for remediation of subsurface waste storage/disposal structures such as underground storage tanks, cribs, caissons, piping, and buried sites. This technology will produce a physically stable structure, wherein contaminated materials are anticipated to be isolated from the environment over hundreds to thousands of years.

Plasma Arc Glass Cap: This technology uses a plasma torch to generate a high heat flux in the vicinity of the disposal site surface, thereby vitrifying the surface soil to create an impermeable glass cap. Depending on how the torch is operated, the cap may be anywhere from 1-6 inches (2.54 to 15.24 cm) deep.

The mobility of the toxic contaminants will be greatly reduced by placing an impermeable glass cap over the site. Moisture from rain and snow melt will be shielded from the waste, eliminating leaching and downward migration of the contaminants. Contaminants will be constrained from migrating upward. This technology can be used with all contaminants and soils that can be vitrified.

Slurry Wall: Slurry walls are subsurface barriers that are used to reduce groundwater flow in unconsolidated earth materials. Slurry wall construction involves excavating a narrow vertical trench through pervious soils, and then backfilling the trench with an engineered material. The backfill material is usually a mixture of soil and bentonite or cement and bentonite. The cement-bentonite slurry initially provides trench support (also prevents high fluid losses to the surrounding soil) and then sets to form an impervious barrier. Some slurry walls also use geomembrane liners to help prevent the migration of contaminants.

Slurry walls can be used to contain most contaminants with a few exceptions. Soil-bentonite slurry walls are not suitable for leachate or contaminated groundwater containing strong acids/bases and alcohols. Also, cement-bentonite slurry walls are not applicable for wastes or leachates containing chlorinated hydrocarbons, organic acids, or acid chlorides. Barrier walls are not totally impermeable to water and can only inhibit the spread of contaminants.

Soil/Cement Wall: The soil/cement wall technology involves fixation, stabilization, and solidification of contaminated soils. Solidification/stabilization agents are blended in situ with the contaminated soils by a multi-axis overlapping hollow stem auger. The product is a monolithic block that extends down to the treatment depth.

This technology is effective on soils that are contaminated with metals and semi-volatile organic compounds. This technology has been used on various construction applications, including soil stabilization and cutoff walls.

Vitrified Barriers: In situ vitrification (ISV) is a thermal treatment technology in which a region of soil volume is melted. This process can also be used to produce vitrified barriers. Upon cooling, the resulting product is a glass and crystalline monolith resembling natural obsidian. The process involves creating a barrier by inserting electrodes in the ground and placing a conductive starter path between them. Soil is melted when an electric potential is applied to the electrodes causing the starter path to heat up above the melting point of the soil. Vitrified walls and floors can be joined as needed to isolate waste sites from transport mechanisms or to totally contain them, if necessary (e.g., for additional in situ treatment). The vitrified soil barrier is extremely leach resistant and possesses about ten times the strength of unreinforced concrete. It is predicted stable over geologic periods of time. It also results in significant volume reduction because no additives are required and the soil is densified in the melting process.

This technology can be used to isolate or contain all contaminant types and can be used on all media states. It can be used to permanently contain a waste site or to temporarily contain a waste site while another method of in situ remediation is applied. There may be a concern in the presence of acids and salts. (see discussion of in situ vitrification).

6.6.3 Physical/Chemical Treatment

Physical/chemical treatments involve physical (heat, freezing, etc.) and/or chemical manipulation of a waste in order to reduce the toxicity and/or volume of the waste. In situ physical/chemical treatments can be used on soils, sludges, slurries, gases, sediments, and water. Contaminants may include metals, organics, radioactive contaminants, inorganics, acids, or bases. Following is a discussion on several physical/chemical treatments, including: dechlorination; electroacoustics; electrokinetics; neutralization; oxidation/reduction; precipitation/flocculation; soil flushing; in situ steam/air stripping; simultaneous injection, extraction, and recharge; and vacuum extraction.

Dechlorination: This process is based on the affinity of alkali metals for chlorine. Polyethylene glycol and some other hazardous chemicals can be used as catalysts for the reaction. The reagent reacts with the chlorinated organic by displacing a chlorine molecule. This chlorine displacement produces a lower toxicity, water soluble material. The reagent can be recovered and recycled after the reaction is complete. For in situ dechlorination, the mixture is typically heated by radio frequency heating or microwave heating to reduce the viscosity of the reagent.

Alkali metal dechlorination is used on contaminated oils and liquid wastes to displace chlorine from chlorinated organic compounds such as polychlorinated biphenyls (PCBs) and dioxins. In situ dechlorination should be used for shallow, uniformly contaminated soils. Conventional agricultural equipment is used to mix the soil and the reagent. If the contaminated soil is deeper than 1 to 2 feet (.3 to .61 m), or if high concentrations are apparent, the soil should be excavated and dechlorinated after it is made into a slurry.

Electroacoustics: Electroacoustic decontamination is used to remediate soils by applying electrical and acoustical fields. The electrical field is used to transport liquids through soils. The acoustic field can enhance the dewatering or leaching of waste such as sludges. Electroacoustic decontamination is effective on soils contaminated by inorganic, organic, and/or heavy metal liquids. Because this technology depends on surface charge to be effective, fine-grained soils are an ideal medium for application.

Electrokinetics: This process is a separation/removal technique for extracting heavy metals and/or organic contaminants from soils and sediments. Electrokinetic soil processing uses electricity to remove/separate organic and inorganic contaminants and radionuclides from the soil. A low direct current is run between an anode and a cathode inserted in a soil mass saturated with deionized water. This results in an acid front at the anode and a base front at the cathode. The acid front advances toward the cathode and eventually flushes across the soil and neutralizes the base. The movement of the front results in desorption of contaminants from the soil. The concurrent mobility of the ions and the advection of pore fluid under the electrical gradients supplies the method to flush contaminants from the soil.

Neutralization: The in situ neutralization process is performed by injecting dilute acids or bases into the ground in order to optimize pH for further treatment, or to neutralize plumes that do not require further treatment. Neutralization is used on liquids, sludges, slurries, and gases contaminated by acidic or alkaline wastes.

Oxidation/Reduction: This process takes advantage of the reactant's oxidation state and chemically transforms it by reduction-oxidation (REDOX). By raising one reactant's oxidation state while lowering the other, the toxicity of many organics and heavy metals can be reduced or destroyed using REDOX reactions. Decreased permeability of soils (due to hydroxide precipitation) or loss of adsorption (due to oxidation/reduction of soil organics) may affect in situ soil treatment. Violent reactions may occur with in situ methods because subsurface injection of reagents and water is required.

This process can be used in situ on soils that are contaminated with cyanide, aldehyde, mercaptans, phenols, benzidine, unsaturated acids, pesticides, benzene, organics, arsenic, iron, manganese, chromium VI, mercury, lead, silver, chlorinated organics, or unsaturated hydrocarbons. Oxidation/reduction may also be used ex situ on water, slurries, and sludges.

Precipitation/Flocculation: Precipitation is a treatment technique that transforms a substance in solution to a solid phase by physical/chemical mechanisms. It involves alteration of the ionic equilibrium to produce insoluble precipitates that can be easily removed by sedimentation or filtration. Typically, flocculating agents are added to cause the precipitate to become agglomerated. The solubility of metal hydroxides and sulfides is greatly affected by pH.

Precipitation may be used as an in situ process to treat aqueous wastes in surface impoundments. In this type of application, lime and flocculants are added directly to the lagoon and mixing, flocculation, and sedimentation occur within the lagoon. Wind and pumping action can provide the energy for mixing in some cases. Contaminants that may be affected by this treatment include zinc, cadmium, chromium, copper, lead, manganese, mercury, phosphate, sulfate, fluoride, arsenic, iron, nickel, and organic fatty acids.

Soil Flushing: The use of soil flushing to remove soil contaminants involves the elutriation of inorganic constituents from soil for recovery and treatment. The site is flooded with the appropriate washing solution, and the elutriate is collected in a series of shallow wellpoints or subsurface drains. The elutriate is then treated and/or recycled back into the site. The technology can introduce potential toxins into the soil system. An effective collection system is required to prevent contaminant migration. Flushing solutions may include water, acidic solutions, basic solutions, chelating agents, and surfactants. Water can be used to extract water-soluble or water-mobile constituents.

Soil flushing and elutriate recovery may be appropriate in situations where chemical oxidizing or reducing agents are used to degrade waste constituents and results in the production of large amounts of oxygenated, mobile, degradated products. In situ soil flushing is effective on sludges, soils, sediments, and other solids contaminated with inorganic corrosives, organic corrosives, oxidizers, halogenated non-volatiles, halogenated volatiles, non-volatile metals, volatile metals, organic cyanides, inorganic cyanides, non-halogenated volatiles, non-halogenated volatiles, PCBs, pesticides, dioxins/furans, oxidizers, and reducers. Chelation is used on liquids and soils contaminated by metals.

In Situ Steam/Air Stripping: Steam/air stripping involves injecting steam or air into the soil beneath a contaminated zone to volatilize and strip organic contaminants. A transportable treatment unit for detoxification is used with this technology and consists of 2 main components—the process tower and process train. Hot air and steam carry the contaminants to the surface where a metal shroud collects the vapors for off-gas treatment and ducts them to the process train for treatment.

In situ steam/air stripping system is effective in reducing the toxicity of soil by removing contaminated organics, such as hydrocarbons and solvents. This system is also commonly used to remove VOCs from ground or surface waters for the purpose of reinjection (for groundwater) or discharge. Soil particle size, initial porosity, chemical concentration, and viscosity do not limit the technology. The compound's vapor pressure and polarity are important in determining how effectively this technology will remove the contaminants.

Simultaneous Injection, Extraction, and Recharge: This process involves the remediation of unsaturated soils by injection of a medium to strip and transport contaminants to an extraction well(s). Water and steam are commonly used media. In unsaturated soil, steam will condense at some distance from the injection point and form a diffuse front consisting of a transient saturated zone with soil permeated by condensing steam on one side and relatively cool, unsaturated soil on the other side. This front is a region of radical contrasts in electromagnetic properties. The placement of injection points and extraction wells are designed to allow injection fronts to consolidate and move the contaminant to strategically located extraction wells. After being transported to the extraction wells, the contaminants are removed and treated.

This technology removes contaminants that can be mobilized by steam or water from unsaturated soil. Highly soluble or volatile contaminants in transmissive soils will be the best application for this technique, and these contaminants are expected to be removed very rapidly.

Vacuum Extraction: Vacuum extraction systems involve the extraction of contaminants from unsaturated soils through air injection. Clean air is injected into the contaminated soil, and a vacuum apparatus is used to extract the air filled with VOCs from recovery or extraction wells. The established air flows are a function of the equipment used and soil characteristics. Spent carbon and contaminated water are residuals of this treatment and further treatment of these residuals is necessary.

Vacuum extraction is used for the treatment of soils, sediments, sludges, and groundwater contaminated with volatile or semi-volatile organic compounds (VOCs or SVOCs) at ambient temperatures. This technology is effective on VOC and SVOC total concentrations ranging from 10 ppb to 100,000 ppm by weight. For effective removal, contaminants should have a Henry's constant of 0.001 or higher. The use of vapor extraction systems is typically limited to permeable unsaturated soils such as sands, gravels, and coarse silts; diffusion rates through dense soils, such as compacted clays, are much lower than through sandy soils. Clayey soils usually lack the conductivity necessary for effective vapor extraction, unless they are first fractured.

6.6.4 Solidification and Stabilization

Solidification and stabilization are treatment processes designed to accomplish one or more of the following: (a) improve the handling and physical characteristics of the waste by producing a solid from a liquid or semi-liquid waste, (b) reduce the solubility of the contaminants in the treated waste, or (c) decrease the exposed surface area across which transfer or loss of contaminants may occur.

While solidification and stabilizing reduce the mobility of a contaminant, the volume of the waste increases slightly, and there is only an incidental effect on toxicity. In addition, the effectiveness of the binders in incorporating organics and acid salts is questionable.

With proper recipe and additives, solidification and stabilization can be applied to virtually all contaminants including organics, inorganics, heavy metals, mixed wastes, and all classes of radioactive wastes. Solidification and stabilization can be applied to refuse, sediment, sludge/slurry, soil, structures, and water.

Waste solidification/stabilization systems discussed below include the lime-fly ash pozzolan systems, organic binding, pozzolan-portland cement systems, sorption, and thermoplastic microencapsulation.

Lime-fly Ash Pozzolan Systems: Lime-fly ash pozzolanic processes use a finely divided, non-crystalline silica in fly ash and the calcium in lime to produce low-strength cement. The solidification/stabilization of the waste is produced by microencapsulation in the pozzolan concrete matrix.

With proper recipe and additives, the lime-fly ash pozzolan process can be applied to inorganics, metals, mixed, low-level, and TRU radioactive wastes; specifically refuse, sediment, sludge/slurry, soil, structures, and water mediums.

Organic Binding: Modified clays can be used to immobilize organic contaminants. Clay particles are platy-shaped minerals that have negative charges on their surfaces as a result of isomorphous substitution. To achieve neutrality in their structure, clay particles attract cationic metals such as Li, Na, Ca, and Mg on their surfaces. Introduction of these organic cations into clays increases the interplanar distance between the clay particles and provides more suitable conditions for bonding of organic contaminants. Other organic binder types are epoxy, polyesters, asphalt, polyolefins, and urea-formaldehyde. Organic binding is useful for soils or sludges contaminated with organic materials.

Pozzolan-Portland Cement Systems: In this process, portland cement and pozzolan materials (i.e., fly ash) are combined to create a high-strength waste and concrete matrix, where solidification/stabilization is achieved through the physical entrapment of waste particles. Fly ash or another pozzolan is often added to the cement to react with free calcium hydroxide and thus improve the strength and chemical resistance of the solidified product. The types of cement used for the solidification can be selected specifically to emphasize a particular cementing reaction, or to enhance cementation (such as sulfate resistance).

Hazardous/toxic waste sites effectively treated by the pozzolan-portland cement process include: (1) heavy metals in metallic or cationic forms, (2) inorganics in anionic form, (3) water-soluble organics, and (4) water-soluble organics. The wastes that can be treated include aqueous solutions, sludges, and contaminated soils.

Sorption: Sorption is the addition of solid adsorbents to soak up and prevent the loss of drainable liquids through the mechanisms of capillary action, surface wetting, and chemical reaction. To prevent undesirable reactions, the absorbent material must be matched to the waste. Zeolite, kaolite, vermiculite, calcite, amorphous entonites silicates, acidic and basic fly ash, and kiln dust are all typical adsorbents. There are also synthetic adsorbents available. Adsorbents can be spiked with scavengers to bind trade metals, flocculating agents, and agents to improve subsequent solidification (cementing) processes.

Sorption can be used to solidify any contaminants in water, liquids other than water, or sludges/slurries. For in situ treatment, the waste can be in the groundwater, surface water, saturated soil, or source term.

Thermoplastic Microencapsulation: Thermoplastic microencapsulation involves blending fine particulate waste with melted asphalt or other matrix. Liquid and volatile phases associated with the wastes are driven off, and the wastes are isolated in a mass of cool hardened asphalt.

6.6.5 Thermal Treatments

Thermal technologies elevate the temperature of the soil to volatilize certain contaminants. Volatized contaminants are captured at the surface, thereby reducing the toxicity of the soil. Thermal treatment can be used to treat most contaminants and can be used in most media states.

The thermal treatments covered below include a high energy corona, radio frequency and electromagnetic heating, and in situ vitrification.

High Energy Corona: Use of a high energy corona is an innovative thermal treatment process that does not require high temperatures or additives. Electrodes/vents are placed in the contaminated soil. Peripheral electrodes/vents are used as air inlets, while a center electrode/vent is used as an off-gas vent. A form of corona develops at higher voltages to generate energetic electrons and robust oxidants from soil gases. The high energy corona technology is used to treat organic contaminated soils, sludges, slurries, and sediments.

Radio Frequency and Electromagnetic Heating: In situ radio frequency (RF) heating is a rapid process that uniformly heats soil without excavation or digging. This process uses electromagnetic wave energy in the range of 45 Hz to well over 10 GHz to heat soil. Exciter and guard electrodes are placed in the ground, and the temperature rise occurs due to ohmic or dielectric heating mechanisms. The RF technology is capable of heating soils to temperatures in excess of 212°F (100°C) (boiling point of water). The gases and vapors formed in the soil are recovered at the surface or through vented electrodes used for the heating process. A vapor containment cover collects volatilized organics for incineration or carbon absorption. This process is also referred to as electromagnetic (EM) heating. The only major difference between RF and EM is in the choice of frequency of the applied power. The EM technology is suitable for heating soils only to the boiling point of water.

RF and EM heating processes are used to treat sludges, solids, soils, and sediments contaminated with volatile and semi-volatile dioxins/furans, pesticides, halogenated volatiles, halogenated non-volatiles, radioactive materials, PCBs, non-volatile metals, volatile metals, non-halogenated non-volatiles, and non-halogenated volatiles. This technology can be used in saturated or unsaturated soil. Both of these technologies have the potential for economic and efficient remediation of soils at hazardous waste sites contaminated with organic compounds.

In Situ Vitrification: In situ vitrification (ISV) involves the electric melting of contaminated soils in place. ISV uses an electrical network typically consisting of 4 electrodes placed in a square pattern and at the desired depth, to electrically heat and melt contaminated soils and solids at temperatures of 2,900 to 3,600°F (1,600 to 2,000°C). ISV destroys organic pollutants by pyrolysis. Inorganic pollutants are immobilized within the vitrified mass, which has properties of glass. Both the organic and inorganic airborne pyrolysis by-products are captured in a hood, which draws the contaminants into an off-gas treatment system that removes particulates and other pollutants of concern.

ISV is effective on aqueous media, organic liquids, sediments, soils, and sludges contaminated with halogenated volatiles, halogenated non-volatiles, non-halogenated volatiles, non-halogenated non-volatiles, pesticides, dioxins/furans, organic cyanides, organic corrosives, volatile metals, non-volatile metals, and PCBs.

On saturated soils or sludges, the initial application of the electric current is needed to reduce the moisture content before the vitrification process can begin. This increases energy consumption and associated costs. Also, sludges must contain a sufficient amount of glass-forming material (non-volatile, non-destructible solids) to produce a molten mass that will destroy or remove organic and immobilize inorganic pollutants. The ISV process, however, has the following limitations: (a) individual void volumes in excess of 150 ft^3 (4.25 m^3); (b) buried metals in excess of 5% of the melt weight or continuous metal occupying 90% of the distance between 2 electrodes; (c) rubble in excess of 10% by weight; and (d) the amount and concentration of combustible organics in the soil or sludge. These limitations must be addressed for each site.

Acids and salts in the soil can also be a concern when using this technology. Acids and salts can cause the soil to have an abnormally high electrical conductivity (hence, a low electrical resistance), which is generally more pronounced as the moisture content of the soil increases. This low resistance will require the application of more electrical energy to the treatment area in order to achieve a vitrified melt. This may also result in a much higher melt temperature.

6.7 PERMEABLE BARRIERS

Subsurface permeable barrier technologies are potentially applicable to existing waste disposal sites. Two types of subsurface barrier systems are described: 1) those that apply to contamination in the unsaturated zone, and 2) those that apply to groundwater and to mobile contamination near the groundwater table. These barriers may be emplaced either horizontally or vertically depending on waste and site characteristics.

Materials for creating permeable subsurface barriers are emplaced using one of 3 basic methods: injection, in situ mechanical mixing, or excavation-insertion. Injection is the emplacement of dissolved reagents or colloidal suspensions into the soil at elevated pressures. In situ mechanical mixing is the physical blending of the soil and the barrier material underground. Excavation-insertion is the removal of a soil volume and adding barrier materials to the space created. These 3 basic methods can be used for the emplacement of both horizontal and vertical barriers.

Major vertical barrier emplacement technologies include trenching-backfilling; slurry trenching; and vertical drilling and injection, including boring (earth auguring), cable tool drilling, rotary drilling, sonic drilling, jetting methods, injection-mixing in drilled holes, and deep soil mixing. Major horizontal barrier emplacement technologies include horizontal drilling, microtunneling, compaction boring, horizontal emplacement, longwall mining, hydraulic fracturing, and jetting methods.

Subsurface barriers may be monitored to ensure the quality of barriers as they are being emplaced and to verify that material and installation specifications are met. Continuous or periodic monitoring over barrier design lifetimes may also be required to verify that the barrier is functioning as expected. The type and degree of quality control is determined by the specific barrier technology, the contaminants present, and the geologic/hydrogeologic conditions present at the site. Three quality control and monitoring methods are applicable to the installation and performance verification of permeable barriers: 1) inference by monitoring the emplacement process, 2) sampling and analysis, and 3) non-destructive monitoring.

Subsurface permeable barriers can be created in situ by emplacing soluble or insoluble materials into the soil around and/or under a contaminated waste site. The chemicals used must reduce the rate of migration of contaminants of concern through the barrier zone.

Subsurface permeable barrier technology is largely unproven, but shows significant promise of temporarily or permanently remediating contaminated waste sites. A potential advantage of this technology over subsurface impermeable barrier technology is less rigorous installation requirements. For illustration, an impermeable barrier must stop the flow of contaminated groundwater. This may require effective sealing between each individual segment of the installed barrier and between the barrier and the aquitard. Achieving a watertight seal under subsurface conditions is difficult, especially in unsaturated zone applications where water advection rates are very low. Low water flow rates may facilitate relatively high flow through barrier joints and other imperfections. Methods to adequately identify and repair leaks in a subsurface environment have yet to be developed. Subsurface permeable barriers, in contrast, are designed to allow water to flow through the barrier. Most subsurface permeable barriers will not cause the buildup of a significant hydraulic head behind them, which otherwise could accelerate flow of water through cracks and areas of high hydraulic conductivity. An imperfectly installed permeable barrier will result in diminished barrier performance, but probably significantly less than impermeable barriers.

Selecting materials for creating effective permeable barriers requires consideration of the contaminants of concern, their concentrations, their speciation, and the physical and chemical conditions of the soil. The contaminants of concern, their concentrations, and their speciation are important for identifying potential chemical additives that may effectively reduce migration rates of the individual contaminants to acceptable levels. Reducing migrations rates can occur by several mechanisms, including precipitation, adsorption, ion exchange, biodegradation, biofixation, and chemical degradation (e.g., hydrolysis). The function of a permeable barrier is to induce or enhance one or more of these mechanisms.

Two basic types of permeable barrier technologies exist: 1) those that apply to contamination in the vadose zone, and 2) those that apply to mobile contamination in the saturated zone. Vadose zone technologies are largely conceptual and undeveloped. These technologies include both vertical and horizontal barrier components. Horizontal barrier components are emplaced under the waste area to intercept meteoric water. Vertical barrier components that intersect the horizontal component may be necessary if horizontal dispersion and/or diffusion of contaminants are occurring.

The vadose zone barrier should be designed with a higher matrix potential than the native soil above, if possible. The higher matrix potential will help distribute water more evenly in the barrier and may be effective for minimizing the potential for channeling. Higher matrix potentials can usually be created by adding barrier materials as very fine solids. Sufficient permeability (hydraulic conductivity) of the barrier must be assured, however, to avoid perching of water on the barrier, resulting in lateral flow without penetrating the barrier.

Saturated zone barriers may be installed in a vertical configuration downgradient of the contaminated groundwater plume. Installing the barrier around the entire plume is unnecessary because the permeability of the barrier ensures that groundwater flow paths will not be significantly altered, i.e., forced around the barrier. Horizontal barriers under the water table may also be considered if significant vertical migration is occurring. The saturated hydraulic conductivity of the barrier materials should be higher than that of the soils, if possible. This will help ensure the groundwater contacts the barrier additives if imperfect mixing of soil and additives occurs. In some cases, it will be unnecessary to key the barrier to the aquitard. This applies when limited vertical dispersion of contaminants in the groundwater has occurred. In those cases, it is only necessary to emplace the vertical barrier a short distance below the maximum depth of contamination.

A potential modification of the permeable barrier concept is to add impermeable barrier components to the design. For example, impermeable barriers constructed on the edges of a groundwater plume will direct the plume toward a downgradient permeable barrier. If the permeable barrier is engineered for ease of replacement, the combined barrier concept offers potential for long-term effectiveness and lower maintenance costs. Impermeable barrier components could include the following developed technologies:

1) Sheet piling
2) Grout curtains
3) Slurry walls
4) Freeze walls

Hydraulic fracturing is widely used in the petroleum industry to stimulate the release and recovery of oil and gas from geologic formations. It has also been at Oak Ridge National Laboratory as a means of disposing radioactive waste grout between layers of bedded shale. This technology injects water or slurry at pressures exceeding the lithostatic pressure of a formation at the bottom of a borehole. The fluid pressure generates fractures that propagate from the borehole. Sand or other propping agents can be introduced into the fracture to hold it open and create a permeable pathway.

This technology holds potential for injection of permeable barrier materials in various geologic media such as consolidated bedrock and glacial tills with fracturing and bedding planes.

Two jetting methods may be applicable for emplacing permeable barriers: kerfing and jet slurrying.

Kerfing is a jetting technology currently used to produce a notch or slot either perpendicular or parallel to the axis of a previously drilled borehole. A potential extension of the kerfing technology is using a high-pressure water jet and an abrasive material to cut a slot along the full length of the borehole. The high-pressure jet is placed in the borehole, where it is moved without rotation along the axis of the hole to create an axial slot. Controlled, partial rotation of the jet may be necessary to cut a slot with a relatively uniform thickness. The jet may advance the slot at a rate of several centimeters per second with a depth of penetration of approximately 3 m. The slot can be filled with permeable barrier material.

Jet slurrying uses a water jet to fragment a formation. The jet is introduced into the borehole and can be rotated 360 degrees. The slurry created will drain through the horizontal borehole, which requires casing to prevent its collapse. The borehole can then be backfilled as the casing is withdrawn. The soil-filled space between 2 non-intersecting, backfilled holes must be drilled and backfilled to create a planar barrier.

The permeable barriers are being designed to operate unattended with minimal maintenance for long periods of time (i.e., years). However, periodic inspections will be required because these enhanced barriers might fail because of cracking. Since the barriers are passive, no power is required for their operation.

6.8 CHEMICAL, LIQUID AND REACTIVE BARRIERS

A project at PNL will develop, test, and evaluate an in situ method for immobilizing inorganics (metals, ions, and radionuclides) and destroying organics (primarily chlorinated hydrocarbons) using chemical or microbiological reduction of both the groundwater and the solid materials within the aquifer to form a permeable treatment barrier. The great majority of the chemically reactive mass in the subsurface system resides in the solid phases, rather than in the groundwater. Therefore, to have a substantial influence on the chemistry of the system, the solid phases should be involved. If changes are made only to the aqueous component, it will quickly re-equilibrate with solid phases.

There are several ways to approach the addition of reagents or nutrients into the subsurface. Three possible approaches include: (1) direct injection into the contaminant plume, (2) injection ahead of the contaminant plume to form a geochemical barrier by reacting with the solid phases, and (3) use of horizontal drilling technology to introduce a gaseous reagent to the contaminant plume. In the second alternative, a reagent or nutrient is injected ahead of the contaminant plume to form a permeable treatment barrier by reacting with the solid phases. The contaminant plume then reacts with the permeable treatment barrier. The second alternative will be used in this project. The basic approach involves a forced gradient, single-well, reactive tracer test. The reagent is pumped into the aquifer in a circle approximately 60 to 100 feet in diameter, allowed to react for 10 to 60 days, and then water containing the reaction by-products and any remaining reagent is pumped back out.

The purpose of a task at INEL is to evaluate the effectiveness of invasive barriers in controlling the migration of contaminants.

The technology involves injecting a latex emulsion and a reactant/coagulant solution through a series of wells into an aquifer. The latex emulsion, with polymeric particles ranging from 200 to 0.2 μm in diameter and having viscosities less then 40 cp reacts with the coagulant solution to form a solid groundwater barrier around or under a contaminated site. This barrier greatly lowers the permeability of selected regions of the aquifer.

The reactant/coagulant solution is composed of polyvalent cations dissolved in water. It is anticipated that these solutions could be injected into subsurface formations using upgradient wells or boreholes. Injected polymer solutions will tend to migrate along preferential flow and highest permeability zones within the subsurface formations prior to coagulation. Groundwater will facilitate mixing of the emulsion and reactant solutions and the solid coagulant formed from the reaction will effectively block the highly transmissive pathways.

The invasive barrier technique is novel because it uses newly formulated barrier material composed of natural rubber and commonly occurring multivalent cations (e.g., Ca^{++}, Mg^{++}) in conjunction with the existing soil structure to form a barrier to groundwater flow. It can potentially form a wide barrier with potentially fewer wells than conventional slurry grouting techniques, and it can potentially place a barrier <u>under</u> a contaminated site without disrupting the site or generating hazardous air emissions.

The injection of materials from the surface which reacted in situ to form a continuous, unreactive, impermeable barrier would add substantially to the ability to control groundwater contamination problems.

Viscous liquid barriers are being investigated at LBL, SNL, and BNL. Viscous liquid barriers are a relatively new class of impermeable barriers. They can be installed as horizontal and/or vertical barriers. The barrier material is emplaced as a low-viscosity liquid which, after emplacement, forms an impermeable high-viscosity barrier under the ambient subsurface conditions. Several classes of materials (e.g., colloidal silica gels, waxes, polysiloxanes and polybutenes) have been or are being developed for use as viscous liquid barriers. The desired viscosity characteristics (i.e., initial low viscosity to facilitate subsurface emplacement and high post-emplacement viscosity) are achieved, for example, by heating the material during emplacement or by promoting polymerization or gellation after emplacement. To meet the DOE need, it will be necessary to emplace the viscous liquid barriers in horizontal (floor) and vertical (wall) configurations (or possibly in V- or bowl-shapes).

The properties of viscous liquids make them amenable to subsurface manipulation techniques that can be applied to facilitate emplacement within the soil. For example, some materials can be made to flow preferentially into heated zones, and electrolyte materials can be made to flow in the direction of a voltage gradient.

Research by Chem-Nuclear Geotech, Inc. is examining the sorption and immobilization capacity of a natural iron mineral for a number of site contaminants, and its use in the formation in situ of a permeable barrier that removes target contaminants, but does not impede groundwater flow.

Ferric oxyhydroxide or hydrated iron oxide is a naturally occurring non-hazardous substance that has sorption affinities for a number of inorganic contaminants found at DOE sites. Such contaminants as uranium, molybdenum, copper, lead, zinc and radium can potentially be removed from groundwater.

The iron is injected as a solution in water. Reaction underground with aquifer mineral alkalinity converts it to the sorbing phase. Precipitation within the aquifer pores coats the rock particles forming a barrier zone around the contaminated area. This coating extracts the contaminants moving with the groundwater and confines them in the barrier zone.

The investigation at LBL is examining liquids which, when injected into the subsurface, produce nearly-inert impermeable barriers through a very large increase in viscosity. Appropriate emplacement of these substances provides an effective containment of the contaminated zone by trapping and immobilizing both the contaminant and the plume.

This project will identify and characterize promising materials and evaluate their containment potential by means of laboratory pilot-scale experiments and field testing and demonstration. The general purpose

TOUGH2™ model, developed at the Lawrence Berkeley Laboratory (LBL) is being modified to simulate barrier fluid behavior and to design experiments.

The first type of barrier fluid under examination belongs to the polybutene family. Polybutenes are chemically and biologically inert, hydrophobic and impermeable to water and gases, and are approved by the Federal Drug Administration for food contact. Their performance is unaffected by the soil and waste type, and is only controlled by their drastic viscosity dependence on temperature.

The second type, colloidal silica, is a silicon-based chemical grout that poses no health hazard, is unaffected by filtration, and is chemically and biologically inert. Its containment performance is controlled by the gelation time, which depends on pH, temperature, the chemistry of the injected suspension, and chemistry and mineralogy of the aquifer porous medium. The third type of barrier fluid is polySiloXane. These fluids are chemically and biologically inert silicon-based polymers used for medical implants. They are mixtures of 2 fluids, are unaffected by the aquifer or waste chemistry, and their containment performance depends on temperature and the ratio of the 2 constituents.

The strong adsorption of many contaminants to soil particles makes physical extraction slow or ineffective. Excavation of contaminated soils and disposal in protected facilities is very expensive. Containment on-site and control of groundwater transport can limit the off-site threat, and may supply a long-term solution.

A barrier containment system that does not require excavation would be a useful groundwater contamination control technique. Formation of a barrier with surface injected components that polymerize or change their viscosity under aquifer temperature and pressure conditions would allow barrier emplacement without excavation. In situations where complete control is necessary, an impermeable barrier is preferred over the sorption barrier.

In some areas aquifer mineralogy or regulatory restrictions may preclude the use of one or another barrier component. A variety of barrier systems must be available to match the range of contaminants and circumstances.

A reactive barrier is an innovative containment technology to prevent the migration of contaminants in a groundwater plume, while allowing water to pass through a treatment barrier. The reactive barrier may be used in conjunction with an impermeable wall when the transverse extent of the plume is broad, in order to direct the contaminated plume toward the reactive barrier that serves as a permeable window through the hydraulic barrier.

A key issue affecting the technical feasibility of reactive barriers is that the reactive material may become exhausted and need periodic replacement. A closely related issue that affects economic feasibility is that an improperly installed reactive barrier will not achieve its design capacity for contaminants and will require more frequent replacement.

Permeable, reactive barriers allow the passage of water while prohibiting the movement of contaminants by employing such agents as chelators (ligands selected for their specificity for a given metal), sorbents, microbes, and others. In Department of Energy (DOE) sites where multiple contaminants are ubiquitous, multicomponent barriers need to be evaluated. Field-scale experiments were constructed using natural sand with the permeable barriers consisting of zeolite + silica gel + sand, bentonite + Al crosslink polyacrylamide + sand, and peat + Al crosslink polyacrylamide + sand. The reactive barriers could be designed (1) to remain in place as permanent or semi-permanent installation; (2) to be removed and replaced periodically, thus serving as a component of the remediation process; and/or (3) to be used as part of the post-closure monitoring system in which the appearance of a contaminant in the barrier would then serve to warn of impending contaminant migration.

Chemical gel barrier systems, based on petroleum technology, are being investigated at LANL.

In-place bioreactors use the capabilities of native bacteria for degrading hazardous organic compounds in a cost-effective, publicly acceptable manner. The capability can be managed to provide prolonged treatment, as well as treatment of relatively short duration. In addition to full-scale site, biodegradation has significant near-term potential as an effective containment strategy. Thus, evaluation of approaches to managing biologic communities on the margins of a site, in combination with other barrier approaches, will provide both significant information for both limitation of contaminant transport and full-site clean up.

6.9 GROUTING

In Situ Grouting: In situ grout injection contains waste material in a solid monolith by mixing it with cement grout, thereby increasing the waste's physical stability and compressive strength, decreasing water intrusion to the waste, and decreasing the leachability of waste constituents. This section discusses the applicability of in situ grout injection for radionuclides.

In general, in situ grout injection can be considered at any site from which wastes cannot be removed, but several characteristics of the soil influence whether the technology will be able to contain waste effectively. These characteristics include void volume, which determines how much grout can be injected into the site; soil pore size, which determines the size of the cement particles that can be injected; and permeability, which determines whether water will flow preferentially around the monolith. Soil with the appropriate characteristics can be treated using a very simple in situ grout injection system.

A pipe is drilled or hammered into the ground where the waste is located. A grout consisting of cement and other dry materials, which can include fly ash or blast furnace slag, then is injected to the waste through the pipe by a pump, conveyor belt, or pneumatically controlled blower. Once all of the voids at a particular depth become saturated, the pipe is raised and more grout is injected. This process continues until the grout forms a rough column extending to the surface from as far as 50 to 60 feet below the surface. A variation on the basic design involves using a pipe with a mixing apparatus that rotates as the grout is injected. This apparatus mixes soil with the grout, creating a distinctly recognizable column of mixed grout and soil. If necessary, a hood can be placed over the system to capture volatile contaminants released during the injection process.

Whichever system is used, the object is to create a solid monolith of adjacent columns that contains the waste. If the permeability of such a monolith is at least 2 orders of magnitude less than that of the host soil, water flows preferentially around the monolith and through the soil. This decreases both water intrusion to the waste and leaching of hazardous constituents from the monolith.

Of the many types of grout available, cement-based grouts are the most common, for several reasons. First, materials for cement-based grouts, such as cement, fly ash, and blast furnace slag, usually are available within 150 miles of any site, making cement-based grouts relatively inexpensive. Second, cement-based grout is a proven material. The construction industry has extensive experience with in situ grouting and has shown that cement-based grouts can withstand extreme natural forces.

There is also a chemical grouting process in which polymer is injected to fill in any cracks or fissures in the host rock at the edge of the contaminated zone. There is a wax-based grout process, permeation grouting, jet grouting, and soil-heating-based grouting.

Soil solidification and stabilization with sodium silicates in combination with concrete is a proven, reliable technology for treating liquid and semi-liquid wastes. Despite its proven effectiveness, using silicates and cement to solidify and stabilize soil has not achieved widespread use. One cause of underutilization is the reputation the process has for being difficult to use.

The ability of the monolith to resist leaching is its most important feature. Cost is another advantage of in situ grout injection. Although the initial capital costs for batch or surface processes often are less than those for in situ processes, the total costs for batch and surface processes, including transportation and disposal, tend to be greater.

Grouts can be formulated to set very quickly. This is an advantage at sites, such as solar ponds, that essentially are open pits. Within a day, previously grouted areas become a platform for further grout injection operations. The injection apparatus also is fairly small and portable, so it can be maneuvered into sites with tight space constraints.

Because the technology operates in situ, process control is relatively poor and it is difficult to verify that the grout actually contained the waste. Rigorous verification involves digging up the perimeter of the grouted area. In addition, in situ grouting does not lend itself to waste retrieval, so it is not a good choice for DOE sites from which wastes may have to be retrieved after 30 to 40 years.

Cement-based grouts have some specific disadvantages. First, injection of a cement grout creates a volume increase—once the grout fills the available voids, it returns to the surface as berm. Second, since cement is particulate, it can flow only to soil pores of sufficient size. The first 2 or 3 injection holes at any site usually are test holes to determine how much grout the soil uptakes. Third, cement-based grouts have limited application.

The Soil Saw is a concept based on the hydraulic erosion principle. It is an in situ technology which uses reciprocating jets of cement grout or bentonite slurry to cut a continuous path through the soil. The result is a homogeneous grout wall of very uniform quality and thickness. The physical properties of the wall can be tailored to create a plastic-like material of high compressive strength concrete.

In the Soil Saw concept, jet grouting nozzles are mounted along a rigid beam that is reciprocated through the soil media producing a sawing action like "a hot knife through butter." The combined sawing, jet slurry grouting action, and the effects of gravity on the (locally) rigid beam results in the construction of a continuous soil/cement slurry wall. This wall can be constructed to depth only limited by the mechanical or hydraulic means to reciprocate the Soil Saw. The process does not require a structurally rigid beam. A Soil Saw demonstration was recently completed at the Savannah River Site.

A project at SNL is examining the potential application of a bentonite/mineral wax formulation, developed in Germany, and an inorganic grout, developed in France, as barrier materials for DOE sites. Because these materials have been used for grouting, bringing them to regulatory and public acceptability within the U.S. should be rapid.

This investigation is examining the compatibility of these barrier formulation within the range of DOE soils and waste types. Technical challenges include lateral permeation of the soils, physical and hydraulic stability of the barrier over time, and the regulatory acceptance of the overall approach and grout materials.

Subsurface barrier emplacement involves putting an impermeable barrier (composed of some kind of grouting material) in below a landfill. It has to be emplaced without disturbing the landfill. There are 2 emplacement methods that are being tested. The first is permeation grouting, which uses a slight pressure to inject the grout and takes advantage of the natural porosity of the soil by letting it flow into the soil. The second is jet grouting by mixing, which takes a drill and rotates while injecting the grout. This intentionally fractures the soil and intermixes it with the grout. These techniques are being investigated at SNL.

Temporary or long-term containment of mobile contaminants from existing waste sites require effective surrounding barriers. Vertical barriers are relatively well-known from standard construction project work, but methods for building horizontal barriers in situ are only now being developed. For old sites, the problem is to place a containment barrier without disturbing the waste. A barrier alternative to cement grout is the Enviro wall concept (interlocking polyethylene panels) developed by Barrier Member Containment Corp.

Horizontal Grouting: Temporary or long-term containment of mobile contaminants from existing land disposal waste sites requires effective surrounding barriers. Vertical barriers (cut-off walls) are relatively well known from standard construction practice, however construction of the type of bottom they require to contain vertical movement of contaminants is less well known. Methods for building horizontal barriers in situ are considered below.

Horizontal barrier placement technology as currently practiced is not highly developed. A search of the barrier industry indicates that no existing/developed technique is as capable as the innovative horizontal grout barrier method promises to be in providing means for vertical containment of pre-existing land disposed materials. The primary competitive technologies are:

1) **Triple rod jet grouting:** a proprietary technique which can be problematic in requiring placement of individual grout disks from directly above and through the waste. The maximum size slab that has been formed to date using this technique is approximately 70' x 170' Verification of the integrity of this type barrier is difficult due to the composite placement method (individual, grouted-in-place, soil cement disks), and due to the lack of direct verification opportunities (the horizontal barrier is formed in situ, buried underground and currently available sensor technology has only limited resolution).

2) **Freeze walls:** which require continuous application of energy to sustain and will function only if the soil moisture is within a fixed range. This method of providing horizontal containment is used in barriers formed and maintained in place through refrigeration using the ground freezing technique. Here, the lower end is closed off by freezing a "V" trench or by freezing a container formed by placing cooling pipes into precision bored holes under the area of concern. Verification of closure is somewhat more positive in the freezing method (in saturated zone) due to measurable effect on the interior water table when closure is reached. Refrigeration is necessary to maintain this

barrier and, moving groundwater can prevent it from forming or breach it.

The innovative horizontal barrier placement technique promises to be capable of placing a barrier layer beneath leaking waste sites, failed storage tanks, uncontrolled dumps and spill sites. With containment achieved, the waste can be held on site, processed in situ or excavated for removal with reduced risk of negative environmental impact.

The process begins by placing 2 roughly parallel directionally drilled holes which curve down from the surface at one end of the area of concern, pass beneath the waste, and return to the surface at the other end. The drill pipes in the holes are attached via a draw bar to a winch or a tractor. At the hole's opposite end, the drill pipes are attached to the front of a jet grouting bar: a device that emplaces grout under very high pressures (10k psi). A grout feed line and a trailing drill pipe are attached to the rear of the jet grouting bar forming a capital "H" shaped arrangement when viewed from above. A high pressure pump, operating at 5-10 thousand psi, feeds a cement grout mix to a horizontal mixing bar. The tractor then pulls on the drill pipes at the top of the "H" and draws the jet grouting bar through the ground along the path of the directionally drilled holes. A typical jet grouting bar is a hydraulically driven injector-mixer which leaves a soil-cement (grout and native soil) slab in its path as it is pulled through the ground and beneath the waste. As the bar moves forward, the high pressure grout erodes the soil surrounding the bar and mixes with it, leaving behind a soil cement slab which cures to form a 10 foot wide, 18 inch thick slab.

Joined slabs, necessary to form an extensive bottom, will be made by using the trailing drill pipe and an additional directionally drilled in-place pipe to make each following pass. It is expected that the use of the trailing pipe ensures slab overlap, a tight seam and a continuous bottom. This process is repeated until a bottom of the required length and width is formed.

The types of grout that may be placed are extensive. The jet grouting bar may be expected to mix and place materials that can be fed to it as a liquid at the required pressures. It should also be noted that conceptually the barrier can twist from the horizontal plane through vertical for special applications simply by varying the arrangement of the directionally drilled guide holes.

The technology is conceptually simple to use. It relies on proven oil field grouting and high pressure pumping techniques and is guided by understood directional drilling methods. The difficulties of operation in rocky ground or other challenging conditions remains to be determined.

The horizontal grout barrier construction process is expected to produce drilling spoil, grout overflow and washout water as secondary waste. Because the path of the holes and barrier does not have to contact the waste, these secondary wastes may not be problematic. The demonstration is taking place at Fernald in conjunction with Halliburton NUS, DOE, EPA, Ohio EPA, and the University of Cincinnati.

Horizontal barrier technology is potentially applicable to containment of a wide range of existing land disposal sites, underground storage tanks, spills and ruptures. Interest has been expressed in this technology as a possible containment method for leaking or deteriorated single wall tanks such as those located at Hanford, WA. Additionally, it may have direct application at the FEMP to isolate units undergoing remediation, such as the waste pits or the K-65 Silos, from the underlying aquifer. It will have widespread application throughout the waste management industry when successfully matured. Potential users abound as this technology would allow for containment of uncontrolled landfills, toxic/hazardous spills and leachates from failed engineered land disposal sites.

6.10 SOIL FREEZING

Frozen Soil Barriers would provide a temporary barrier to quickly halt the migration of contaminant plumes or would permit construction of large equipment for in situ treatment.

An investigation by Martin Marietta, addresses the feasibility of frozen soil barriers (ground freezing technology) as a means of containing hazardous and radionuclide-contaminated soil in a non-arid setting. Because ground freezing has long been a civil engineering technique for ground control, water entry control, etc., this project is essentially a new application of an established technology. A series of holes are drilled and refrigerant is circulated, freezing the soil around the holes such that a confined volume is created, thereby preventing contaminant migration.

Another project at the DOE Grand Junction Projects Office addresses the feasibility of using frozen soil barriers (ground freezing technology) to contain hazardous and radionuclide-contaminated soil in an arid setting.

Many of DOE's contaminant sources (e.g., landfills, dry wells, evaporation ponds, etc.) are located in arid climates and are typically far above the natural groundwater level. Frozen soil barriers are thought to be useful in providing containment at these sites. However, most experience with their hydraulic performance is associated with natural, fully saturated environments. Under arid conditions, performance may be affected by the need to first create full saturation, (i.e., achieve near-zero air porosity), then maintain this condition under the frozen state. This project will examine potential performance factors arising from arid site conditions and evaluate specific measures to mitigate or minimize adverse effects.

CRYOCELL® (RKK, Ltd.) is a frozen soil barrier that completely contains waste migration to the soil or isolates a contaminated area during an in situ remediation program. The CRYOCELL® design involves installing freeze pipes in an array outside and beneath the contaminated zone to completely surround the waste source or groundwater plume. Standard well drilling equipment is used to drill or drive the freeze pipes into place. Once installed, the array of pipes is connected to a freeze plant by a distributive manifold. The pipes carry a cooled brine in a completely closed system, which freezes the entire inner volume between the pipes, and the adjacent earth to the outside of the row(s) of pipes.

The barrier thickness and temperature may be varied to suit site conditions. RKK, Ltd. (RKK), reports that barriers can be established at depths of 1,000 feet or more and may vary in thickness from 15 to 50 feet.

6.11 ELECTROKINETICS

In electrokinetic remediation, a DC voltage is applied across electrodes that are emplaced in the soil that is to be treated. The resulting electric field induces motion of the liquid, dissolved ions, and possibly colloid-sized particles suspended in the liquid. The contaminants are moved by 3 processes, namely electroosmosis, electromigration, and electrophoresis.

The relative contribution of each mechanism—electroosmosis, electromigration, and electrophoresis—depends on the physical and chemical properties of the soil matrix, the contaminants, and the liquid. In cases where these processes are slow or concentration gradients are high, simple concentration diffusion may also affect the overall transport process.

In the electroosmotic purging process, a purge solution which is introduced at one of the electrodes serves to enhance the efficiency of the process.

Electrokinetics is a relatively new remediation technology that uses low-level direct current on the order of mA/cm^2 of cross-sectional area between electrodes placed in the ground in an open flow arrangement. This arrangement allows processing or pore fluid to flow into or out of the porous medium. The low-level direct current results in physico-chemical and hydrological changes in the soil mass, leading to species transport by coupled and uncoupled conduction phenomena.

Electrokinetic remediation is incompletely understood because of the complexity of parameters and their interactions that occur when one applies a direct current between buried electrodes in contaminated soil. Contaminants can move through soil by 3 different processes induced by the applied field: electroosmosis, electrophoresis, and electromigration. Electrolysis reactions that occur at the electrodes induce pH changes that can affect contaminant speciation and solubility. Contaminant mobility can also be influenced by soil permeability and the degree of water saturation.

The primary advantage of this technology is the potential for many in situ applications. Electrokinetics has several potential applications in waste management. Besides enhancing chemical migration, the technique can be employed in implementing electrokinetic flow barriers; diverting plumes; detecting leaks; and injecting chemicals, grouts, microorganisms, and nutrients to subsurface deposits.

The fact that the technique requires a conducting pore fluid in a soil mass could be considered a shortcoming, particularly at sites where there are concerns about introducing an external fluid into the soil. In addition, the technique has been demonstrated to be successful at electrode spacings of only 6 to 10 m. Large-scale applications will require that several electrodes be placed across the site.

Pilot-scale demonstration of electrokinetic removal of uranium from contaminated soil will be accomplished in a project by Hazwrap/Martin Marietta. Site selection and treatability studies will precede the pilot test, and a full-scale field test at a site to be determined is envisioned following evaluation of the pilot scale results. Removal efficiency, control of added fluids, contaminant recovery and disposal, power consumption, mass balance, and control of soil pH must all be evaluated to assure that this process is viable. Technology advances made by Russian scientists in this area of environmental remediation will be used as much as possible. The selected site should be such as to allow easy permitting for testing, be representative of the uranium problems throughout the DOE, and be accessible to industry, regulatory agencies, and academia.

The Electro-Klean™ (Electrokinetics, Inc.) electrokinetic soil process separates and extracts heavy metals and organic contaminants from soils. Electro-Klean™ can be applied in situ or ex situ, and uses direct currents with electrodes placed on each side of the contaminated soil mass. Conditioning fluids such as suitable acids may be used for electrode (cathode) depolarization to enhance the process.

An acid front migrates towards the negative electrode (cathode) and contaminants are extracted through electroosmosis (EO) and electromigration (EM). The concurrent mobility of the ions and pore fluid decontaminates the soil mass. The EO and EM supplement or replace conventional pump-and-treat technologies.

6.12 HYDRAULIC CAGE

A hydraulic cage is an engineered system to passively control geohydrological gradients over the long-term in and round either a controlled placement of hazardous materials or an existing contaminated region. The cage is constructed by drilling a series of boreholes around the region of interest and enhancing the hydraulic conductivity of the rock between them. The boreholes can also be used for pre-construction characterization and post-construction monitoring, as well as for dewatering during construction (for underground waste placement), or for groundwater removal during remediation.

If properly constructed in appropriate conditions, the hydraulic cage can minimize the hydraulic gradient across a site and, thus effectively provide containment to advective aqueous contaminant transport. However, complete containment can probably not be achieved (due to the difficulty in developing continuity between holes) and such a passive system may clog up in the future. At the extreme, the hydraulic cage may exacerbate the problem under certain conditions by providing a fast pathway for contaminant transport.

For initial waste disposal, would supplement other engineered barriers; that is, no reasonable alternative is available. For containment of contaminated site, would replace grouting. May enhance other types of engineered containment/disposal systems (concrete, metal, glass, etc., waste receptacles) by providing a redundant measure of protection (more robust system). May also improve the efficiency of remediation technologies such as pump and treat, bioremediation, and circulation pumping by slowing the flux of contaminated groundwater.

May not be applicable in formations containing large-scale discontinuities (faults, fracture zones, etc.) or where thermal or osmotic gradients are present. Also, is generally not applicable at shallow depths, in soils, in unsaturated zone, or where gradients are near vertical, because other technologies are preferred.

May be difficult to verify completeness of hydraulic cage and the absence of fast pathways intersecting the cage. Also, may be difficult to ensure long-term performance (that is, no clogging). Counter to the standard philosophy of a low-permeability barrier. However, should be acceptable as a supplement to other engineered barriers.

Previous attempts to construct conductive zones between adjacent boreholes have had limited success. Should be acceptable as a supplement to engineered barriers, but may not be acceptable for providing containment by itself.

6.13 IN SITU CHEMICAL TREATMENT

A project at PNL will develop, test, and evaluate an in situ method for immobilizing inorganics (metals, ions, and radionuclides) and destroying organics (primarily chlorinated hydrocarbons) using chemical or microbiological reduction of both the groundwater and the solid materials within the aquifer to form a permeable treatment barrier. The great majority of the chemically reactive mass in the subsurface system resides in the solid phases, rather than in the groundwater. Therefore, to have a substantial influence on the chemistry of the system, the solid phases should be involved. If changes are made only to the aqueous component, it will quickly reequilibrate with the solid phases.

There are several ways to approach the addition of reagents or nutrients into the subsurface. Three possible approaches include: (1) direct injection into the contaminant plume, (2) injection ahead of the contaminant plume to form a geochemical barrier by reacting with the solid phases, and (3) use of horizontal drilling technology to introduce a gaseous reagent to the contaminant plume. In the second alternative, a reagent or nutrient is injected ahead of the contaminant plume to form a permeable treatment barrier by reacting with the solid phases. The contaminant plume then reacts with the permeable treatment. The second alternative will be used in this project.

An unconfined aquifer is usually an oxidizing environment; therefore, most of the contaminants that are mobile in the aquifer are those that are mobile under oxidizing conditions. If the redox potential of the aquifer can be made reducing, then a variety of contaminants could be treated. Chromate could be immobilized by reduction to highly insoluble chromium hydroxide or iron chromium hydroxide solid solution. This case is particularly favorable since chromium is not easily reoxidized under ambient environmental conditions. In addition, uranium and technetium could be reduced to less soluble forms. Laboratory studies have shown that carbon tetrachloride and other chlorinated solvents can be degraded by microbes if the redox potential is reduced to the point where nitrate acts as an electron acceptor in place of oxygen.

A project at Westinghouse Hanford is testing the feasibility of treating unsaturated soils by injection of reactive gases. Dilute mixtures of hydrogen sulfide in air or nitrogen will be used to treat soils contaminated with heavy metals, while chromate or uranium contaminated soils are being treated with hydrogen sulfide and sulfur dioxide gas mixtures diluted by inert gases. Initial testing activities are using clean soils that have been artificially contaminated with hexavalent chromium, uranium, and other selected metals or radionuclides. Clean soils from several DOE sites are being used in this testing phase to verify that the approach is applicable to a variety of soil types, and to better evaluate the impact of gas concentrations and residence time on performance of other heavy metals. An objective of this activity will be to elucidate the chemical interaction between groundwater solutions, aquifer sediments, contaminants, and treatment agents.

6.14 BIOLOGICAL APPROACHES

DOE is evaluating biotechnological processes for their potential to decontaminate uranium-contaminated soils. Use of microorganisms (bacteria and fungi) to catalyze the uranium extraction process is being investigated. This method is similar in many respects to those already used by the commercial scale heap and in situ leach processes for uranium and copper extractions. The microorganisms involved in the leaching processes are indigenous and appear in nature wherever favorable living conditions exist: thiobacillus ferrooxidans are naturally occurring acidophilic iron oxidizing bacteria that function as an electron pump, oxidize iron, and in turn, oxidize uranium. In nature, thiobacillus ferrooxidans acidifies its environment to very low pH (sometimes as low as pH 1.5). This ability may be used as part of the clean up process, or it may be accelerated by addition of sulfuric acid. In this experimental system, the pH is achieved and maintained through titration of the soil with sulfuric acid.

Fungal metabolism is also being considered for uranium extraction; fungi-produced compounds that complex or chelate uranium, and fungal mycelia that accumulate uranium directly. Penicillium simplicissimum and spergillis niger are non-pathogenic, naturally occurring fungi that are capable of utilizing low-value carbon sources. Current experiments are utilizing 2 approaches for uranium extraction. In one set of experiments the fungi and soil are incubated together, and in the other set, depleted media is extracted from the fungi and applied to the soil. The media and cell mass are then analyzed for uranium, content. Preliminary results have been encouraging. It is likely that manipulation of culture conditions will significantly improve extraction. Specific evaluation parameters that are being assessed include:

1) The metabolic alteration of contaminant chemistry;
2) The microbial generation of acids;
3) The use of chelators or specific lixivants in conjunction with microbes for their potential contribution to solubilization and extraction;
4) The retrievability of the leachate;
5) The residual contaminant concentrations in the soil; and
6) The potential for treating the resulting contaminated leachate.

The uranium content of some contaminated soils at Fernald is very low compared to uranium contents of ores used for yellowcake production. This situation has created a need for evaluation of new technological possibilities for the treatment of large volumes of contaminated soils containing low or trace concentrations of uranium. One of these possibilities is the biotechnological approach.

Results indicate bacteria would likely be effective in improving uranium extraction in soils with high percentage of tetravalent uranium or low iron content (the opposite of that found at Fernald).

Researchers at the U.S. Geological Survey in Reston, Virginia, have shown unequivocally that bacteria can directly reduce soluble uranium (VI) to insoluble uranium (IV).

Under proper conditions micro-organisms bring about dissolution or immobilization of radionuclides and toxic metals by one or more of the following mechanisms: (1) oxidation-reduction reactions that affect solubility; (2) changes in pH and Eh that affect the valence or ionic state; (3) solubilization and leaching of certain elements by microbial metabolites or decomposition products such as organic acid metabolites, chelation, or production of specific sequestering agents; (4) volatilization due to alkylation reactions (biomethylation); (5) immobilization leading to formation of stable minerals or bioaccumulation by microbial biomass.

Microorganisms solubilize various metals and radionuclides from ores, soils, and fossil and nuclear energy wastes by production of mineral acids, organic acids, and oxidizing agents. Treatment of certain types of DOE radioactive wastes by heterotrophic microbial action offers great promise. However, fundamental information at the mechanistic level, in particular with the actinides and selected fission products, is very scanty and warrants basic research that can lead to the development of reliable treatment methods.

Basic research studies should be performed to determine the mechanism of oxidation-reduction reactions catalyzed by aerobic and anaerobic microbes. Enzymes involved in oxidation or reduction of elements leading to solubilization or precipitation of radionuclides and toxic metals under both aerobic and anaerobic conditions need to be isolated and characterized. Particular attention should be given to the enzymes (reductase) involved in reduction of actinides and fission products.

Mechanisms whereby microbes stabilize (i.e., immobilize) radionuclides need to be studies and understood. Singled out for special study should be microbial reactions involving (1) sulfate reduction and formation of insoluble metal sulfides; (2) formation of stable minerals as the result of precipitation reactions or redistribution of solubilized elements with the stable mineral phases in the waste matrix; and (3) biosorption by microbial biomass and biopolymers.

The overall basic research program on microbial waste pretreatment technology should also include studies to isolate and characterize novel microbial metabolites including chelating agents that can be used to solubilize and/or selectively complex radionuclides in wastes.

A project at PNL will develop, test, and evaluate an in situ method for immobilizing inorganics (metals, ions, and radionuclides) and destroying organics (primarily chlorinated hydrocarbons) using chemical or microbiological reduction of both the groundwater and the solid materials within the aquifer to form a permeable treatment barrier. The great majority of the chemically reactive mass in the subsurface system resides in the solid phases, rather than in the groundwater. Therefore, to have a substantial influence on the chemistry of the system, the solid phases should be involved. If changes are made only to the aqueous component, it will quickly reequilibrate with the solid phases.

6.15 IN SITU VITRIFICATION

The in situ vitrification (ISV) process fixes fission products and immobilizes or destroys mixtures of hazardous chemicals in soils. This technology can be applied to radionuclides, heavy metals, and hazardous organic-contaminated soil.

ISV is the conversion of contaminated soil into a durable glass and crystalline waste form through melting the soil by joule heating. Contaminants are destroyed by or immobilized in molten glass (melted soil). Soil is melted by electrical energy from electrodes that are placed in the ground. Off-gas from this process is treated by conventional off-gas treatment methods.

This technology has a number of benefits. Specifically, ISV may safely immobilize or destroy both radioactive and hazardous chemicals before they impact the groundwater or other ecosystems. It is applicable to soils contaminated with fission products, transuranics, hazardous metals, and hazardous organics. It reduces risk to the public by immobilizing or destroying radioactive and hazardous materials in the soil. Finally, in situ treatment poses a lower potential risk to workers than traditional treatments because contaminants are not brought to the surface. This technology, however, has not yet been demonstrated at depths beyond 20 feet.

The ISV technology can be applied to a wide range of soil types and contaminants. Melt depths of approximately 5 meters are considered the practical limit for most sites at this time. However, additional research is being conducted to ultimately achieve melt depths of up to 10 meters. There are no practical limits for inorganic contaminants; current processing systems are designed to process up to 8 wt. percent organics based on heat loading consideration. High moisture soils can generally be processed, but saturated soils with free flowing groundwater would require the use of methods to minimize groundwater recharge. With use of electrode feeding technology (vertically moveable electrodes), inclusions such as scrap metals and buried piping can be processed without concern of electrical short circuits.

Laboratory-scale experiments at PNL have demonstrated the following: (1) a subsurface ISV melt can be initiated and maintained, resulting in a horizontal, planar, glass block; (2) the downward growth of a vertical ISV melt can be directed and controlled such that enhanced melt rate and limited outward growth is achieved, resulting in a vertical, planar, glass block; and (3) a vertical ISV melt can be vitrified to a subsurface horizontal ISV block, forming a bond that joins them into one continuous formation. The results from these experiments demonstrate the feasibility of generating vitrified underground barriers beside, beneath, and/or around a waste site.

Since its development, ISV has been tested more than 190 times at various scales, including bench, engineering, pilot, and full-scale. It has been used to treat a wide range of hazardous materials, including heavy metals, organics, and radioactive materials.

The PNL/Geosafe in situ vitrification (ISV) process uses an electric current to melt soil or sludge at extremely high temperatures (1,600°C to 2,000°C), thus destroying organic pollutants by pyrolysis. Inorganic pollutants are incorporated within the vitrified mass, which has glass properties. Water vapor and organic pyrolysis by-products are captured in a hood, which draws the contaminants into an off-gas treatment system that removes particulates and other pollutants.

The vitrification process begins by inserting large electrodes into contaminated zones containing sufficient soil to support the formation of a melt. An array (usually square) of four electrodes is placed to the desired treatment depth in the volume to be treated. Because soil typically has low electrical conductivity, flaked graphite and glass frit are placed on the soil surface between the electrodes to provide a starter path for electric current. The electric current passes through the electrodes and begins to melt soil at the surface. As power is applied, the melt continues to grow downward, at a rate of 1 to 2 inches/hr. The large-scale ISV system melts soil at a rate of 4 to 6 tons/hr.

The mobile ISV system is mounted on 3 semitrailers. Electric power is usually taken from a utility distribution system at transmission voltages of 12.5 or 13.8 kilovolts. Power also may be generated on-site by a diesel generator. The electrical supply system has an isolated ground circuit to provide appropriate operational safety.

Air flow through the hood is controlled to maintain a negative pressure. An ample supply of air provides excess oxygen for combustion of any pyrolysis products and organic vapors from the treatment volume. Off-gases are treated by quenching, pH controlled scrubbing, dewatering (mist elimination), heating (for dewpoint control), particulate filtration, and activated carbon adsorption.

Individual settings (placement of electrodes) may grow to encompass a total melt mass of 1,000 tons and a maximum width of 35 feet. Single-setting depths as great as 25 feet are considered possible. Depths exceeding 19 feet have been achieved with existing large-scale ISV equipment. Adjacent settings can be positioned to fuse to each other and to completely process the desired volume at a site. Stacked settings to reach deep contamination are also possible. Void volume present in particulate materials (20 to 40 percent for typical soils) is removed during processing, reducing the waste volume.

The ISV process can be used to destroy or remove organics and to immobilize inorganics in contaminated soils or sludges. In saturated soils or sludges, water is driven off at the 100°C isotherm moving in advance of the melt. Water removal increases energy consumption and associated costs. Also, sludges must contain a sufficient amount of glass-forming material (non-volatile, non-destructible solids) to produce a molten mass that will destroy or remove organic pollutants and immobilize inorganic pollutants. The ISV process is limited by (1) individual void volumes in excess of 150 ft^3, (2) rubble exceeding 20 percent by weight, and (3) combustible organics in the soil or sludge exceeding 5 to 10 percent, depending on the heat value.

ISV is applicable to soils containing radionuclides, transuranics, fission products, organic chemicals, metals and inorganic chemicals, and mixed waste. Amenability and achievable depth may be limited by the presence of rock or gravel layers where heat transfer is less efficient.

It is extremely effective in immobilizing radionuclides, including transuranics and fission products. Criticality limits are conservatively placed at 30-kg plutonium/setting. Typically, there is no volatilization of Sr-90, Am-241, Pu-239/240, and measurements indicate greater than 99.993% retention of these isotopes. Rare earth tracers, Ce, La, and Nd, were used as surrogates for transuranic isotopes in an Oak Ridge National Laboratory (ORNL) field test with greater than 99.9995wt% retention in the melt. Cesium is more volatile than most radionuclides and has been measured with volatilization of 0.029% up to 2.4 wt% of Cs-137 in some cases.

In Situ Vitrification (ISV) is a patented thermal treatment process. The technology was originally developed by Pacific Northwest Laboratory, operated by Battelle Memorial Institute, and has been undergoing testing and development since 1980. A majority of the development work was performed by the U.S. Department of Energy, however, significant work also has been done for various private and other government sponsors. The technology has been licensed exclusively to Geosafe Corporation for the purpose of commercial applications of hazardous and radioactive waste remediation.

This technology has been demonstrated at a variety of sites, including Geosafe's test site in Kirkland, Washington, and the DOE's Hanford Nuclear Reservation in Richland, Washington, Oak Ridge National Laboratory in Oak Ridge, Tennessee, and Idaho National Engineering Laboratory in Idaho Falls, Idaho.

DOE plans to turn a waste pit containing radioactive wastes into a 1,000-pound mass of glass. Using In Situ Vitrification an obsidian-like glass will be created from waste in the pit.

6.16 IMPOUNDMENTS

Impoundments are areas where contaminated waters have been discharged into ponds or diked areas that prevent runoff. The soil in impoundments often can be treated with technologies discussed earlier, but the localized concentration in impoundments requires special technologies. These special technologies, not the more general soil treatment technologies, are discussed below.

Containment and Isolation:

1) *Capping* - This technology involves the placement of a hydraulic barrier over and around the contaminated impoundment.
2) *Drains* - French drains, biopolymer drains, and horizontal wells aid in hydraulic isolation of the impoundments.
3) *Walls/Barriers* - Grout Curtains, slurry walls, sheet pile walls, clay bentonite additives and cryogenic barriers were considered for hydraulic isolation of the impoundments.

Treatment:

1) *Physical Separation* - Soil washing and in situ soil flushing were considered for removal of metals and radionuclides from impoundments. Sludge drying was also considered for removal of water from impoundment sediments as a pretreatment for storage/disposal.
2) *Fixation* - In situ fixation using silicates, cement or polymers and ISV were considered for immobilizing contaminants in impoundment sediments.

Retrieval:

1) *Excavation* - Mechanical excavation, vacuum loading, and dredging to retrieve contaminated impoundment sediments were considered.

Note that some technologies (sludge drying and soil washing) are ex situ treatment technologies that require mechanical excavation or vacuum loaders (technologies listed for waste retrieval), prior to processing. Similarly, cryogenic barriers are an in situ treatment that retain the contamination temporarily (perhaps for relatively long times) until radioactivity has decayed or until permanent treatment plans are complete.

Several technologies aim to restrict the entry of water into the impoundment area. These include slurry walls, sheet-pile walls, grout curtains, french drains, capping, and biopolymer drains. These may be considered permanent or temporary solutions, depending upon the lifetime of the retention and the lifetime of the contaminant. These technologies can also be useful for preventing pollutant migration while the contaminated region is being treated.

Other impoundment technologies are in situ treatment for permanent solutions to the problems. These technologies may be of more interest for impoundments than for most soil contamination because impoundments are more likely to contain high local concentrations of contaminants that can be immobilized by solidification of relatively small volumes of soil.

6.17 LANDFILL STABILIZATION

The Department of Energy (DOE) selected the Savannah River Site (SRS) to lead 2 focus groups—landfill stabilization and contaminated plumes—in researching and implementing technologies to clean up national landfills and contamination in groundwater and soils. Other areas targeted under the program are mixed waste, high-level waste tanks and decontamination and decommissioning.

Annually, about $50 million should be earmarked for the landfill stabilization and contaminated plumes groups during the next 3 or 4 years. The teams primarily will focus demonstrations at 3 major DOE sites: Idaho, Savannah River, SC, and Richland, WA.

7

Thermal Processes

The importance of thermal processes is related to the volume reduction that can be obtained in destroying the non-radioactive portions of low-level and mixed waste, allowing disposal of the radioactive portion in a much smaller volume.

Several thermal processing techniques such incineration, supercritical water oxidation, plasma destruction, and oxidation in molten salt media are applicable to the treatment of a variety of DOE wastes. Incineration processes are well developed. Although thermal techniques are broadly applicable to the destruction of the organic components of almost any kind of waste, their applicable to DOE wastes is now restricted primarily to the incineration of low-level waste. Reliable, essentially fail-safe equipment and processes; better monitoring of effluents; better process control; and small units suitable for processing wastes locally without the requirement for off-site transportation are all necessary to obtain public acceptance of thermal processing of wastes. In addition, the broadest application of thermal processing will require advanced designs that allow efficient treatment of non-conventional materials such as soils heavily contaminated with hazardous or mixed wastes.

An integrated systems engineering approach is being developed by EG and G, Idaho, Inc. for uniform comparison of widely varying thermal treatment technologies proposed for management of contact-handled mixed low-level waste (MLLW) currently stored in the U.S. Department of Energy complex. Ten different systems encompassing several incineration design options are studied. All subsystems, including facilities, equipment, and methods needed for integration of each of the 10 systems are identified. Typical subsystems needed for complete treatment of MLLW are incoming waste receiving and preparation (characterization, sorting, sizing, and separation), thermal treatment, air pollution control, primary and secondary stabilization, metal decontamination, metal melting, mercury recovery, lead recovery, and special waste and aqueous waste treatment. The evaluation is performed by developing a pre-conceptual design package and planning life-cycle cost (PLCC) estimates for each system.

7.1 INCINERATION

Incineration is a very versatile process. Many devices and techniques are available that work with combustible materials in almost any physical form. Volume reductions greater than 100 are routinely realized when burning low-level waste; burning liquids that are not contaminated with inorganic materials leaves almost no solid residue. The performance of modern incinerators operated within their design envelopes is excellent; organic materials are converted quantitively to CO_2 and H_2O with essentially no production of toxic by-products. Transients associated with changes in feed, mechanical problems, etc., do occur, however, and can result in operation outside the envelope of tested conditions with presently unpredictable results.

Future research efforts should be devoted to:

1) Reaction mechanisms and rates.
2) Fluid mechanics.
3) Monitoring and control.

Incineration serves several purposes as a management strategy: (1) it destroys some hazardous materials by breaking them down into simpler chemical forms, (2) it eliminates liquids in waste that otherwise complicate waste management, (3) it decreases the volume of waste, and (4) it may generate usable energy. Incineration currently is a critical component in DOE's strategy for managing low-level radioactive and mixed wastes.

The incinerator at Chalk River Nuclear Laboratories (CRNL) has achieved a volume reduction of about 170:1 (on as-received volume basis) for miscellaneous combustible uncompacted trash generated at the laboratories. Incineration of baled waste has also been successfully implemented at the CRNL incinerator. Radiochemical analysis of the resultant ash from the CRNL incinerator has shown that Co-60 and Cs-137 account for about 12 and 8% respectively of the total activity in the ash, whereas other radionuclides (Sb-125, Cs-134, Ru-106, Ce-144, Ag-100m, Ce-141, Ru-103, Nb-95, Zr-95 and Zn-65) account for the remaining 80%. The activity in the ash is typically about 720 μCi/kg. Because of the radioactive decay of the shorter-lived radionuclides, after 2.5 years in storage, Co-60 and Cs-137 jointly account for about 85% (Co-60 52% and Cs-137 33%) of the total activity remaining in the ash; the other radionuclides account for about 15%. The activity remaining in the ash at this time is about 150 μCi/kg. Thus, the ash can be stored to allow a significant part of the radioactivity to decay.

Incineration is both effective and technically feasible as evidenced by typical volume reduction factors, before final ash immobilization and packaging, of 30:1 to 100:1. Even after final packaging, the net volume reduction is still

2 to 5 times greater than competing technologies such as super-compaction. Incineration as an alternative to direct shallow-land burial of LLW at licensed radioactive waste sites also has the benefits of providing a very limited and monitored release of radionuclides to the environment and of providing a waste that is readily stabilized, which minimizes long-term ground residence and leaching by rain and groundwater.

The larger incinerators designed to burn low-level/mixed wastes are usually: 1) rotary kilns; 2) controlled air, dual chamber; and 3) fluidized bed.

The Mixed Waste Management Facility (MWMF) (Lawrence Livermore, CA) is a national demonstration test bed that will be used to evaluate, at pilot scale, emerging technologies for the effective treatment of low-level radioactive, organic mixed wastes. The primary and initial goal will be to demonstrate technologies that have the potential to effectively treat a selection of organic-based mixed waste streams, currently in storage within the DOE, that list incineration as the best demonstrated available technology (BDAT).

7.1.1 SEG Incinerator

SEG operates the world's largest radioactive waste incinerator. SEG's incinerator is an automated, controlled-air incinerator capable of burning 1,000 lb of waste per hour. It is located in Oak Ridge.

Treatable Wastes: The following radioactive materials are incinerated at the SEG operation:

1) Dry active wastes, such as paper, plastic, wood, cloth, rubber, canvas, fiberglass, and charcoal.
2) Ion exchange resins used to polish condensate from nuclear power plants.
3) Animal carcasses from scientific—but not medical—research.
4) Sewer sludges and lubricating oils that have become contaminated with radioactive materials.
5) High efficiency particulate air (HEPA) filters.

Other materials, including metals, explosives, flammable liquids, shock-sensitive materials, or polyvinyl chloride (PVC), might not be suitable for incineration at SEG. In addition, large pieces of metal, such as sections of pipe, cannot be incinerated, because they can jam the augers that slowly propel ashes from the charging area to the discharge area of the incinerator. Items smaller than a 10-in. crescent wrench do not interfere with the action of the augers.

The incinerator has 3 chambers—the primary combustion chamber, secondary combustion chamber, and retention chamber—each with its own burner and thermostat. The total residence time for gases, from the dumping of waste materials into the primary combustion chamber to the emission of flue gases from the retention chamber, is about 3 seconds.

Draft fans, air supply fans, gas monitors, opacity detectors, HEPA filters, negative air-pressure controllers, and an emergency power source are among the redundant features that can improve the safety of an incinerator. The most important feature is an emergency power source. SEG has a 300-kilowatt diesel backup generator, capable of carrying the entire incinerator load when outside power is lost.

Since SEG's incinerator is used to process radioactive wastes, it must be operated under a vacuum. SEG's primary combustion chamber is operated at -0.5 in. H_2O, while the vacuum at the suction of the ID fan is -30.0 in. H_2O. The difference between these is the differential pressure that occurs across the scrubber, baghouse, boiler, and HEPA systems.

Advantages: Incineration produces a waste form that is dense and easy to transport, and takes up relatively little space when buried. Incineration has been shown to yield varying volume reduction factors (VRFs): commonly 4 to 40 for most types of compressible dry active wastes and combustible solids, and greater than 100 for liquids and most plastics. SEG also operates a supercompactor, which exerts up to 10 million lb of pressure on the filled ash box and can produce further VRFs of 2 to 5.

The annual permissible dose equivalent release limit from the SEG site is 10 mrem, but actual releases tend to be much lower. In 1991, the SEG incinerator processed 5.3 million lb of radioactive wastes, exposing the nearest resident to an estimated dose of 0.027 mrem for the year, compared to natural background levels of approximately 150 mrem/year.

Limitations: The primary disadvantage of incineration is that it can produce toxic ash that requires further processing prior to disposal. This is a particular concern for incineration of radioactive waste, which yields waste residues that have much higher radionuclide concentrations than does the original waste stream. As a result, containers or bins of ash from the incineration of radioactive waste may have high external radiation exposure rates. When radiation exposure levels are expected to be high, personnel interaction with equipment and ash bins should be minimized. Ash collection bins and other ash handling equipment also might need to be shielded.

Incineration produces 3 types of ash: hearth ash, which is discharged from the primary chamber during combustion; fly ash, which gets stripped from the flue gas in the baghouse; and boiler ash, which gets stripped from the flue gas in the boiler. Hearth ash from a incinerator operated at the proper temperature usually passes EPA's Toxicity Characteristic Leaching Procedure (TCLP). Fly and boiler ash always are characteristic because of the presence of lead which emanates from the primary chamber and passes from the incinerator in fine aerosol form.

Ash that passes TCLP testing can be compacted immediately and shipped for burial, while ash that fails TCLP testing must be solidified by concrete or epoxy into a monolithic waste form by mixing it with a hardener and fixer base material and allowing it to harden. Once hardened, the waste form is sampled and retested. If the sample passes, the waste form may be buried;

if it fails, the waste must be reprocessed. To date, SEG has not experienced a TCLP failure of its stabilized fly ash waste form.

Another disadvantage to incineration is that the operation of wet scrubbers generates salt that must be removed. SEG uses a quick dry dewatering system in which salt drums are decanted into larger drums that contain filtering systems. A vacuum then is applied to draw the water out of the salt mixture. The remaining salt is not hazardous and can be disposed of accordingly. SEG currently is developing a spray dryer to provide a one-step drying process for the salt slurry.

7.2 MOLTEN METAL/SALT PROCESSES

High-temperature molten metal-molten salt processes are extensively used in the United States and United Kingdom for plutonium metallurgical/chemical operations. Such processes are also being extensively studies for potential use in recovery and recycle of actinides in irradiated integral fast reactor (IFR) fuel. High-temperature non-aqueous processes also are being studied for potential use in oxidation of solid combustible waste from plutonium recovery, recycle, and/or purification operations, and for pretreatment of various DOE residues and sludges.

Many of the process steps in high-temperature schemes are based upon partitioning of solutes such as actinide and either daughter or fission products between 2 immersible liquid phases; some involve 2 liquid phases plus a third solid phase. Most of the extraction steps use a molten metal and a molten salt; the partitioning behavior of the solutes depends upon oxidation-reduction reactions.

Other high-temperature, non-aqueous schemes involve either gas-solid or gas-liquid reaction interfaces. The products of these reactions may be either solids, gases, or a combination of gases and solids.

Pyrochemical processes, used extensively in plutonium metal fabrication and processing operations, are carried out at high temperatures (typically 700-1000 °C) in a molten salt medium; molten metals and molten alloys can be, and often are, used. As the solubilities of various reactants and products are usually lower in molten salts than in aqueous solutions, large volumes of salts are necessary for reactions to proceed. These large volumes of salts, generally chloride-based, contain plutonium and other actinides and constitute a large amount of DOE radioactive waste. Existing pyrochemical processes must be modified and new processes must be devised to allow meaningful reduction in pyrochemical process wastes.

The molten salt process for waste destruction provides for oxidation of organic material within a fluidized bed of molten salt at ambient pressures and temperatures of about 700-900 °C. The salt is a mixture of alkali and alkaline earth halides and carbonates. Liquid or pulverized solid wastes are injected into the melt along with oxygen diluted with carbon dioxide. The sparging produces an expanded foam (50% liquid) of large aggregate surface area which exhibits catalytic properties for oxidation. The alkaline nature of the melt prevents formation of acidic gases (HF, HCl, SO_2, etc.) by forming solutions of the corresponding neutral salts.

Potentially fugitive particulates and volatile substances are retained by the melt by a combination of particulate wetting and encapsulation, and by dissolution and solvation. The salt is eventually exhausted by either neutralization or by accumulation of ash. The exhausted salt is withdrawn and separated by aqueous processing into a mixed-waste ash (for subsequent refinement or immobilization and burial) and a LLW salt (if radioactive isotopes are present). Any unreacted carbonates may be returned to the melt.

The Molten Salt Oxidation (MSO) Process is carried out in a highly reactive oxidizing and catalytic medium. It uses a sparged bed of turbulent molten salt such as sodium carbonate at 800 °C to 1,000 °C with waste and air introduced beneath the surface of the molten salt. Generally, the heat of oxidation of the waste keeps the salt molten. The off-gas, containing carbon dioxide, steam, nitrogen, and un-reacted oxygen is cleaned of particulates by passing the gas through standard filters before discharging to the atmosphere.

MSO has a high treatment potential for radioactive and hazardous forms of high-heating liquids (organic solvents, waste oils), low-heating value liquids (high-halogen content organic liquids), other wastes (pesticides, herbicides, PCBs, chemical warfare agents, explosives, propellants, infectious wastes), and gases (VOCs and acid gases). By virtue of the latter, MSO could replace conventional wet-scrubbers as a superior dry-scrubber system for use with incinerators. The typical residence time is 2 seconds for the treatment of wastes by the MSO Process.

Wastes containing heavy metals are converted to oxides and retained in the melt. Organic solids and other combustible materials are destroyed, but MSO is not suitable for direct treatment of inert solids, such as soils and rubble. However, MSO can treat the extracted residuals of commercially available soils pretreatment technologies such as vapor extraction, solvent extraction, thermal desorption, and base-catalyzed dechlorination. Carbon has been destroyed in all of the process demonstrations, including graphite oxidation and coal gasification.

Ash and the reaction products of acid gases and salt are retained in the molten salt. The MSO Process has been tested at 900 °C for the destruction of solid combustible waste-bearing plutonium at TRU levels (>100 mCi/g). Measurable amounts of plutonium downstream of the oxidizer have shown that 99.9% of the plutonium remains in the melt.

The final waste form is a product of the spent salt disposal or recycle subsystem. In the destruction of chlorinated waste compounds, the melt becomes unreactive as the salt converts to approximately 90% sodium chloride (NaCl). The NaCl can be discarded unless it is extracted from the disposable salt by ion exchange chemistry coupled with biosorption techniques. Otherwise, when the salt is reusable, but contains ash and possibly metal products, conventional dissolution and fractional filtration techniques with radionuclide extraction apply.

The Office of Technology Development (OTD) of DOE organized a technical review panel to study Molten Salt Oxidation (MSO) for mixed low-level waste (MLLW). They concluded that although it appears to have capabilities to effectively and efficiently treat only a limited portion of DOE's MLLW, they recommended proceeding with a design.

7.3 FLUIDIZED BED UNIT

Rocky Flats has a serious mixed waste problem. One solution under study is to use a catalytic fluidized bed unit (FBU) to destroy the combustible portion of the mixed waste. The fluidized bed thermal treatment program at Rocky Flats is building on knowledge gained over 20 years of successful development activity. The FBU has numerous technical advantages over other thermal technologies to treat Rocky Flats' mixed waste, the largest being the lower temperature 700 °C versus 1,000 °C which reduces acid corrosion and mechanical failures and obviates the need for ceramic lining. Successful demonstrations have taken place on bench, pilot, and full-scale tests using radioactive mixed wastes. The program is approaching implementation and licensing of a production-scale fluidized bed system for the safe treatment of mixed waste.

7.4 PLASMA TORCH INCINERATION

The plasma torch incineration process is an experimental technology currently under development. The concept is similar to that of an electric arc welder and can generate temperatures up to 5,500 °C. The intended use of the plasma torch is for disposal of liquid chemical wastes, in particular hazardous organic compounds and solvents from Environmental Protection Agency (EPA) Superfund sites; conceivably, this process could also be adapted to handle liquid low-level wastes. It offers the complete destruction of the organic component of mixed wastes.

7.5 SUPERCRITICAL WATER OXIDATION (SCWO)

Supercritical Water Oxidation (SCWO) technology holds promise for treating a portion of DOE's mixed waste. The process involves bringing together organic waste, water, and an oxidant (such as air, oxygen, etc.) to temperatures and pressures above the critical point of water (357 °C, 22.1 MPa). Under these conditions, organics in the waste are destroyed to levels of over 99.99%. The resulting effluents, which consist primarily of water and carbon dioxide, are relatively benign. In contrast to incineration, SCWO can easily be designed as a full containment process with no release into the atmosphere. In contrast to wet air oxidation, SCWO can achieve the high destruction efficiencies for hazardous wastes such as polychlorinated biphenyls (PCBs) or dioxins. In comparison to plasma treatment methods and incineration, SCWO processes achieve high organic destruction efficiencies at much lower temperatures and without NO_x production. When compared to other thermal treatment technologies, SCWO can process wastes with low concentrations of organics.

SCWO experiments are carried out at temperatures up to 800 °C and pressures as high as 300 atmospheres. Present reactors suffer corrosion, erosion, and embrittlement from exposure to mixtures containing water, acids, chlorides, hydrogen, oxygen, and various inorganic solids. The ultimate utility of the SCWO process will be strongly coupled to the success of research aimed at finding new materials that will withstand its extremely hostile environment.

There are 2 general schemes. The first is one in which the effluent is cooled and effluent treatment is primarily done in the liquid phase. In the second scheme, most treatment is performed with the effluent in the gas phase.

The SCWO Program will construct and demonstrate hazardous and mixed waste pilot-scale SCWO units. The program will be conducted in 2 phases; Phase 1 - the Hazardous Waste Pilot Plant (HWPP) Demonstration, and Phase 2 - the Mixed Waste Pilot Plant (MWPP) Demonstration. The goals of the HWPP Demonstration are to demonstrate the technical viability and cost effectiveness of SCWO technology for treating DOE hazardous/surrogate mixed wastes, to provide the necessary design, operational, environmental, and safety data to evaluate the feasibility of a MWPP demonstration as well as to serve as the design basis for the MWPP. The HWPP is comprised of a test bed system which provides feed pressurization, separation, pressure letdown, and waste disposal systems. The test bed provides the support systems required for testing alternative reactor designs and components to evaluate their ability to resolve the technical issues facing SCWO application to mixed waste. The hazardous waste SCWO unit was scheduled to start testing in the spring of 1995.

The Idaho National Engineering Laboratory (INEC) is the lead laboratory for this program.

7.6 CATALYZED ELECTROCHEMICAL PLUTONIUM OXIDE DISSOLUTION (CEPOD) PROCESS

This is an oxidation process that dissolves plutonium oxide and other metal oxides by increasing their oxidation state. The process electrolytically generates silver ions in a nitric acid solution. Because these silver ions are powerful oxidants, they oxidize plutonium oxide and other metal oxides. This dissolution process allows the separation of plutonium from other constituents in wastes so that the total amount of material requiring treatment can be dramatically reduced.

7.7 MEDIATED ELECTROCHEMICAL OXIDATION

In the Mediated Electrochemical Oxidation process, an oxidizing metal ion (such as silver (II), cobalt, (III) or cerium) is generated at the anode of an electrochemical cell containing an acidic solution. The oxidizing metal then attacks and destroys the organic components of the waste.

It is mentioned here because Mediated Electrochemical Oxidation was originally developed to dissolve an insoluble form of plutonium oxide. Later, the ability to achieve high-destruction efficiencies for organic contaminants was demonstrated along with the effective dissolution of metals. The process operates at near-ambient temperatures and pressures using an acidic solution.

Evaluation of Mediated Electrochemical Oxidation for Rocky Flats is being conducted at LLNL and by PNL.

7.8 PYROLYSIS

Pyrolysis converts portions of municipal solid wastes, hazardous wastes, and special wastes such as tires, medical wastes, and even old landfills into solid carbon and a liquid or gaseous hydrocarbon stream. Pyrolysis heats a carbonaceous waste stream typically to 290-900 °C in the absence of oxygen, and reduces the volume of waste by 90% and its weight by 75%. In principle, pyrolysis could be used to treat mixed low-level wastes (MLLW) at DOE sites.

7.9 EVAPORATORS AND EVAPORATIVE CRYSTALLIZERS

Considering that evaporator technologies are controlled by physical and chemical characteristics of the waste streams and not by their radioactivity, almost any type of evaporation technology can be applied to LLW consistent with keeping radiation exposures "as low as reasonably achievable" (ALARA). Evaporators are used extensively in association with the nuclear power plant industry. They are typically used for treatment of relatively large volumes of liquids.

When separating a solution of salts in water, the water can be vaporized from the solution without salt removal because, for all practical purposes, salts are non-volatile under normal operating conditions. Loss of water by evaporation leaves behind a more concentrated solution of radioactive material (often called sludge or evaporator bottoms), thereby reducing the volume of radioactive liquid waste requiring disposal. Evaporator bottoms account for 700 to 7,000 ft^3/year of waste from a typical nuclear power generation station. The vaporized water can be condensed and reused in process applications or, in many cases, can be discharged.

In general, evaporators are capable of producing concentrations of the treated effluent of up to 12 weight % for boric acid LLW and 25 weight % for sodium sulfate. On the other hand, crystalline systems produce slurries of sodium sulfate up to a 50 weight % concentration (50% water and 50% salt).

Following evaporation, the concentrated liquid or slurry waste may undergo additional drying to further reduce waste volume. The waste must then be solidified, encapsulated, or in some manner treated prior to disposal. The extruder-evaporator unit produces a solidified waste material, but other evaporator systems require post-evaporative treatment. It is also possible that the evaporated water may still not be of sufficient quality for direct discharge to the environment, particularly if organics are present in the waste stream.

Just as evaporators can be used to reduce large volumes of liquid LLW, they can also be used to reduce volumes of dilute liquid mixed waste, provided the hazardous component(s) of the liquid are not volatile and remain with the evaporator bottoms for further treatment and disposal.

The Evaporator-Crystallizer operation is important at the Hanford site. Liquid radioactive waste and mixed waste currently undergo evaporation in the Evaporator-Crystallizer Facility. Approximately 5 to 10 Mgal of waste volume reduction are achieved annually during normal operations.

Waste concentration has reduced the storage space requirements for DSTs (Double-Shelled Tanks) by more than 100 Mgal. The 242-A Evaporator-Crystallizer is the cornerstone of waste management's treatment facilities in that it maximizes the use of available DST space and minimizes the need to construct additional DSTs.

Oak Ridge has studied the use of a wiped film evaporator for concentrating Melton Valley storage tank low-level wastes.

8

Radionuclides In Water

Radionuclides can occur in groundwater, surface water, process waste streams, and in storage tanks. This chapter discusses both high-level and low-level wastes, with the exception of high-level wastes in underground storage tanks which are discussed in chapter 13.

8.1 GROUNDWATER EXTRACTION

Groundwater can become contaminated from numerous sources. At the Nuclear Weapons Complex, sources include accidents and spills; intentional introduction of waste into the ground (cribs, surface impoundments, underground injection wells, landfills); and failure of containment methods (underground storage tanks).

Groundwater contamination is very site-specific in terms of the contaminants present and their behavior. Groundwater contamination is such a difficult problem to characterize and clean up because the environment is not uniform. In general, the less uniform the environment (such as fractured limestone at Oak Ridge or the presence of clay lenses at Savannah River), the more difficult it is to characterize contamination problems and clean them up. Some contaminants will be easier to find and clean up than others. For example, those contaminants that move with water are easier to find than those that do not.

Contaminants at the Weapons Complex include radionuclides, heavy metals, nitrates, and organic contaminants. Often these are present as complex mixtures that affect the mobility and fate of individual contaminants in the subsurface. Contaminants also behave in different ways, depending on the characteristics of a site. As contaminants move through the ground to an aquifer, many processes occur that affect the amount or concentration of the contamination by the time it reaches a receptor of concern such as a well or surface water. The processes may also affect the performance of remediation techniques. Many of these processes, however, are not well understood.

Some contaminants adsorb onto soil particles in the unsaturated zone or onto the aquifer media, thereby slowing their movement and possibly preventing groundwater contamination. Contaminants may also form or adsorb onto colloidal particles, which allows them to move with, or faster than, the average groundwater flow. Flow can result from an apparently unrelated force, such as the flow of water and contaminants due to a thermal or electrical gradient instead of the expected hydraulic gradient. Chemical reactions and biotransformation may occur, possibly changing the toxicity or mobility of contaminants. Some contaminants dissolve and move with the water; some are in the gas phase; other are non-aqueous phase liquids; some are more dense than water and may move in a direction different from groundwater; others may be less dense than water and float on top of it.

Contaminants that dissolve in water can often be extracted from groundwater and cleaned up with pump and treat techniques. This is the most commonly used procedure to clean up contaminated groundwater. Pump and treat can successfully remove great quantities of contaminants; however, the approach often takes much longer than originally planned to reduce contaminants to desired levels. Pumping can often be an effective way to prevent the spread of groundwater contamination and even reduce the size of a contaminated plume, but in some cases, it may not be possible to restore aquifers by pump and treat methods.EPA recognizes that, with current technologies, complete groundwater restoration may not be practicable in some circumstances, such as highly contaminated zones near the source of contamination that remain contaminated at levels preventing beneficial use. Long-term containment, natural attenuation, wellhead treatment or alternate water supply, and institutional controls to restrict water use may be necessary rather than attempting to restore an aquifer to health-based standards.

Because contaminated groundwater is so difficult to clean up it is especially important to prevent contamination from occurring in the first place and to prevent it from spreading further once it has occurred.

Extraction of groundwater for treatment is currently the primary method of groundwater remediation. Technologies to extract contaminated groundwater for treatment have limitations that make it difficult to predict the amount of time required to remove sufficient concentrations of contaminants. Limitations include adsorptive partitioning of contaminants between the aquifer and aquifer materials, and diffusion of contaminants into the small pores of the aquifer materials, which increase the amount of time required for remediation; aquifer heterogeneity, which makes it difficult to control groundwater flow; and residual contaminant sources in the soil or in a non-aqueous phase, which replace the dissolved contaminants as they are removed. In some cases, when sources of contamination cannot be eliminated, it may be necessary to operate pump and treat systems for long periods to achieve the desired reductions in contaminant concentrations.

There are 2 basic methods for extracting groundwater: pumping systems and passive systems. Both are based on 2 assumptions: 1) that it is possible to design a system that will withdraw all the contaminated water (this can be a problem in aquifers of low transmissivity, which do not release much water to wells, or in aquifers that have zones of low permeability, such as clay lenses); and 2) that the contaminants will come out of the aquifer with the water (this can be a problem if contaminants are sorbed onto aquifer materials or are present in a non-aqueous phase). Non-aqueous phase contaminants may be either more or less dense than the groundwater. When dense non-aqueous phase liquids (e.g., TCE and some other solvents) are present, they may be difficult to locate, and aquifer restoration may be judged impossible. When less dense,non-aqueous phase liquids (e.g., many petroleum products) are present, prospects for clean up are improved by the use of additional restoration techniques such as vapor extraction or bioremediation.

Pumping systems, or wells, can extract or divert groundwater at virtually any depth. The system should be optimized to remove contaminated groundwater, while extracting only a limited volume of uncontaminated water.

Passive interceptor systems can be excavated to a depth below the water table with the possible placement of a pipe to collect contaminated water or to lower the water table beneath a contamination source. These systems are relatively inexpensive to install, have low operating costs because flow is by gravity, and provide a means for leachate collection without impermeable liners. Although these systems can be more effective than wells for extracting water from some lower permeability materials, they are not suited to all low permeability conditions. They are limited in depth to the capabilities of trenching equipment (about 100 feet) and require continuous and careful monitoring to ensure adequate leachate collection.

DOE's plumes are contaminants including VOCs (volatile organic compounds) and other organic compounds, inorganic compounds, heavy metals, tritium, and radionuclides found in soil, surface water, and groundwater. Plumes can come from aqueous solutions leaked from underground pipes, volatile liquids that have vaporized in the subsurface and migrated into the soil, airborne releases deposited on the soil surface by wind and precipitation, wells used for underground injection of wastes, and waste disposal areas with contaminants that are spread by water moving through the site. In many cases, DOE's contaminant plumes were created by waste handling and discharge practices that used to be acceptable, but no longer are because of improved understanding of contaminant fate and associated risks. The plumes have also resulted from unintentional leaks and spills. The department has contaminated thousands of acres of land and associated groundwater, surface water, and soils with hazardous materials, which are often found combined with radioactive materials. Land within the DOE complex contains more than 4,800 individual release sites.

With such a large problem to tackle, the Department of Energy has named contaminant plume containment one of its 5 focus areas. The plume focus area must identify and develop cost-effective solutions to contain DOE's contaminant plumes on site, remediate groundwater aquifers to useable conditions, and treat or remove contaminants from soil above aquifers to prevent continuing groundwater contamination.

Six sites have initiated some sort of remediation process for removing and treating contaminated groundwater from certain areas. These involve pump and treat systems alone, or with French drains or interceptor trenches. Treatment consists of air stripping, ultraviolet light exposure, physical-chemical treatment, and ozonation.

All weapon sites in non-arid locations (i.e., those that have a net positive water balance) either have confirmed or suspected surface water contamination. This results from several factors, such as contaminated groundwater discharge to surface water, point source outfalls, and non-point source discharge to surface water (due to precipitation on contaminated soil and subsequent erosion of soil particles to surface water). Some arid sites also have surface water contamination.

Surface treatment techniques developed for water or wastewater are available for most contaminants. However, treatment systems for extracted water must be designed to deal specifically with the mixtures of contaminants and varying concentrations present at a site. Combinations of treatment processes may be required, and there is little experience designing systems to handle the mixtures of organics, radionuclides, and inorganics that may be present. Some processes are not applicable to the low concentrations of contaminants in question. Measures are required to discharge treated water back to the subsurface, to surface water, or to further treatment at a treatment plant. Some processes also generate residuals that must be handled as hazardous waste.

The volume of secondary waste that is generated in the clean up of a groundwater or plant effluent is very important to the economic viability of a process. This is especially true when the secondary waste must be treated as a hazardous waste or mixed hazardous/radioactive waste and disposed of in accordance with applicable State and Federal laws. Therefore, the testing of these processes is not only to evaluate their capability to reduce contaminants to MCL, but to help estimate the amount of secondary waste that will be generated during processing.

The secondary waste volume for each process that is tested will be one of the process characteristics used to determine the process with the best capability to economically decontaminate groundwater.

Processes for removal of contaminants from groundwater and surface waters can be carried out in either of 2 general methods: in situ or aboveground. Some of the technologies considered have been practiced widely in the chemical process industries and can be designed with confidence for aboveground application. When such technologies (e.g., ion exchange) are considered for in situ application, uncertainties arise concerning the control of flow distribution and rate and monitoring concentrations and associating them with flow rates so that mass balances are closed properly. No experience is available to indicate efficiencies that may be expected when such materials must be replaced.

The Westinghouse Hanford Company has been testing various technologies for decontaminating groundwaters and liquid effluents. The technologies are iron co-precipitation/filtration, supported liquid membranes, and reverse osmosis. The processes were tested to determine their capability to remove uranium, chromium, nitrates, and technetium. All processes removed contaminants to less than maximum contaminant limits. The secondary waste volumes were estimated for each process. The supported liquid membranes secondary waste volume was the smallest, followed by iron co-precipitation, and the largest volume was created by the reverse osmosis process.

An investigation at PNL is examining a number of substances that can immobilize chemical and radionuclide contaminants in groundwater beneath waste sites. Substances and processes under investigation include adsorption of chlorinated hydrocarbons from groundwater using a variety of organic materials, reductants to destroy chlorinated hydrocarbons and induce precipitation of various metals and oxyanions, and zeolites to sequester mobile metals.

In situ remediation of contaminants at disposal sites would be significantly enhanced if techniques existed that could target mobile substances without restricting groundwater movement. Control of groundwater requires limiting surface water penetration and vadose zone movement. Errors in control can lead to release of contaminated water.

Magnetic Separation: Selective adsorption of radioactive/heavy metals from groundwater on magnetically separable particles has been demonstrated in bench–scale experiments. Field demonstration is planned by the Resource Recovery project at the Berkeley Pit in Butte, Montana, and at Savannah River. Selentec has developed the MAG*SEP process.

8.2 AQUEOUS-BASED SEPARATION PROCESSES

Metal Ion Specific Ligand Design: The key ingredient in most aqueous-based separation processes is the use of some type of metal ion complexant. Metal ion specific ligands may be incorporated in solvent extractants, ion exchange resins, liquid membranes, and reagents used for precipitation processes or may be bonded or sorbed on substrates. There is a need for continued basic research in the design of metal ion specific ligands, particularly in computer-aided design. Appropriate criteria that are to be incorporated in the ligands need to be better defined, e.g., improved specificity, together with ease of recovery, solubility, and chemical and radiolytic stability. Design should also attempt to avoid the use of chemicals that will produce hazardous or toxic wastes.

An important facet of this area of research is the design of new solvent extractants that will be the key ingredients in new separation schemes. Metal ion specific ligand design should focus on the transuranic elements, actinide/lanthanide separations, and fission products such as Se-79, Sr-90, Zr-93, Pd-107, Sn-126, Cs-137, and Sm-151, even though considerable progress has already been made for separation processes for some of the latter isotopes.

The separation of useful materials (e.g., Ru, Rh, Pd) from high-level waste is another area where research should be focused.

Syntheses of Metal Ion Specific Compounds: The difficulty in the syntheses of new classes of reagents frequently hinders the exploitation of these new compounds for separation processes. This is particularly true in the preparation of organophosphorus reagents and in the preparation of individual isomers of certain macrocyclic reagents. Even when synthetic routes are available, the cost of the final product militates against their use in commercial separation processes. Therefore, there is a need for research in new approaches to the preparation of metal specific reagents that will allow the facile preparation of a variety of potentially useful compounds in aqueous-based separation schemes.

New Separation Techniques: Most separations processes used in the nuclear fuel cycle involve either solvent extraction or ion exchange. Although both of these techniques are well established and equipment used in plant-scale operations has been in existence for many years, there is still a need for improved methods of performing both of these unit operations. For example, basic research is needed in the use of electrostatic fields to rapidly mix and separate 2 immiscible phases and in the calculation and measurement of axial and radial dispersion achieved in continuous chromatographic and radial flow chromatographic modes of operation.

New separations techniques such as emulsion and supported liquid membranes and those based on ligands bonded to or sorbed on inert substrates need to be more completely investigated from the standpoint of the underlying basic physicochemical processes. A number of electrochemical phenomena, known for many years, but not fully understood, are also in need of basic research. This group of electrochemical methods includes electrodeposition, electrodialysis, electroosmosis, and electrophoresis.

New Reagents for Dissolution of Sludges and Residues: Chemical pretreatment of wastes by aqueous-based separations techniques requires the dissolution of large volumes of intractable sludge from waste storage tanks or the dissolution of residues from thermal treatment of wastes. New reagents or new formulations of existing reagents will be needed to facilitate the dissolution process. Basic research in the design and synthesis of new aqueous-soluble complexing reagents specifically tailored for metal oxide dissolution is needed. These reagents may require both a strong metal ion complexing group and a reducing functionality incorporated into the same molecule or may involve mixtures of complexing agents and reducing reagents. Alternative, chemically stable complexing agents together with oxidizing agents may be an important area for research.

Thermodynamics of Aqueous and Organic Species Relevant to Aqueous Pretreatment Process: To understand and predict partitioning behavior in organic and/or aqueous systems, the standard Gibbs free energies of formation and activity-concentration data are needed for the species and compounds of interest. A program to provide the missing thermodynamic data for plutonium, other actinides, and major fission products needs to be reinstituted.

8.3 ION EXCHANGE

This section discusses the use of ion exchange for low-level wastes, uranium, and radium. Ion exchange for high-level tank wastes is also discussed in chapter 13.

Uranium can be a cation, neutral, or an anion depending on the pH of the water. In water with a pH less than 5, uranium is a cation; in water with a pH between 5 and 7, uranium is neutral; in water with a pH greater than 7, uranium is an anion. As a result, ion exchange for uranium may involve either cation exchange or anion exchange. The pH of the water also affects the uranium removal efficiency of iron coagulation. Iron coagulation is very efficient at pHs near 6 and near 9; the treatment is not efficient, however, at pHs between 7 and 8 or below 5. When alum is used as a coagulant, the removal pattern is similar to that of iron coagulation. The uranium removal efficiency of lime softening and anion exchange depends on the presence of naturally occurring elements in the water. There is an impact of magnesium levels on the effectiveness of lime softening for uranium removal, as well as an effect of sulfate levels on uranium removal by ion exchange.

As with uranium, the effectiveness of ion exchange for radium removal depends on the presence of other elements, such as barium, calcium, and magnesium, in the water being treated. These elements may be preferred to radium in the resin's selectivity sequence.

Even if radium is highly preferred by a particular cation resin, the final percentage of radium removed will depend on the selectivity sequence of the resin and other elements present in the water.

Water with more than one radioactive contaminant may require more than one treatment process. For example, radium usually is treated by cation exchange with sodium, and uranium usually is treated by anion exchange with chloride. Water contaminated with radium *and* uranium can be treated by a mixture of cation resin and anion resin.

Treatment efficacy can depend on the source of the water being treated. A treatment appropriate for contaminated groundwater often will not be appropriate for contaminated surface water. Surface waters that are high in turbidity will foul ion exchange media, reverse osmosis membranes, or GAC. These methods can be used only if surface water is pretreated to achieve groundwater turbidity levels. Lime softening can be used for both ground and surface waters without pretreatment, though it might be more costly for surface water. Coagulation/filtration treatment is designed to remove turbidity and therefore is used only on surface waters.

Leaching used in extraction of uranium and other minerals is a non-selective process resulting in the dissolution of elements in addition to the desired constituents. Ion exchange is one process used for concentrating the desired constituents from the leached solutions. The resin ion exchange technique involves the interchange of ions between the aqueous solution and a solid resin. This provides for a highly selective and quantitative method for recovery of uranium and radium. The process of removing dissolved ions from solution by an ion exchange resin is usually termed adsorption in the uranium industry.

There are several resins available for extraction of both radium and uranium. For uranium extraction by ion exchangers, strong and intermediate base anionic resins are loaded from either sulfuric acid or a carbonate leach feed solution. The loaded resin is stripped with a chloride, nitrate, bicarbonate, or an ammonium sulfate-sulfuric acid solution to remove the captured uranium. These resins are semi-rigid gels prepared as spherical beads. Radium can be extracted by using synthetic zeolites.

The total amount of uranium that may be adsorbed is a function of the quantity of anionic complex in solution. Two to five pounds of U_3O_8 can be captured for each cubic foot of resin. Higher capacity is not possible because of competition for ion sites in the resin by other anions present.

The ion exchange process is, in most plants, a semi-continuous series of operations integrating the adsorption and elution steps with various stages of washing, resin regeneration, etc.

There are 3 types of ion exchange systems: fixed bed, moving bed, and resin-in-pulp.

Ion exchange is an excellent and economic method for removing very fine radioactive contaminants from liquids. In the absence of ion exchange equipment, more expensive ultrafiltration or solvent extraction techniques are used. Ion exchange is less sensitive to the volume or grade of liquor than the solvent extraction techniques. Ion exchange has been extensively used in cleaning radioactive contaminants from nuclear power plant streams, providing a valuable database for the development of ion exchange equipment to clean contaminated soils.

In using ion exchange, impurities in the liquor can overload the ion exchange resins. Trace metals such as molybdenum, vanadium, radium, and sulfate in the leached liquor can poison the resin, reducing its life.

The use of ion exchange has been documented in a number of applications. These include:

1) Decontamination of uranium mill processing water and water pumped from the mine. Ion exchange also has been used to remove radium from uranium mill tailings.
2) The Mining Science Laboratory in Canada has demonstrated ion exchange extraction as a means of cleaning the leach liquor from tailings for uranium, thorium, and radium.
3) Extraction of uranium in several plants in the U.S.
4) An alkaline leaching process in which ion exchange is used to extract the impurities and produce a high grade liquor for precipitation and recovery of uranium.

A problem with ion exchange has been that the ion exchanging materials clump together and then break down into particles that can no longer function as ion exchangers. Collins and his team at Oak Ridge have used a process called internal gelation to make inorganic ion exchangers in the form of small, stable, porous beads that resist clumping. The beads are made of hydrous oxides of metals such as aluminum, iron, titanium, or zirconium. Highly radioactive contaminants permeate the beads, and the isotopes are concentrated.

The AlgaSORB/ion exchange treatment process can remove low concentrations of heavy metal ions from contaminated groundwater. Compared to ion exchange resins, an advantage of the immobilized algal biomass resins is that they are capable of producing effluent metal ion concentrations in the low part per billion range, even in the presence of high concentrations of hardness salts, such as those of calcium or magnesium.

Improvements in the specificity of solid ion-exchange materials may be possible by the use of polymeric materials containing metal ion specific ligands. Such materials, for example, might contain pendant ligands prepared by functionalization of commercial polymers or by polymerization of functionalized monomers, or might contain coordinating groups within the polymer backbone. There is a need for basic research to determine how to design and fabricate new, much more specific, ion-exchange materials.

Inorganic ion-exchange materials show several advantages over conventional ion exchange resins made from organic polymer matrices. Foremost among these advantages are greater chemical and radiolytic stability and, in some cases, greater metal ion specificity. However, ion-exchange rates are frequently slow, and particle size integrity is usually poor with inorganic materials. Inorganic ion exchangers could be very useful in chemical pretreatment and in waste-minimization processes if their major disadvantages could be solved. Therefore, there is a need for basic research in the design and fabrication of inorganic ion-exchange materials.

8.4 SOLVENT EXTRACTION

Solvent extraction is an efficient method for separating uranium on a commercial scale. There are no commercial solvent extraction processes to extract radium or thorium. The solvent extraction, as applied to uranium extraction plants, consists of a two-step process. In the first step, termed "extraction," the dissolved uranium is transferred from the feed solution (or aqueous phase) into the organic or solvent phase. The second step, called "stripping," recovers the purified and concentrated uranium product into a second aqueous phase after which the barren organic is recycled back to the extraction step. The aqueous and organic solutions flow continuously and countercurrently to each other through the required number of contacting stages in the extraction and stripping portions of the circuit. The uranium is recovered from the second aqueous solution by precipitation.

The extraction of metal from the aqueous solution and its eventual transfer to another aqueous solution (the strip liquid) involves the use of various reagents (extractants, diluents, and modifiers) and requires a suitable vessel to bring about intimate contacts between the different liquids. The extractants are the reagents in the solvent that extract the metal ions. Extractants that are used in recovery of uranium from acid leach solutions are alkylphosphoric acid, amines, tri-n-butyl phosphate (TBP) and trioctyl phosphine oxide (TOPO).

The diluents comprise the bulk of solvent and are inert ingredients whose principal function is to act as carrier for the relatively small amount of extractant. Kerosene is the most commonly used diluent, although other organics such as fuel oil, toluene, and paraffins are also used. The most commonly used modifiers for increasing the solubility of the extracted species are long chain alcohols such as isodecanol.

Solvent extraction is the preferred technology for extracting uranium from acid leach liquor circuits. However, it has not proved feasible to apply solvent extraction to carbonate leach liquors or to slurries containing appreciable amounts of solids.

Since solvent extraction technology involves only liquid-liquid contacts, it is readily adaptable to other systems and can be performed as a continuous operation. Solvent extraction is also readily adaptable to efficient and economical automatic continuous operation. Other advantages of solvent extraction are better selectivity and greater versatility than ion exchange.

The main disadvantage of solvent extraction is that the feed solution must be essentially free of solids. It has not proved economically feasible to apply solvent extraction to carbonate leach liquors. Emulsion formation in solvent circuits causes trouble. The small loss of solvent to tailings is not only costly, but may be a source of stream pollution. Solvent reagents are also very costly. The solvent extraction process is more sensitive to the volume and grade of liquor than the ion exchange process. Molybdenum is strongly extracted by amines and builds up in the amine, acting as poison.

The TRUEX solvent extraction process is discussed in chapter 13.

8.5 PRECIPITATION AND CO-PRECIPITATION

Precipitation and co-precipitation have been used in some extraction schemes to separate uranium from the leach liquor. All currently operated uranium extraction plants, with the exception of a few using a carbonate leaching circuit, employ precipitation to recover the uranium from the solvent extraction stripping liquor or from the ion exchange eluate. Precipitation could be used directly to extract the radionuclide from the water and inorganic salt extraction pregnant liquor.

Precipitation and co-precipitation involve a difficult, cumbersome, and costly operation requiring complex chemical separation. Close control of operating conditions is required. The pH must be monitored and controlled to have better product recovery. The precipitation procedure has not been adaptable to automatic control, and most plants currently operate on manual.

Precipitation of heavy metals and radionuclides can be enhanced by various reagents, iron hydroxide and aluminum hydroxide being the most common.

Iron co-precipitation is a process that is used in the Uranium Mill Tailing Remedial Action (UMTRA) program to remove radium, uranium, and other contaminants from the surface runoff wastes generated during remedial action. It is also used at the Oak Ridge Y-12 Plant to remove uranium from nitrate-containing wastes.

Iron is added to the stream and then precipitated with the contaminants when the pH of the solution is raised by the addition of lime or sodium hydroxide.

Once the precipitation has occurred, the contaminant-containing solids must be separated from the water. This can be done using micro-filtration as at the UMTRA site at Lakeview, Oregon, or by settling as used at the Oak Ridge Y-12 Plant. Co-precipitation is a process that removes metal ions, however, it will not remove nitrate ions, which are a serious contamination problem in some of the groundwaters at Hanford.

Another technique is the tetraphenyl borate precipitation process.

The Techtron Environmental, Inc. process is a combined chemical precipitation, physical separation, and binding process. This technology removes heavy metals and radionuclides from contaminated waters. The process combines the proprietary RHM-1000 powder, as well as a complex mixture of oxides, silicates, and other reactive binding agents, with a contaminated water stream. Selectively enhanced complexing and sorption processes form flocculants and colloids, which are removed through precipitation and physical filtration. The pH, mixing dynamics, processing rates, and powder constituents are optimized through chemical modeling studies and laboratory tests. The contaminants are concentrated in a stabilized filter and precipitate sludge, which is then dewatered. The dewatered sludge meets toxicity characteristic leaching procedure criteria and may, depending on the contaminant, be classified as nonhazardous.

Hanford is studying the precipitation of plutonium from acidic solutions using magnesium oxide. Plutonium (IV) is only marginally soluble in alkaline solution. Precipitation of plutonium using sodium or potassium hydroxide to neutralize acidic solutions produces a gelatinous solid that is difficult to filter and an endpoint that is difficult to control. If the pH of the solution is too high, additional species precipitate producing an increased volume of solids separated. The use of magnesium oxide as a reagent has advantages. It is added as a solid (volume of liquid waste produced is minimized), the pH is self-limiting (pH does not exceed about 8.5), and the solids precipitated are more granular (larger particle size) than those produced using KOH or NaOH. Following precipitation, the raffinate is expected to meet criteria for disposal to tank farms. The solid will be heated in a furnace to dry it and convert any hydroxide salts to the oxide form. The material will be cooled in a desiccator. The material is expected to meet vault storage criteria.

8.6 FILTRATION/MEMBRANES

8.6.1 Reverse Osmosis

Reverse osmosis is a separation technology used for very difficult separations (i.e., salt from water) and can produce highly purified water. However, the rejection of salts is a function of the salt concentration in the feed. In order for reverse osmosis to compete as a process for cleaning groundwater, it must produce a small secondary waste. This means that the feed will become very concentrated in dissolved solids, and the percent rejected may decrease. The goal is to reduce the contaminants to below MCL. These levels are very low; therefore, a small decrease in rejection of a contaminant may cause that contaminant to exceed the drinking water standards in the permeate.

The combined disadvantage and advantage of reverse osmosis is that it removes all of the ions present. This is a disadvantage because the secondary waste volume is increased by ions, such as sodium and calcium, that do not need to be removed. It is an advantage because ions, such as nitrate, are also removed. Most of the contaminated groundwaters contain nitrate in excess of the MCL.

8.6.2 Membrane Microfiltration

The membrane microfiltration system uses an automatic pressure filter, combined with a special Tyvek filter material (Tyvek T-980) made of spun-bound olefin. The filter material is a thin, durable plastic fabric with tiny openings (about 1 ten-millionth of a meter in diameter) that allow water, other liquid, and soil particles smaller than the openings to flow through. Solids in the liquid stream that are too large accumulate on the filter and can be easily collected for disposal.

This treatment can be applied to hazardous waste suspensions, particularly liquid heavy metal- and cyanide-bearing wastes; groundwater contaminated with heavy metals; constituents such as landfill leachate; and process wastewaters containing uranium. The technology is best suited for treating wastes with solid concentrations of less than 5,000 ppm. At higher concentrations, the cake capacity and handling become limiting factors. The system can treat any type of solids, including inorganics, organics, and oily wastes, with a wide variety of particle sizes. Moreover, the system is capable of treating liquid wastes containing volatile organics because the unit is enclosed.

This technology was demonstrated at the Palmerton Zinc Superfund site in Palmerton, Pennsylvania. The shallow aquifer at the site, contaminated with dissolved heavy metals—such as cadmium, lead, and zinc—was selected as the feed waste.

8.6.3 Liquid Membranes

Liquid membrane technology is being considered for treating contaminated groundwater at the Department of Energy's Hanford, WA, laboratory. This clean up process, developed at Argonne National Laboratory, would require 2 liquid membranes working in tandem—one to remove uranium salts from groundwater and the other to capture technetium, nitrates, and chromates. The liquid membranes are thin, porous, plastic films that are impregnated with an appropriate ligating reagent. Cyanex 272 (2,4,4-trimethylpentylphosphinic acid) is used to bind uranium whereas the other contaminants are trapped by Amberlite LA-2. The salts are then stripped off the film and the films recycled. The developers claim that the treatment not only cleans the groundwater, but reduces the amount of contaminated liquid for disposal by more than a thousand-fold.

The Supported Liquid Membrane (SLM) technology has been investigated on a laboratory scale for the past 20 years and on a very limited pilot scale for the past 10 years. There are 2 types of liquid membranes: emulsion membranes and SLM.

SLM have several advantages over conventional solvent extraction and water treatment technologies as follows:

1) Very low solvent requirements.
2) Improved selectivity.
3) Reduced cross contamination.
4) Recovery of species present in low concentrations.
5) Low secondary waste volumes.
6) Nitrate removal.

The SLM development at Westinghouse Hanford is a joint effort with the Argonne National Laboratory (ANL) and has focused on the problem of removing uranium and nitrate from groundwater. An extraction system for removing uranium has been identified and tested at ANL.

8.6.4 Hollow Fiber Membrane

Highly radioactive wastes as well as electroplating shop waste streams contain multiple heavy metals. There is need for developing novel membrane-based synergistic solvent extraction technologies to remove individually the heavy metals and concentrate them for recycling and reuse.

The feasibility, efficiency and usefulness of individual metal removal from a mixed waste stream by a novel hollow fiber membrane-based synergistic extraction technique will be investigated. The researcher will employ a module having 2 separate sets of microporous hollow fibers, one having an acidic organic extractant in the bore and the other having a basic extractant. Metals present as cations will be extracted from the mixed waste flowing in the shell-side of the device into the acidic extractant stream while metals present as anions will be extracted into the basic extractant stream synergistically. For fractionation of individual cations, modules in series with an aqueous raffinate pH control will be used. Base extraction into water will be studied to concentrate and recover individual metals.

8.6.5 Colloid Sorption Filter

The Filter Flow Technology, Inc. (FFT) Colloid Polishing Filter Method (CPFM) was tested as a transportable, trailer mounted, system that uses sorption and chemical complexing phenomena to remove heavy metals and non-tritium radionuclides from water.

The colloid sorption filter is a "polishing" filtration process that removes inorganic heavy metals and non-tritium radionuclides from industrial wastewater and groundwater. The filter unit employs inorganic, insoluble beads/particles (Filter Flow-1000) contained in a dynamic, flow-through configuration resembling a filter plate. The pollutants are removed from the water via sorption, chemical complexing, and hydroxide precipitation. By employing site-specific optimization of the water chemistry prior to filtration, the methodology removes heavy metal and radionuclide ions, colloids, and colloidal aggregates.

A three-step process is used to achieve heavy metal and radionuclide removal. First, water is treated chemically to optimize formation of colloids and colloidal aggregates. Second, a prefilter removes the larger particles and solids. Third, the filter bed removes the contaminants to the compliance standard desired. By controlling the water chemistry, water flux rate, and bed volume, the methodology can be used to remove heavy metals and radionuclides in a few to several hundred gpm.

The methodology has applications for heavy metal and radionuclide remediation from pond water, tank water, groundwater, or for in-line industrial wastewater treatment systems. The technology also has been successful in removing natural occurring radioactive materials (NORM), man-made low level radioactive wastes (LLRW) and transuranic (TRU) pollutants from groundwater and wastewater.

8.7 POLYMER EXTRACTION SYSTEMS

There is an urgent need for alternative technologies for treatment of radioactive waste water to meet regulatory limits, decrease disposal costs, and minimize waste. In particular, this technology would address the need to replace sludge-intensive precipitation methods at LANL TA-50 and at the RFP, and to reduce the TRU wastes from batch processes for recovery of plutonium.

Selective separation and preconcentration techniques are required to analyze increasingly lower concentrations of elements often at levels below the detection limit. The use of water-soluble chelating polymers combined with ultrafiltration is an effective method for selectively removing metal ions from dilute aqueous solutions on both an analytical and process scale. New polymer materials can provide a cost-effective replacement for sludge-intensive precipitation treatments and yield effluents that meet more stringent discharge requirements. New waste treatment facilities using this technology could be downsized relative to facilities using precipitation/flocculation, considerably reducing capital costs.

Investigation of chelators containing multiple hydroxamic acid functional groups has continued to yield new compounds with improved selectivity for binding actinide ions relative to potential competing metal ions. A systematic series of compounds has been investigated and a number of these chelators have shown a strong preference for binding tetravalent plutonium and thorium over trivalent ions. These new compounds show considerable promise for yielding extraction systems with improved selectivity for actinide metal ions over potential interfering ions, such as iron and aluminum.

Preconcentration of actinides uses water-soluble chelating polymers to selectively retain the metal ions of interest while the unbound metal ions are removed with the bulk of the aqueous solution as the permeate by membrane ultrafiltration. Water-soluble polymers have been evaluated for selective retention of americium (III) and plutonium (III) from dilute aqueous solutions high in salt content that simulate waste streams from the TA-50 treatment plant at LANL. The PEI phosphoric acid was chosen over other experimental polymers because of its high solubility over a wide pH range, ease of synthesis, high selectivity for actinides over the metal ions, and the ability to bind actinides at low pH.

Another project will develop a series of polymer supported, ion-specific, extraction systems for removing actinides and other hazardous metal ions from waste water streams. The work is initially focused on the metal contaminants (especially plutonium and americium) in waste streams at the Waste Treatment Facility at LANL TA-50 and at the Rocky Flats Plant (RFP). Reducing the concentration of a target metal to extremely low levels will require that the chelating system have a high-binding strength for that ion, while also having a high-selectivity for the target ion in the presence of competing cations. To this end, the work involves testing and selection of ligands with the required selectivity and binding constant, incorporation of the chosen ligands into polymeric structures, evaluation of the separation properties (capacity, recycle and long-term stability) of the supported ligands, and a complete engineering assessment of the polymer systems in combination with complementary and competing technologies. Chelating ligands under consideration include polyhydroxamates, bis(acylpyrazolones), malonamides and water soluble polymers for ultrafiltration.

Another is to investigate the use of polymer-supported pendant ligand technology in the removal of toxic metal ions from DOE waste streams. Polymer pendant ligands are organic ligands with metal-ion removal capabilities attached to the modified surfaces of 3%, 10% or 20% cross-linked divinylbenzenepolystyrene beads. The metal-ion removal step usually occurs through an ion exchange or binding phenomena, and consequently regeneration and reuse of the beads is achievable. The research objectives of this project are to prepare the polymer supported ligands, to evaluate the ligands for selectivity with respect to the metal ions of interest, to study rates of removal of metal ions in order to determine residence times necessary for demonstration experiments, and to define regeneration and reuse procedures. The work will initially focus on the waters in the Berkeley Pit, and the metal ions targeted for removal from the pH ~2.6 solution are Cu, Zn, Mg, Mn, Al, Fe, Cd, Ni and Ca. The first 6 of these metals are of economic importance and represent ~$720 million of projected recovery value.

Another project will implement and demonstrate Bradtec's Mag*SEP™ technology for in situ groundwater treatment. The MAG*SEP™ technology uses specially designed particles to selectively adsorb contaminants from effluent water or groundwater. The technology can recover low levels of radioactive and/or inorganic hazardous contamination (in the ppb range) while leaving non-radioactive non-hazardous species unaffected.

The selective adsorption particles are composites manufactured in one of 2 forms. The particles can range in size from 1 to 15 microns, have a magnetic core, a polymer coating for durability, and either a "functionalized" resin coating or selective seed materials embedded in the polymer coating.

In treating contaminated water, the particles are injected into the water where they adsorb the contamination. Because the particles are small, and adsorption is a surface phenomenon only, the adsorption kinetics are very rapid (typically less than one minute). The particles are then recovered from the water using a magnetic filter. The magnetic core gives the particle a very high magnetic susceptibility. Also, because the contamination is chemically bound to the particles, adsorbing non-magnetic contaminants can be removed from water with high decontamination factors. Once the particles have been recovered on the magnetic filter, the filter is backwashed, the particles regenerated (much in the same manner as ion exchange resin is regenerated), the contaminants recovered (for recycle or treatment), and the particles reused.

The technology can be applied in situ for the recovery of radionuclides, heavy metals, and nitrates from groundwater. For in situ treatment, a "filter wall" is installed to prevent the groundwater from moving beyond the filter wall, except by passing through it.

8.8 FREEZE CRYSTALLIZATION

Freeze crystallization technology is capable of concentrating liquid effluents and separating organic and inorganic contaminants by removing the bulk of the water. All freeze crystallization processes are based on the difference in component concentrations between solid and liquid phases that are in equilibrium. As an aqueous solution is cooled, ice usually crystallizes as a pure material, and dissolved components of the stream are concentrated into a reduced volume. It could be applicable to low-level mixed wastes.

Freeze crystallization can be used to decontaminate fluids containing inorganics, organics (including volatile organics), heavy metals, and radionuclides. Freeze crystallization is a flexible process that can be designed to adjust to the needs of the application so that it can operate at high efficiency. Potential benefits of freeze crystallization technology over conventional treatment and concentration technologies include:

1) High decontamination factors and high waste volume reduction factors.
2) More efficient partitioning of volatile and semi-volatile components as compared to that of evaporation/crystallization and membrane technologies.
3) The process is a low temperature, low pressure process, and is intrinsically safe. It is highly energy efficient, removing heat

rather than adding it. Also, heat exchangers can be used in the process to recover the cooling value in the melt and concentrate streams.

4) No additives are needed.
5) Potential for salt recovery and purification.

8.9 MICROBIAL PROCESSES

Microorganisms play a major role in the transformations of organic and inorganic compounds in nature. DOE mixed wastes contain a variety of organic and inorganic compounds. Radionuclides are present in several forms, including elemental, ionic, oxides, carbonates, naturally occurring minerals, and organic complexes. The organic compounds include chlorinated solvents as well as various types of aliphatic and aromatic compounds.

Microbial processes have not been fully exploited in the treatment of nuclear wastes. Microbial treatment of such wastes can result (1) in the removal, recovery, and stabilization of radionuclides; (2) in the biodegradation of organic constituents to innocuous products, and (3) in the overall reduction of the volume of such wastes for disposal. Fundamental information is lacking on specific microbial processes and the biochemical mechanisms involved in the transformation of radionuclides and toxic metals in wastes.

Under proper conditions, microorganisms bring about dissolution or immobilization of radionuclides and toxic metals by one or more of the following mechanisms: (1) oxidation-reduction reactions that affect solubility; (2) changes in pH and Eh that affect the valence or ionic state; (3) solubilization and leaching of certain elements by microbial metabolites or decomposition products such as organic acid metabolites, chelation, or production of specific sequestering agents; (4) volatilization due to alkylation reactions (biomethylation); (5) immobilization leading to formation of stable minerals or bioaccumulation by microbial biomass. Microorganisms solubilize various metals and radionuclides from ores, soils, and fossil and nuclear energy wastes by production of mineral acids, organic acids, and oxidizing agents. Treatment of certain types of DOE radioactive wastes by heterotrophic microbial action offers great promise. However, fundamental information at the mechanistic level, in particular with the actinides and selected fission products, is very scanty and warrants basic research that can lead to the development of reliable treatment methods.

Basic research studies should be performed to determine the mechanism of oxidation-reduction reactions catalyzed by aerobic and anaerobic microbes. Enzymes involved in oxidation or reduction of elements leading to solubilization or precipitation of radionuclides and toxic metals under both aerobic and anaerobic conditions need to be isolated and characterized. Particular attention should be given to the enzymes (reductases) involved in reduction of actinides and fission products.

Mechanisms whereby microbes stabilize (i.e., immobilize) radionuclides need to be studied and understood. Singled out for special study should be microbial reactions involving (1) sulfate reduction and formation of insoluble metal sulfides; (2) formation of stable minerals as the result of precipitation reactions or redistribution of solubilized elements with the stable mineral phases in the waste matrix; and (3) biosorption by microbial biomass and biopolymers.

The overall basic research program on microbial waste pretreatment technology should also include studies to isolate and characterize novel microbial metabolites including chelating agents that can be used to solubilize and/or selectively complex radionuclides in wastes.

Technologies are needed for the treatment of wastewater contaminated with low concentration of uranium. Current technologies for uranium removal include precipitation with alkali, adsorption onto activated carbon or other inorganic resins, and extraction with tributyl phosphate or other materials. These technologies generate unacceptable secondary wastes, e.g., large amounts of sludge and/or mixed waste. None of these technologies is suitable for treatment of dilute waste streams, and processes based on these technologies are expensive.

Various strains of bacteria, yeast, fungi, and algae have been screened for their ability to extract uranium from contaminated water containing low concentrations of this heavy metal. Certain species of *Pseudomonas* have been identified as the optimal biological material for the binding of uranium from acidic water (i.e., pH less than 3). Isolated microorganisms in solution could bind uranium and reduce uranium concentrations from 10 ppm to 0.35 ppm. Studies are underway to find the best microbial uranium binder and the best binding conditions. Polyacrylamide will be tested for use as a matrix material in which to immobilize the biomass and form permeable beads.

In previous work *Pseudomonas aeroginosa* was found to be the best biomass for the binding of uranium at alkaline pH (pH 8.8). Heat-killed *Pseudomonas aeroginosa* immobilized within a matrix of calcium alginate reduced uranium levels in a simulated wastewater from 10 ppm to 6.8 ppb. Over 350 column volumes were successfully treated before breakthrough occurred, and bound uranium was quantitatively removed from the column by treatment with 5 column volumes of 0.15M nitric acid.

A proposed process utilizes biosorbents (sorptive biomass, or biological material) immobilized in permeable beads that are, in turn, contained within flow-through bioreactor systems. Systems will be operated in a continuous or semi-continuous mode, and will be operated on-site as a pump-and-treat methodology. The system will achieve waste fixation and volume reduction. Uranium concentrations will be reduced from ppm to ppb levels.

Uranium is particularly difficult to remove from water because it's highly soluble. But when metabolized by specific microbes, the compound is converted to an insoluble sediment that can be skimmed from water streams, as reported by Lovely of the USGS in Reston, VA.

The bacteria may also be capable of removing other radioactive metals, such as plutonium and technetium and possibly chromium. And the bacteria doesn't appear to be harmful to humans or animals.

Enzymatic uranium reduction by *Desulfovibrio desulfuricans* readily removed uranium from solution in a batch system or when *D. desulfuricans* was separated from the bulk of the uranium-containing water by a semi-permeable membrane. Uranium reduction continued at concentrations as high as 24 mM. Of a variety of potentially inhibiting anions and metals evaluated, only high concentrations of copper inhibited uranium reduction. Freeze-dried cells, stored aerobically, reduced uranium as fast as fresh cells. *D. desulfuricans* reduced uranium in pH 4 and pH 7.4 mine drainage waters and in uranium-containing groundwaters from a contaminated Department of Energy site. Enzymatic uranium reduction has several potential advantages over other bioprocessing techniques for uranium removal, the most important of which are as follows: the ability to precipitate uranium that is in the form of a uranyl carbonate complex; high capacity for uranium removal per cell; the formation of a compact, relatively pure, uranium precipitate.

Uranium Biosorption: Use of beaded bacterial biomass for adsorption of uranium from groundwater as a selective adsorption process for low levels of uranium is being evaluated.

Another project is to demonstrate the feasibility of using plants (both terrestrial and aquatic) to remediate soils, sediments, and surface waters contaminated by heavy metals and radionuclides.

Considerable heavy metal contamination exists in soils and groundwater across the DOE complex, and much of this contamination is of low concentration. For such low levels of contamination in a relatively large quantity of soil and water, removal and storage or remote treatment (such as incineration, for soil) become extremely expensive. The bioremediation technology proposed could be less expensive than soil removal and treatment given the areal extent and topography of the sites under consideration, the problems associated with process-generated fugitive dust emission, and the investment of energy and money in the soil-moving or water-pumping and treatment processes. Moreover, in situ technology may receive regulatory acceptance more easily than ex situ treatments. Taking advantage of the natural ability of plants to take up metals is indeed an inexpensive and publicly appealing method by which remediation of low-level heavy-metal/radionuclide contamination can occur.

MT International, a U.S. corporation, has established a joint venture agreement, American-Ukraine Biotech JV (AUB), with the Central Scientific Research Laboratory of Comprehensive Processing of Plant Raw Material of The Ukrainian Academy of Agrarian Sciences. The Ukrainian Academy of

Agrarian Sciences and Berevetnik Scientific Research Institute have conducted large scale oil remediation, implementing a biomass processing system, near the radioactive Chernobyl site.

The technology will be tested initially using plant material grown on heavy metal contaminated soils in the area of the Silver Bow Creek Superfund site (near Butte, Montana) and on Berkeley Pit water. Other plant biomass specimens derived from contaminated DOE facilities or from other sites will also be subjected to testing and evaluation in the initial feasibility study.

A small, bench-scale FST system will be built and operated at a site in Butte, Montana. Following successful demonstration and evaluation of the bench-scale process system, a larger, field-scale, mobile FST system will be tested at a DOE demonstration site (to be selected).

8.10 TRITIUM REMOVAL

Tritium, a radioactive isotope of hydrogen, is used to enhance the performance of nuclear weapons and is a necessary component for all weapons in the stockpile. Tritium decays at 5.5 percent per year and must be replaced periodically as long as the Nation relies on a nuclear deterrent. Currently, there is no capability to produce tritium within the Complex, yet projections require that new tritium be available by approximately 2011. DOE is proposing to build a facility to produce tritium for the next 40 years.

Tritium supply deals with the production of new tritium in either a reactor or an accelerator by irradiating target materials with neutrons and the subsequent extraction of the tritium in pure form for its use in nuclear weapons. Tritium recycling consists of recovering residual tritium from weapons components, purifying it, and refilling weapons components with both recovered and new tritium when it becomes available.

Alternatives for new tritium supply and recycling facilities consist of 4 different tritium supply technologies and 5 locations. The 4 technologies being evaluated to provide a new supply of tritium are Heavy Water Reactor (HWR), Modular High Temperature Gas-Cooled Reactor (MHTGR), Advanced Light Water Reactor (ALWR), and Accelerator Production of Tritium (APT). Both small (600 MWe) and Large (1300 MWe) ALWRs will be evaluated, as well as a phased approach for the APT.

DOE will also include an analysis of the MHTGR and ALWR technologies for tritium production together with plutonium disposition and steam/electricity production.

The 5 candidate sites are: the Idaho National Engineering Laboratory (INEL), Nevada Test Site (NTS), Oak Ridge Reservation (ORR), the Pantex Plant, and the Savannah River Site (SRS).

DOE will also study the possibility of using one or more commercial light water reactors for tritium production as a contingency in the event of a national emergency. This contingency option would permit the commercial reactor to continue electricity generation. They will also analyze an existing commercial light water reactor that would be purchased for tritium production and withdrawn from commercial electricity production.

Tritium removal from both wastewaters and groundwaters presents particularly difficult problems. To remove tritium from water, an isotope separation process is needed. There are no highly specific adsorbents for tritium or tritiated water, and the separation factors available are sufficiently low that many separation stages are needed. Although the separation factors for separation of hydrogen isotopes are among the highest available for any isotope separation, they are still low compared to those often available for removing most other inorganic components from wastes or groundwater. This makes the capital and operating costs for high throughput systems very high; most isotope separation systems operate with much lower throughputs than those often needed for treating major wastes and groundwater flows.

Innovative hydrogen isotope separation methods are needed to extend the range of conditions for which it is practical to treat wastes for tritium removal. Since several methods have been developed and used in the past, it is necessary that any R&D on tritium removal be sufficiently innovative and potentially better than the methods available presently. The current "standard" technology may be the hydrophobic water-hydrogen catalytic process being employed by the Canadian nuclear program for removing tritium from heavy water coolant in their reactors. This would be a suitable standard for comparing any new tritium separation concept. Although there would be some merit in using the waste or groundwater in its existing form as feed to a tritium separation system, special feed preparation such as demineralization, evaporation, or even electrolysis may be needed.

8.11 TREATMENT TRAINS

One DOE project is to develop and demonstrate an improved ex situ treatment process for removing groundwater contaminants.

A combination of process steps will be used consisting of sequential chemical conditioning, microfiltration and dewatering by low-temperature evaporation and/or filter pressing to achieve high contaminant removal efficiencies.

The experimental program will focus on performing screening tests to identify key process variables. Key variables to be evaluated include: type and composition of waste influent to the process; precipitation conditions; type of ion exchange/adsorbent material, their concentration and treatment condition; type and concentration of non-contaminant metals; type and concentration of leaching agent for soil washing; and conditions to solidify secondary waste produced by the process.

The conditioning of the contaminated water by a sequential addition of chemicals and adsorption/ion exchange materials produces a poly-disperse system of size enlarged complexes of the contaminants in 3 distinct configurations; water soluble metal complexes, insoluble metal precipitation complexes and contaminant-bearing particles of ion exchange and adsorbent materials. Waste volume is reduced by dewatering of the polydisperse system by cross-flow microfiltration, followed by low-temperature evaporation. The bulk of the filtrate is discharged if it meets the specified target water quality, or is recycled.

Overall test results revealed that a three-step chemical treatment/microfiltration sequence combined with a final dewatering step is optimal for effective treatment of mixed waste having diverse physico-chemical properties.

Key chemical treatment steps include: pH adjustment by lime addition combined with zeolite heavy metals; sequential addition of natural zeolite ion exchange/adsorbent powder to remove radionuclides, and; sequential addition of powdered activated carbon with or without zeolite powder as a polishing step to remove organics and residual radionuclides.

The use of $FeCl_3$ as the leachant to extract strontium-90 from the contaminated soil showed a synergistic effect: excellent strontium-90 removals from the soil at low concentrations at room temperature; and effective in the removal of metal ions in the first step of the laced treatment by co-precipitation and adsorption-scavenging.

Process economics were assessed for treatment plants of 2 and 300 gpm throughput capacities. Installation costs are estimated at $275,000 for the 2 gpm plant, and at $4 million for the 300 gpm plant, while annual operating costs are estimated at $418,000 and $12 million for the 2 gpm and 300 gpm plants respectively, including secondary waste disposal and capital recovery costs.

A simplified, single-stage version of the process was successfully implemented to treat groundwater and surface water contaminated with Strontium-90 at the Chalk River Laboratories (CRL) site in Chalk River, Ontario, Canada.

8.12 LOW-LEVEL WASTE TANKS

Remedial alternatives considered for the inactive low-level waste tanks include removal of the tank contents, in situ treatment via fixation or sorption, removal of the tank shells, hydrogeologic containment or isolation of the tanks, and treatment of the tank contents via chemical methods.

1) *Removal of the tank contents (liquid alone, or both liquid and sludge)* - Specific technologies included in this alternative are mechanical excavation, pneumatic removal, and hydraulic removal (sluicing/slurry pumping). The use of cryogenics to freeze and thereby facilitate removal of the contents has been

suggested; however, this technology was not pursued further because of concern over potential for cracking or breaching the tank shells.

2) *In situ treatment of tank contents* - Specific fixation technologies include cement grouting, silicate-based stabilization, vitrification, chemical grouting, and thermoplastic encapsulation; in situ sorption is also possible as a physical separation treatment.
3) *Removal of the tank shells (Gunite and steel)*
4) *Hydrogeologic containment or isolation of the tanks using walls or barriers or combinations of capping and drains* - Examples of wall or barrier technologies are grout curtains, slurry walls, sheet-pile walls, and cryogenic barriers. Examples of capping are intruder barriers and multilayer caps, and examples of drains are french drains and horizontal wells.
5) *Treatment via chemical methods* - Examples of chemical treatments are dissolution, neutralization, and radiochemical separation and recovery.

For many of these alternatives, access to the inside of the tanks with special equipment will be necessary. However, due to the radiation hazard inside the tanks, the remediation equipment that enters the tanks must be operated remotely. The robotics and automation program addresses the different types of equipment that can be used remotely in hazardous environments.

With regard to tank contents removal (particularly sludge removal), the principal science and technology needs are the development of retrieval and transport devices or systems. Retrieval equipment must be capable of removing hard salt cake, sludges of varying consistencies, tank hardware, and foreign objects. Conveyance systems must be capable of handling dislodged or mobilized wastes and tank hardware, and, in some cases, it must be capable of transporting the slurry over long distances with negligible solids separation or in-line clogging. Special emphasis is being placed in the Integrated Demonstration projects of DOE's Office of Technology Development on robotics-integrated retrieval equipment (e.g., articulated and remotely operated mechanical arms). Because the equipment must be versatile, this work is technically difficult. If scheduled implementation time is short, large expenditures in terms of dollars and manpower will be needed for development.

Another principal science and technology need is for a detailed characterization of the waste matrix that will determine the suitability of the solidifying agent for specific tanks. This is a necessary first step in development of in situ fixation methods. Better understanding of immobilization mechanisms and the chemistry of these solidified waste forms can lead to improved performance and better predictions about their durability. It is important also to note that high confidence in the long-term effectiveness of these fixation techniques is not universal. More careful demonstrations and operations are needed to create a higher level of confidence in the long-term performance of the fixation techniques. Long-term assessments of the ability of the fixation technologies to retain and contain the contamination would be useful. The assessments would address, for example, whether the cement grout and silicate-based formulations maintain the pH necessary to prevent migration and whether the polymer grout and thermoplastic formulations retain their stability in high-radiation environments.

One low-level separations investigation involves a three-phased development approach corresponding to the types of UST waste to be treated: supernate, salt cake, and sludge. The first phase will focus on removing key constituents for supernate using ion exchange, calcination and other methods, and methods yet to be identified for removing selected radionuclides. The second phase will focus on treating salt cake by dissolution and will develop methods for separating solids and liquids. Lastly, sludge treatment will be developed in conjunction with the Efficient Separations and Processing Integrated Program.

To support the separation technologies, compact processing units (CPU) will be developed using a modular or distributed processing concept. These CPUs are an alternative to a large, permanent facility and are currently being considered. The ion exchange technologies developed by the Savannah River National Laboratory will be evaluated for incorporation into the first fieldable CPU. The organic and nitrate destruction technologies will be initiated later.

Technologies for treating sludges developed by the Oak Ridge National Laboratory will be demonstrated and validated using the transuranium extraction (TRUEX) model. Sludge from the Melton Valley waste tanks will be washed, the supernate passed through ion exchange columns containing the resorcinol-formaldehyde resin in development at Savannah River. The sludge will be treated with a TRUEX process, an the results will be compared to the predictive model for TRUEX, supported by the Argonne National Laboratory.

The LLW form development will focus on testing 2 alternatives to the current disposal form for low-level waste (grout): nitrate to ammonia and ceramic (NAC) and polyethylene. The NAC process destroys nitrates and produces a ceramic LLW form in one process. The resulting ceramic can be sintered, which would destroy all organics by the high heat added during the final phase. The polyethylene process takes a dry waste stream and encapsulates it into a solid polyethylene matrix that can be extruded into the desired form.

Centralized Treatment: Improved centralized treatment methods are needed in the management of liquid low-level waste (LLLW) at Oak Ridge National Laboratory (ORNL). LLLW, which usually contains radioactive contaminants at concentrations up to millicurie-per-liter levels, has accumulated in underground storage tanks for over 10 years and has reached a volume of over 350,000 gallons. These wastes have been collected since 1984 and are a complex mixture of wastes from past nuclear energy research activities.

The waste is a highly alkaline 4-5 M sodium nitrate solution with smaller amounts of other salts. This type of waste will continue to be generated as a consequence of future ORNL research programs. Future LLLW (referred to as newly generated LLLW or NGLLLW) is expected to a highly alkaline solution of sodium carbonate and sodium hydroxide with a smaller concentration of sodium nitrate. New treatment facilities are needed to improve the manner in which these wastes are managed. These facilities must be capable of separating and reducing the volume of radioactive contaminants to small stable waste forms.

Treated liquids must meet criteria for either discharge to the environment or solidification for on-site disposal. Laboratory testing was performed using simulated waste solutions prepared using the available characterization information as a basis. Testing was conducted to evaluate various methods for selective removal of the major contaminants. The major contaminants requiring removal from Melton Valley Storage Tank liquids are Sr-90 and Cs-137. Principal contaminants in NGLLLW are Sr-90, Cs-137, and Ru-106. Strontium removal testing began with literature studies and scoping tests with several ion-exchange materials and sorbents.

9

Decontamination, Decommissioning, and Recycling

The aging of the DOE Complex's facilities, along with the reduction in nuclear weapons production, have resulted in 1,200 facilities being shut down at multiple field locations that require deactivation, decommissioning and disposition. The inventory of permanent structures that require deactivation is projected to grow to about 20,000 permanent and temporary buildings and structures. Current clean up processes tend to expose workers to radioactive and hazardous substances. They are labor-intensive, and expensive, producing an unacceptable large volume of secondary waste. Additionally, in the near term, high costs for shutdown, stabilization, surveillance, and maintenance will absorb the majority of dollars budgeted for decommissioning. Similar problems exist for the commercial nuclear power industry, other government facilities, and commercial facilities.

The Decontamination and Decommissioning (D&D) of Facilities Integrated Demonstration (ID) was initiated by DOE in October, 1993, to develop and to demonstrate improved processes for:

1) Facility shutdown and stabilization.
2) Surveillance and maintenance.
3) Sampling, imaging and characterization.
4) Decontamination and dismantlement.
5) Recycling and disposition.

The near-term goals of the D&D ID include: demonstrating the capability to decontaminate surface and contaminated concrete by December 1997; and providing accepted alternative improved processes to Key Decision Documents for the 1196 Decommissioning Characterizations/Assessments and subsequent Decommissioning Projects.

The 5 primary areas crosscut into 10 secondary areas:

1) Contaminated concrete decontamination and dismantlement.
2) Contaminated metal decontamination, dismantlement and recycle.
3) Fuel reprocessing facilities.

4) Research and production reactor facilities.
5) Reactor fuel storage pool facilities.
6) Hot cells research and development.
7) Gaseous diffusion plants.
8) Plutonium processing facilities.
9) Uranium processing facilities.
10) Lithium processing facilities.

Selection of technologies for development and use in D&D facilities is a very difficult task. Multidimensional trade-offs are required relative to the availability, complexity, and cost of technologies that appear applicable, but have not been demonstrated and/or accepted. Also, the variety of facilities makes selection of generic technologies difficult. For example, technologies applicable to hot cells may have limited usefulness for reactor pools.

Among the high-cost items is the dependency on labor intensive techniques and the necessary protective clothing. Experience indicates that 30-50% of the cost can be attributed to reduced crew efficiency due to protective clothing and equipment required. This cost can be offset by automated dismantlement tooling and sophisticated delivery systems, thereby reducing manual labor and improving protection to D&D workers.

Some manner of manual work will always be required regardless of the degree of automation used in the dismantlement tasks. To this degree, the development of improved (cheaper, recyclable, lighter, more convenient, etc.) protective equipment is certainly justified.

New, more efficient, versions of clothing and protective equipment have the potential of saving vast sums of money and labor. Respirators, face masks, air packs, cloth suits, airtight suits, gloves, shoe covers, laboratory coats, air conditioned and filtered equipment cabs, etc., need improved designs to reduce cost in a massive dismantlement project such as a diffusion plant and its support facilities.

Decontamination of metal components with induced radiation, such as reactor cores, is generally not feasible. Metal refining technologies will be ineffective in most cases because the induced radiation comes from atoms that are isotopes of and chemically identical to the bulk of the metal atoms. In principal, transmutation could be used, but only if the radioactive isotope had a larger cross section than the base metals. In any event, this would be a prohibitively expensive operation. In general, these components will have to be removed and disposed of in their radioactive state. Concrete with induced radiation can be removed, leaving clean concrete for concrete that can be decontaminated. Similarly, equipment that is partially contaminated with induced radiation is decontaminated by cutting out only the parts with induced radiation.

Reusable equipment and hardware that is contaminated only on the surface, can be decontaminated, generally with various cleaning fluids. The big advantage of decontamination techniques is that, after decontamination, the equipment and hardware can be released for unrestricted use. This is even more important in the case of high-capital-cost equipment. The main disadvantage is the generation of liquid LLW that must then be appropriately managed. High-pressure-water-jetting techniques for decontamination are relatively inexpensive. For specialized decontamination applications, proprietary technologies are available, such as CAN-DECON (CANDU-Decontamination), CORD (Chemical Oxidation Reduction Decontamination), and LOMI (Low Oxidation State Metal Ion Reagents). Drycleaning techniques can also be used, e.g., method using Freon ($C_2F_3Cl_3$) and employing agitation by ultrasonic waves. Mechanical decontamination techniques include manual cleaning, vacuum cleaning, grinding, and machining.

Major initiatives are underway within DOE to recycle large volumes of scrap material generated during clean up of the DOE Weapons Complex. These recycling initiatives are driven not only by the desire to conserve natural resources, but also by the recognition that shallow level burial is not a politically acceptable option. The Fernald facility is in the vanguard of a number of major DOE recycling efforts. These early efforts have brought issues to light that can have a major impact on the ability of Fernald and other major DOE sites to expand recycling efforts in the future. Some of these issues are; secondary waste deposition, title to material and radioactive contaminants, mixed waste generated during recycling, special nuclear material possession limits, cost benefit, transportation of waste to processing facilities, release criteria, and uses for beneficially reused products.

Hazards associated with contaminated hot cells, canyons, glove boxes, and reactor facilities at DOE sites include radiation, radiological contamination of equipment to be removed, and hazardous chemicals associated with the processes performed at the facilities. Because of these hazards, deactivation, S&M, and ultimate D&D will be performed remotely. D&D operations include disassembly of process equipment, cutting pipes, size reduction of equipment to be removed, transport of pipe and equipment out of the facilities, decontamination of equipment before removal from a facility, and decontamination of floors, walls, and remaining equipment in facilities to be refurbished. Robotics may also be needed to dismantle the facility structure. Hardened robotics systems for facility D&D can provide the capability to accomplish these operations safely with workers away from the work site.

Throughout the DOE Complex, there are numerous facilities identified for D&D with piping that has been placed on the contaminated list because of the internal contamination risk. Much of this piping is inaccessible since it is buried in concrete or it runs through hot cells. Currently, there are no robotics/remote systems capable of characterizing pipe in the one to three-inch inside diameter range. Characterization of this piping is essential before, during, and after D&D activities. By identifying those sections of piping not contaminated can greatly reduce the amount of material sent to waste handling facilities or the amount of waste generated performing unneeded decontamination.

Also, throughout the DOE Complex, there are numerous facilities identified for D&D that have been placed on the contaminated list because of the risk of internal contamination from duct work, much of which is inaccessible due to its location in concrete or running through hot cells. Duct work characterization is extremely difficult because of varying size and direction of travel. Characterization of duct work is essential before, during, and after D&D activities. Identifying portions of ducts that are not contaminated greatly reduces the amount of material sent to waste handling facilities and decontamination. Conventional methods have been applied to duct work with some success, but at the risk of human exposure to high levels of contamination. Limited capability remote duct work characterization systems are commercially available. A robotics/remote duct characterization system with extended travel capability is needed that can perform chemical and radiological contaminant characterization and select hot spot decontamination or partial duct work dismantlement.

The robotics program (Selective Equipment Removal System) is being jointly developed by ORNL, SNL, and PNL, and will entail interaction with the University of Tennessee and Carnegie Mellon University.

The Small Pipe Characterization System, and the Internal Duct Characterization System, are being developed at INEL, in coordination with Oak Ridge.

Costs can be extremely high. An example is the recent dismantlement and decommissioning at a plutonium processing facility at the Savannah River Site (SRS). Excluding the eventual TRU waste processing cost, the project has cost about 12.5 million dollars to completely D&D approximately 5,000 square feet of floor area of the plutonium facility, or $2,500 per square foot of plutonium facility. Although the elapsed time for the clean up was 9 years, the effort was intermittent and is estimated to have required about 5 years of continuous effort.

9.1 TECHNOLOGIES UNDER DEVELOPMENT BY DOE

Massive containment:

1) High-pressure abrasive water jet.
2) Diamond wire cutting.
3) Deep microwave demolition.

Reactor pools:

1) High-pressure abrasive water jet.
2) Demolition compounds.
3) Microwave demolition.

Buried tanks:

1) High-pressure abrasive water jet.
2) Explosive cutting.
3) Demolition compounds.

Structural-only concrete:

1) Grappler.
2) Diamond wire cutting.
3) Conventional jack hammer, headache ball, etc.
4) High-pressure abrasive water jet.
5) Demolition compounds.

Steel structures - high-level contamination:

1) Laser cutting.
2) Abrasive water jet.
3) Explosive cutting.
4) Plasma arc cutting.
5) Oxygen cutting.
6) Grapple and massive mobile shearing.
7) Conventional disassembly.

Steel structures - low-level contamination:

1) Abrasive water jet.
2) Laser cutting.
3) Plasma arc cutting.
4) Conventional disassembly.
5) Grapple and massive mobile shearing.
6) Oxygen cutting.

Asbestos removal:

1) Conventional/automated with vacuum system.
2) Laser cutting.
3) CO_2 blasting.
4) High-pressure abrasive water jet.

Major equipment removal:

1) Plasma arc cutting.
2) Advanced automatic fixtures, bug-o, etc.
3) Advanced laser cutting.
4) High-pressure water jet.
5) Mechanical saws.

Entombment:

1) Encasement - subsurface waste storage with void reduction.
2) Engineered storage - for very "hot" items, future disposal.
3) Sensors for monitoring entombment integrity.
4) Permanent entombment.
5) Capping - above ground environmental barrier.

9.2 TECHNOLOGIES

9.2.1 Bulk Decontamination Methods

1) Solvent extraction.
2) Incineration.
3) Dry heat.
4) Chemical leaching.
5) Catalytic extraction process.
6) Vacuum (low pressure).
7) Transmutation.

Solvent extraction removes organics from bulk materials simply by dissolution. The technology could be applied as a "factory style" process in either a batch or continuous fashion. For waste-volume reduction, separation of the contaminants from the solvent and recycling of the solvent would be needed. Also, the solvent would have to be removed from the decontaminated material by a method such as heating or vacuum. The chemical leaching technology would be applied in a manner similar to that of solvent extraction. The chemical leaching and biological technologies are essentially the same as described under chemical surface cleaning. Solvent extraction and chemical leaching are well-established technologies in current use.

Incineration of various materials is accepted by the EPA. The incineration of contaminated combustible building materials is common in the nuclear industry. The waste generated will depend upon the design of the incinerator and the ash content of the material being burned. The fly ash and smoke particles have to be contained with particular care.

The technology of dry heat (e.g., calcination) would be used to remove volatiles such as oils and PCBs. Treatment of off-gases to remove the volatized contaminants would be necessary.

The catalytic extraction process is similar to smelt purification in that a furnace is used to form a molten metal bath. In the catalytic extraction process, the bath is used not only to purify metal but also, at the high-operating temperatures, to effectively destroy hazardous-organic materials such as aromatic hydrocarbons and PCBs. Also, mixing is better in the catalytic extraction process because material is introduced from the bottom of the process; control of the oxidation potential is better with the catalytic extraction process because oxygen, rather than air, is injected into the furnace. The process has not been used commercially to treat scrap contaminated with radionuclides. Waste generated could include radioactive slag and wet scrubber solutions.

Very limited use of vacuum (i.e., low pressure to remove volatile compounds) has been made for bulk decontamination. Some information has been obtained on a proprietary vacuum-based process for removing water soluble Cr^{+6} from transite sheeting surfaces. Vacuum can also be used to remove contamination from porous materials by boiling solvents out of the material if the solvent foams out and the solvent is collected. Because of their low-vapor pressures, the technique would probably be rather ineffective for most oils and PCBs.

Transmutation appears to be the only technology capable of eliminating induced radiation, which results from exposure of material to high-neutron fluxes. The neutron fluxes, such as those near the core of a fission reactor, often result in the change of some of the atoms in the exposed material to radioactive species that are chemically identical isotopes of the other unchanged atoms in the material. Thus, the radioactive isotope cannot be removed from the base material by chemical means. (If the induced radiation comes from different elements from the bulk of the material, chemical methods such as metal refining techniques could be used to decontaminate the material.) Transmutation might be used to convert radioactive isotopes into non-radioactive ones by bombarding the radioactive species with neutrons or alpha particles. This technique could only be used if particles could be found that have energies such that the cross-sections of the radioactive isotopes were much larger than those of the non-radioactive isotopes. Otherwise, this technique might generate as much, or more, radioactivity than it eliminated. Even then, the practicality of this technology is doubtful.

9.2.2 Surface Cleaning Methods

1) Compressed air cryogenic CO_2 blasting.
2) High-pressure water.
3) Superheated water.
4) Water flushing.
5) Steam cleaning.
6) Hand brushing.
7) Automated brushing.
8) Sponge blasting.
9) Hot air stripping.
10) Dry heat (roasting).
11) Solvent washing to remove radiological contamination.
12) Solvent washing to remove organics.
13) Strippable coatings.
14) Vacuum cleaning.
15) Ultrasonic cleaning.

Surface cleaning methods remove the fixed and/or loose contamination without disturbing the surface of the substrate. They should be particularly effective for loose contamination. Several technologies that are described in the mechanical surface removal methods group may also be listed in this group in their less aggressive modes. These include scraping, ultrahigh pressure water, shot blasting, grit blasting, centrifugal cryogenic CO_2 blasting, and plastic pellet blasting. Depending upon the operating conditions or the particular substrate being treated, these technologies can achieve decontamination without affecting the substrate surface. Other blasting technologies in this generic group are compressed air cryogenic CO_2 blasting, sponge blasting, high pressure water, superheated water, hot water, and steam.

In this CO_2 blasting method, the cryogenic pellets are accelerated by compressed air rather than by centrifugal force. The sponge blasting technology decontaminates by blasting surfaces with various grades of patented, water-based urethane foam particles using 110 psig air as the propellant. The foam is absorptive and can be used either dry or wet to treat a variety of surface contaminants such as oils, greases, lead compounds, chemicals, and radionuclides. A non-aggressive grade of foam is used for surface cleaning on sensitive or otherwise critical surfaces. Aggressive grades, which are impregnated with abrasives, are capable of removing tightly adherent materials such as paints, protective coatings, and rust and can roughen concrete and metallic surfaces, if desired. High-pressure (500-1,000 psi) water sprays are used extensively in the nuclear industry for smearable contamination. Decontamination factors of 1.5-10 are typically obtained, depending upon the amount of fixed contamination. Superheated water (300-400°F and several hundred psi) machines are commercially available. Their use is usually limited to floors. Commercial equipment using steam and hot water jets for decontamination is also available.

Manual decontamination methods of brushing and scrubbing will remove loose contamination, but are ineffective for fixed contamination. Also, they have the same undesirable features of the manual methods in the mechanical surface removal group. Hand grinding, brushing, and vacuuming will continue to find applications in areas of negligible radiation because of their very low capital cost. These technologies are very effective for some applications. Remotized versions of these technologies are needed in higher-radiation or -contamination areas.

Solvent washing is typically used for organic contamination such as oils and PCBs. Loose or smearable contamination may also be removed by this method. To minimize waste, solvent recovery is needed. All low-boiling organic solvents are Resource Conservation and Recovery Act (RCRA) hazardous by the rules and increase the cost of operation.

The strippable coating technology involves "painting" a surface with, for example, a water-based organic polymer. When dry, the coating can be removed, along with the contaminants captured by the coating, mechanically, chemically, or, in the case of "auto-release" coatings, by vacuuming. Decontamination factors are reported to range from as low as 3 to 5 up to several hundred, depending upon how tightly the contaminants are bound to the surface.

The technology of dry heat (e.g., roasting) would be used to remove volatiles such as mercury, tritium oils, and PCBs.

The use of vacuum cleaners and their filters is accepted and widely used for removing and collecting loose contamination at gaseous diffusion plants. This method is often used to clean up during and after grinding, honing, scraping, and other surface removal methods. Vacuuming might be effective for removing loose contamination that has settled on girders, other non-vertical surfaces, cracks, etc.

Ultrasonic cleaning has been used for many years to remove surface contamination from relatively small metal parts. Use of this technology on a large scale would likely require separation of the contaminants from the cleaning fluid so that the fluid could be recycled.

9.2.3 Chemical Surface Cleaning Methods

1) Chemical foams.
2) Chemical gels.
3) Organic acid treatment.
4) Fluoboric acid treatment.
5) Inorganic acid treatment.
6) Detergent treatment.
7) REDOX treatment.
8) Chelation treatment.
9) Electropolishing
10) Gas phase decontamination.
11) UV/ozone (UV light activation).
12) Electromigration.
13) Biological.
14) Volatilization/Low Temperature Thermal Desorption (LTTD).
15) Supercritical Carbon Dioxide Extraction (SCDE).

Technologies comprising chemical foams, chemical gels, acid treatments, caustic treatments, and redox treatments involve reaction with the contaminant to form a species that is dissolved in the cleaning solution and thereby removed from the substrate. Depending upon the strength of the chemical solution, physical degradation of the surface can also occur. Foams and gels are used to enhance the performance of the chemicals and to reduce subsequent waste volumes by holding the decontamination chemical in better contact with the contaminated surface. These methods of application can be used for in situ decontamination of exterior surfaces. The other solutions maybe applied in dipping baths, in loop systems, or through spraying. Recovery would be a problem in the use of the spraying technique for decontamination of exterior surfaces unless decontamination is done in a spray booth with suitable critically safe drains. Spray booths similar to car washes are used at gaseous diffusion plants to decontaminate the interior and exterior of disassembled equipment. For waste volume reduction, regeneration and recycling of the chemical is necessary. Depending upon the strength or reactivity of the chemical, decontamination factors range from as low as 5 to 10 up to several hundred.

Fluoboric acid effectively decontaminates concrete and metals by removing up to a few millimeters of the surface. Fluosilicic acid can also decontaminate concrete, brick, and similar masonry surfaces. The acid can be electrolytically regenerated and recycled. One treatment option is neutralization and precipitation with solidification in cement. Another treatment is ion exchange with solidification of the ion exchange resin or mineral in cement. The optimum materials for solidification of these wastes needs to be determined. Treatment systems that permit recycle of decontamination agents are needed for fluoboric acid and all liquid-based systems.

Electropolishing selectivity removes contaminant metal ions from a metal surface. This method, using an electrolyte and electricity, has usually been applied on small areas. For waste-volume reduction, separation of the contaminant from the electrolyte and recycling of the electrolyte would be needed.

Gas-phase decontamination uses a strong fluorinating gas to convert uranium fluoride and oxyfluoride deposits inside gas-tight equipment to uranium hexafluoride gas, which could then be recovered with chemical or cold traps. Lab studies have shown that over 99.9% uranium removal can be obtained by reaction at room temperature. However, thorium and protactinium fluorides and most of the fluorides of elements formed in the decay chains of U-235 and U-238 are not volatile. Treatment of the off-gases will generate wastes from liquid scrubbers and dry-chemical traps when using current technology.

Plasma etching and fluorination technologies may be used to enhance performance of gas-phase decontamination. The plasmas of fluorinating gas can be created to promote room temperature chemical reactions, converting uranium fluoride and oxyfluoride deposits into uranium hexafluoride gas. A research project would be needed to perform a feasibility test in the gas-phase facility.

The UV/ozone and electromigration technologies are in the evolving technology stage. In the use of UV/ozone to remove oils, greases, and solvents, UV light produces ozone from oxygen and also excites and/or dissociates contaminant molecules. These reactive species then produce CO_2 and water vapor (and HCl in the case of chlorinated solvents). Electromigration uses an electric potential to cause migration of metal ions from a surface into an electrolyte. The technology may be applied in a localized area or as a "factory style" process in batch fashion to contaminated areas. For waste volume reduction, separation of the contaminant from the electrolyte and recycle of the electrolyte would be needed.

Some of the wastes at the Rocky Flats Plant (RFP) consist of bulk items with contaminated surfaces. Technologies to clean these surfaces would allow the separation of the hazardous and radioactive components for treatment and disposal as land disposal restrictions compliant mixed or hazardous and low-level waste. Such technologies are expected to treat not only debris and solid wastes, but to play significant roles in the decontamination and decommissioning operations. The Surface Organic Contaminant Removal Program consists of 2 main technology subtasks: (1) Volatilization/Low Temperature Thermal Desorption (LTTD), and (2) Supercritical Carbon Dioxide Extraction (SCDE).

LTTD is a volatilization technique that has been a successful treatment for removing organic contamination from soils. Its applicability to combustible mixed wastes remains to be determined. In this process, waste is fed into a heating unit where it is heated to temperatures less than 600°F. The actual heating unit could be a rotary kiln, a calcination unit, or a fluidized bed unit. Heated nitrogen or another inert carrier gas is swept through the heating unit and carries the volatized organics into a scrubber system. Particulates are separated from the gases and a condenser recovers the organics for disposal or further treatment. No combustion occurs because of the low operating temperatures of the LTTD units, and the resulting waste form is a dry waste.

SCDE is a process that employs a flowing, noncombustible, nontoxic, environmentally-safe fluid as a solvent. This process takes advantage of the enhanced ability of carbon dioxide to dissolve organic contaminants once it has been heated and compressed above 90°F and 1080 psig. In waste clean up applications, SCDE is used to dissolve the hazardous components and extract them from the substrate material. By lowering the temperature and pressure, the contaminants can be precipitated from the solution to allow separation and recycling of the carbon dioxide. This process would also produce a dry residual waste form which can be treated as radioactive, rather than mixed, waste.

9.2.4 Mechanical Surface Removal Methods

1) Ultrahigh-pressure water.
2) Shot blasting.
3) Scabblers/scarifiers.
4) Grit blasting.
5) Centrifugal cryogenic CO_2 blasting.
6) Ice blasting.
7) Supercritical CO_2 blasting.
8) Plastic pellet blasting.
9) Hand grinding, honing, scraping.
10) Automated grinding.
11) Metal milling.
12) Concrete milling.
13) Explosive.

Mechanical surface removal methods are good for decontaminating concrete and exterior metal items. Mechanical scabblers are available and have been used extensively for decontaminating concrete surfaces. No secondary waste streams other than the removed concrete surface layer are produced. Mechanical scabblers have problems with uniform application on irregular and non-horizontal surfaces. However, this problem is typical of other technologies for concrete decontamination and is one of the areas for development.

Shot blasters and grit blasters are also available and have been used extensively for decontaminating concrete and metal surfaces. The volume of waste shot created is a reasonably small fraction of the waste created from removing a concrete surface layer. The relative proportion of waste grit is larger. Grit blasting is more effective than shot blasting for decontaminating metal surfaces. The proportion of waste shot or grit created is greater for metal surfaces than for concrete surfaces. More durable blasting materials that resist abrasion and can be recycled through the process more times for waste minimization would improve the utility of these technologies. Shot and grit blasting are not recommended inside hot cells.

Ultrahigh-pressure water blasting is also available and accepted for concrete and metal decontamination. The depth of the concrete surface that is removed can be controlled by varying the water pressure and unit speed. A water recycle system may be available for some environments, but generally will have to be developed and demonstrated for each specific application.

Supercritical CO_2 blasting, which is under development commercially, should act very much like ultrahigh-pressure water except that no water will remain that must be treated. Development of this technology should be completed. Commercial cryogenic CO_2 pellet blasting has been used to decontaminate tools at nuclear power plants. More abrasive high-speed pellets from a centrifugal cryogenic CO_2 pellet blasting system under development may be needed to remove many contaminants at ORNL. This technology should be effective in decontaminating metal and concrete surfaces.

Ice blasting offers the advantage that the pellet melts to form water rather than becoming a waste. The contaminated waste water formed by the melting ice particles must then be treated for discharge in commercially available evaporators. Remote-controlled operation of the cleaning head is desirable for some applications and will be required for others (e.g., decontamination of hot cells). Consequently, the adaptation of the equipment to a robotics control system would be necessary.

Mechanical scarifiers and scabblers are available for removing a surface layers of contaminated concrete. Scarifier heads usually have several carbide tips; multiple-head machines are used for large-surface floor or wall applications. Scabblers use moving chains, saws, and steel bars or rollers. The method is effective, but is slow and can generate airborne contaminants.

Manual methods such as scraping tend to be labor intensive and slow and might result in excessive worker exposure to radiation or contamination. Grinding, honing, and milling machines are commercially available for decontaminating surfaces. These machines would be suitable for regular surfaces. The explosive technique involves the selective detonation of small explosive charges in the surface of concrete to shatter the surface layer.

Since the technologies in this generic group remove the surface of the substrate, 100% decontamination should result if the depth of penetration of the contamination is less than the depth of surface removal and if back-contamination can be prevented. For blasting technologies, the waste generated will consist of the particular blasting medium and the contaminated layer of substrate that is removed. If the blasting medium can be separated, the waste generated by the technologies in this generic group will consist primarily of the detached substrate layer. These techniques will also produce a small volume of contaminated filters. Recycle systems for ultrahigh-pressure water will probably produce small volumes of contaminated ion exchange media and other components such as filter cakes or sludges.

9.2.5 Thermal Surface Removal Methods

1) Microwave scabbling.
2) Plasma torch.
3) Laser heating.
4) Laser etching and ablating.
5) Plasma surface cleaning.
6) Plasma etching/fluorination.
7) Flashlamp cleaning.

For surface removal of concrete, the microwave scabbling and plasma torch methods flash the hydration water in the concrete to steam. The resulting volume expansion fractures the concrete surface. Removal of the debris captures most of the contamination.

The plasma torch, flashlamp, and laser heating methods remove the surface layers of other substrates by supplying enough heat to the surface to cause vaporization. The laser ablation method removes surface layers by photochemical and photothermal processes. The vaporized materials and contaminants may then be captured by vacuum removal with a cover-gas stream, thus effecting decontamination. The plasma torch method will produce a significantly higher gas-flow rate than the laser method. Because of the unidirectional nature of these technologies, application to complex geometries might be a problem. Burn-through could also occur when treating walls.

In plasma cleaning, glow discharges might be used either to clean the inside surfaces of vessels or, in reactive plasma cleaning, to produce chemically reactive species that clean contaminated surfaces by processes of chemisorption, reaction, and desorption or gasification. For example, with an oxygen plasma, oxygen atoms, ozone, and UV photons convert hydrocarbons to CO_2 and water vapor, and a fluorine plasma may provide for separation and recovery of valuable uranium by converting it to UF_6. For the latter, the low volatility of UF_6 would need to be taken into account in the system design. Because of the relatively low melting point of metal fluorides, burn through with aggressive fluorine plasmas would have to be prevented. This could be done by monitoring the production of UF_6 and other off-gases and by controlling the plasma discharge parameters. The UF_6 gas can be recovered and the excess fluorine reused. Plasma cleaning is routinely used to clean the inside of fusion devices before the main plasma is lighted. A pollution control methodology for the effluent gas from the plasma treatment step would be needed.

For the technologies in this generic group that remove the surface layer of the substrate, 100% decontamination should result if a sufficient depth of surface is removed. Thus, control of the depth of surface removal is a critical factor. The waste generated will consist primarily of the detached substrate layer.

The thermal surface removal technologies produce smokes that have smaller particle sizes than the dusts resulting from mechanical surface removal. The smokes from thermal surface removal will be harder to remove by filtration than the dusts from mechanical surface removal.

9.2.6 Metal Refining Methods

1) Smelt purification.
2) Electrorefining.
3) Leach/electrowinning.

Smelt purification involves adding oxidizing fluxes to scrap metal that will react with impurities when the metal is heated. The contaminants are removed in the slag that forms and floats to the top of the molten metal. Lab-scale and large-scale studies of smelt purification of radioactive metals have been made. Good results were obtained with the easily oxidized elements like uranium, but reportedly the more noble technetium is not removed by this technology. Estimates of capital and operating costs for smelting 90,000 tons of DOE scrap metal have been developed.

Electrorefining is a well-established, commercial technology, but is much less established for decontaminating radioactive metal. Electrorefining uses an electrolysis cell that contains an anode made from contaminated metal and deposits a cathode of pure metal. A voltage applied between the cathode and anode causes metal to ionize from the anode and to plate out on the cathode. Lab-scale tests that electrorefined contaminated nickel resulted in plate out of technetium with the nickel on the cathode. The electrolytic waste solutions from this process will be characterized as mixed wastes. Decontamination and recycle of the electrolytic solutions will be needed. Studies to determine the electropotentials, current densities, and concentrations to use might make this method successful. At the thermodynamic limit, this process can separate all elements adequately.

In the leach/electrowinning process, impure metal is dissolved, the solution is passed through a purification step, and purified metal is then plated out on a cathode in an electrolytic cell. This process, is a well-established, commercial method for producing nickel from ore, but is not used commercially to refine contaminated nickel metal. As in the case of electrorefining, decontamination and recycle of the electrolytic solutions will be needed. Technetium can be removed by ion exchange with the impure metal solution.

9.2.7 Other Techniques

Forge Hammer: The forge hammer is a pretreatment technology designed to rapidly rubblize nuclear weapon components. This rubblization process allows for easier chemical characterization and subsequent treatment, or separation of metals. However, there is a desire, for waste minimization reasons, to ensure that dissimilar materials are not rubblized. Therefore, it will include an interlock system using barcoding to identify components, and prevent creating mixed waste by not rubblizing components containing radioactive material and heavy metals. Remote operation and dust control systems have been installed to address worker safety and health problems.

Water Jet: It is advantageous to separate waste streams as early as possible in a process to minimize the amount of a particular material (i.e., waste minimization) for subsequent processing and to keep one material from contaminating other recovery or processing steps. Standard cutting procedures, like bandsaws or shears, when used on nuclear weapon components, are generally difficult to use, are slow, and are not very accurate (i.e., wide cutting margins are required to account for errors or failures of the cutting devices), and, therefore, do not minimize waste. To overcome these advantages, DOE is demonstrating the use of water jet cutting techniques. Specifically, water jet cutting can cut through thick, heterogeneous materials quickly, does not appreciably heat up the material being cut (important for potting that may evolve carcinogens when heat up), is essentially vibration free, and can cut intricate patterns within thousandths of an inch of a target material. In addition, because the cutting medium is water and garnet (i.e., sand), clean up is straightforward (i.e., mechanical filtration).

This technology has been demonstrated in the program for the removal of thermal batteries from potted components and for the demilitarization of a parachute. For example, it took just a few minutes to remove the thermal batteries and to cut through a 12 inch diameter section of nylon/Kevlar parachute. Water jet cutting lends itself to small, enclosed systems that recycle the water used for cutting and could be used for explosive or radioactive materials removal where containment boundaries are needed.

Chemical Depotting: Chemical depotting is an advanced treatment technology being used to declassify protected volume parts and classified electronic weapon components containing recyclable metals and either radioactive gap tubes or other types of hazardous components. The process dissolves or attacks a wide range of organic adhesives, sealants, coatings and potting compounds (encapsulants), is non-corrosive and biodegradable. The process is being used at the Kansas City Plant to facilitate removal of radioactive gap tubes from units which presently would constitute mixed waste.

9.3 CONCRETE

It is very easy to clean large concrete slabs and surfaces. It is then easy to survey the decontaminated item and show that it is clean. Clean material can then be recycled. If the concrete is broken up, it becomes harder to decontaminate and virtually impossible to prove that it is clean. Proving that an item is clean is the key feature in recycling. There are a lot of good commercial decontamination methods available. Most people are aware of this aspect of the decontamination equation, but the other part of the equation is proving that the material is clean based upon the release criteria.

Many of the decontamination methodologies that are available commercially require large capital outlays to build and get them to a size that makes sense from an economic point of view. The amount of activity that can be performed on site with mobile operations or temporary units is limited by the economics that drive that technology. Of course, the economics include the size of the cleanup/decontamination activity, the duration of the job, and the mobility of the equipment.

Materials that are too hot to clean economically are volumetrically reduced. In this process the material is repacked, the voids (e.g., the insides of pipes), are filled with other hot waste material (e.g., rubble), and then the entire mass is compacted to reduce its original volume. In some cases, it has been possible to reduce the volume to 1/36 of its original value. This compacted mass is ultimately sent to such places as Barnwell for burial.

In the opinion of Scientific Ecology Group (SEG), it is best to take the entire contaminated concrete structure and rubblize it on the assumption that the concrete can be separated from the rebar. It is better to not separate non-contaminated concrete from contaminated concrete. Thus, if a whole building was rubblized, the volumetric radiation levels would be less due to the amount of uncontaminated concrete that exists within the building, and the fact that the radioactivity tends to concentrate in the dust developed from the rubblizing and not in the aggregate. This is a "dilution is the solution to pollution" tactic. SEG has found it very difficult to prove "cleanliness" of concrete. They have found the concrete slabs they handled were contaminated 7 inches down in a 12 inch slab. SEG's experience shows that it is more economic to assume all of a concrete structure is contaminated, and use the material in such in-house products as waste boxes.

Hanford has taken the lead in the area of using waste to dispose of waste. The idea is to use the concrete, ground up, as aggregate. Along with contaminated purge waters, soil washing sludges and related wastes, it would be processed using a dedicated grout batch plant. The contaminated grouting then would be injected into already contaminated underground burial boxes that are collapsing, or to fill spaces between the drums and boxes in burial grounds, to prevent subsidence.

Methods that did not excessively destroy the concrete surface are preferable to scabbling, especially if one planned to reuse a building being decontaminated. In situ methods were seen as being more economic than off-site. There is no one right way to decontaminate concrete. What method(s) are used should be based in part on the history of the contaminated concrete and a thorough characterization of the contaminants present. Concrete exposed to liquid contaminates can be especially difficult to clean because of the ability of liquids to penetrate the concrete surface and move readily into cracks and crevices. Effective and economical decontamination of concrete may require several methods in concert. The use of a penetrating liquid can bring contamination to the surface of concrete.

If the Federal government wants concrete to be recycled they may have to mandate its use because of the present cost advantage toward burial. There are few suggestions for alternative uses for recyclable concrete within the DOE complex besides SEG's use of recycled concrete for waste boxes. Making the concrete into caps for waste dumps, using it as mine fill, or as a source of aggregate is possible. Reuse has the potential of leading to future problems in liability.

Microwave Heating: Both Harwell in the U.K. and Oak Ridge have been testing a microwave heating process. In the Oak Ridge process, the microwave energy is directed at the concrete surface, using a waveguide applicator. The concrete and free water in the matrix are heated, producing thermal-and steam-pressure-induced mechanical stresses that cause the surface to burst. The resulting concrete particles are small enough to be removed by a vacuum system, but large enough not to qualify as "dust." Less than 1 percent of the debris is small enough to pose an airborne contamination hazard. Two microwave generators were used—a 6 kilowatt, 2.45 gigahertz generator and a 10 kilowatt, 10.6 gigahertz generator.

Electro-Kinetic Decontamination: Another technology is electro-kinetic decontamination, which is based upon a power supply being placed across the reinforcing bars or pipes in the concrete while laying a mesh or a grid on top with an electrolyte or a gel. Some preliminary experiments that have been conducted both within the DOE system and in private industry and they are very encouraging. Isotron at Hanford was successful in handling concrete contaminated with uranium. It is unknown whether or not the electro-kinetic technique can successfully remove technetium.

The process consists of (1) application of a complexant (extractant) to the contaminated concrete, (2) application of a formulated stripping coating, (3) application of a DC potential across the concrete, and (4) removal of the strippable coating which should contain the contaminants, and (5) disposal/treatment of the reduced volume of contaminated coating.

Abrasive Technologies: Glass beads, dry ice, water ice, rice hulls, and grit blasting are all forms of abrasive technologies. Sometimes a wire brush is used on the insides of pipes and other hard to reach places. Dry ice has been used on decontaminating tools and is also used in decontamination projects at nuclear power plants. An exciting new development for the decontamination arsenal is the reuse of your own waste products in the decon cycle. Quadrex is currently investigating the use of sludge materials, generated by chemical cleaning processes, as an abrasive in the blast cleaning processes for metals. By using waste by-products in decontamination processes, Quadrex eliminates part of their waste stream.

Electro-Hydraulic Scabbling (EHS): In conjunction with Textron the EHS system is being developed. In the EHS system, the concrete surface is scrubbed in a two-dimensional traverse of the EHS head positioned by a control arm. Concrete rubble from the EHS enclosure is retained in a tank, with the scrub water recirculated to the EHS head. Real-time on-line elemental analysis ICP-ES technique will control progress of the decontamination. For ease of operation at any desirable location, the EHS system will be mounted on a carriage.

Improved Cutting Methods: Technologies need to be developed and demonstrated for improved cutting methods for the dismantlement of concrete structures. Conventional techniques have difficulty confining and segregating contaminated portions of concrete rubble and resulting dust. Additional problems include worker exposure to high radiation fields, industrial safety hazards, and generation of secondary waste. Development needs include improved methods of concrete cutting and size reduction. These methods include, but are not limited to, laser ablation, water jet, explosive cutting, expansive grout, liquid gas cutting, and diamond saws.

Laser Surface Cleaning: Methods in current use such as chemical paint stripping, cryogenic cracking, and pellet, sand or water blasting add mass to the waste resulting from removal. Laser based technology now being explored will add negligible mass to the waste stream. In addition, the laser technology has the potential for reducing worker health risk associated with the removal process and subsequent waste handling to comply with the ALARA principle, and further reducing the duration and cost of removal.

High Pressure Water: The proposed decontamination system includes a dry vacuum cleaning with HEPA filter, dust collection, foam cleaning agent, low-pressure surface rinsing, and surface concrete removal with high-pressure water. The separation system provides coarse solids screening, oil and grease collection, fine solids removal, and organic removal by activated carbon.

TechXTract™ Process: This process (EET, Inc.) is designed to treat porous solid materials contaminated with PCBs; toxic hydrocarbons; heavy metals, including lead and arsenic; and radionuclides. By extracting the contaminants from the surface, the materials can be left in place, reused, or recycled. After treatment, the contaminants are concentrated in a small volume of liquid waste. The chemicals penetrate through pores and capillaries, and electrotrachnical bonds holding contaminants to substrate are attacked and broken.

9.4 TRANSITE

Transite is composed of 50% asbestos fibers, along with Portland cement and fine silica sand. Even if you can remediate the transite, you are still stuck with the problem of the asbestos. Asbestos is a hazardous waste. Do you treat the contamination, the asbestos, both or neither? What methods are available to treat transite? What products can be made from recycled transite and/or asbestos? Both industry and DOE agreed that a major problem with transite/asbestos is the lack of end-users for the waste.

DOE perceives that there is a huge transite problem facing them. Fernald has 409 cubic yards of transite, 23 cubic yards alone in Plant 7. This amount is small when compared to the amount at Oak Ridge's K-25 facility. That facility is paneled entirely with transite, even the interior compartments are made from it. DOE feels that there are no clearly defined treatment techniques and/or transite derived products available to utilize the transite waste being generated within the complex. FERMCO, and to a lesser extent the rest of the complex want to find a way of solving this problem immediately.

Decontamination techniques include all of those used on metals. For example, strippable coatings, liquid abrasives, blasting with CO_2 pellets or ice pellets, high-pressure water with applied surfactant, steam cleaning, liquid abrasive blasting, and abrasive and impregnated foam media (sponge, or sponge jet unit). Also, painting or coating the transite before disassembly is suggested to fix the radioactive contamination while the transite was being moved to a decontamination site.

Besides the technical problems of removing, decontaminating and disposing of transite, there are the even more difficult problems of (a) public perception (they think of asbestos in the same way as radioactivity), (b) issues of liability, and (c) inability of finding insurance companies willing to issue worker compensation or product liability for asbestos products such as transite.

Several methods for using transite processes, where it is a feed material includes: (a) slagging material added to an iron ore reduction process, wherein the temperatures would be high enough to decompose the asbestos; (b) grinding up the transite such that it can be added to clays in the manufacture of bricks and then firing the bricks in kilns hot enough to destroy the asbestos, and (c) dissolving the cement and asbestos and using the resulting sand in a vitrification process.

Direct uses for transite included a role in shielding materials, and as caps over waste pits at DOE facilities. Two problems with the latter however, is that (1) the transite might become friable with the release of asbestos because of environmental factors similar to what happens to transite used as roofing material, and (2) there may, at some future date, be a ruling that the transite must be exhumed for further treatment/disposal while carrying forth all requisite liabilities. Other direct application of transite include its use in trickle fitters, and structural parts of an artificial "reef" for fish.

The demonstration project—Minimum Additive Waste Stabilization (MAWS)—is intended to combine various waste stream present at Fernald, in order to obtain a synergistic waste. This synergistic waste is turned into glass. Transite contains silicon. Silicon is necessary in the manufacturing of glass. Thus using transite in the MAWS project is a synergistic use of that waste.

9.5 METALS

Beneficial reuse of Radioactive Scrap Metal (RSM) begins by determining whether to melt, refine the scrap, or surface decontaminate it. RSM, which has been melted, can ultimately be used to produce slightly contaminated products for reuse within the DOE complex, while decontaminated RSM can be free released and become a source of revenue. The choice of path is dependent on the final objective. Some advocate decontamination as the "path of choice." It makes little or no sense to use virgin metal for a product, when the process, etc. will ultimately make it radioactive. It is far better to go with scrap that is already contaminated. If the RSM can be used to produce a product usable on site, then melt refinement is the route to go. Process history of the scrap is another factor to be weighed when determining whether to go with decon or melt. Adverse publicity also plays a part in determining which path to take. The institutional, legal, and political problems that occur if a piece of RSM should be accidently free released is enough to make DOE sites act very conservatively. Because of the need to prove it is sufficiently clean, free release of decontaminated scrap can be a costly process. Thus, even though there is a market for decontaminated RSM, most of the DOE complex is opting to look at reuse rather than release. The Federal regulations currently do not encourage free release. There are 2 options for RSM decontamination: (a) melt refining, and (b) surface decontamination.

Based upon preliminary studies, melting technologies that are considered to be viable for treatment of mixed waste and thus justify further examination included:

1) Fossil fuel melting (bath smelting, basic oxygen processes; process heat from carbon combustion with injected oxygen);
2) Electric arc melting;
3) Induction melting;
4) Electron beam melting;

5) Vacuum arc remelting;
6) Electro-slag melting.

Currently, several melt refining processes are available. The first is melting by induction furnace. This capability is readily available. Further, the process has been proven in Japan and Europe. The process is especially good at handling depleted uranium and uranium isotopes, and achieving very high decontamination factors. There is no reason that the 700 tons of uranium contaminated at RSM at Fernald could not be sent to private industry and successfully decontaminated via induction melting. The second option is Electro-Slag Remelt. Because of a need to achieve higher decontamination levels than are currently possible by private industry, INEL has research and development programs in Induction Furnace and Electro-Slag Remelt with Montana Tech and the Oregon Graduate Institute. INEL's objective is to develop slag chemistry technology relating to the transport of radioactivity and the decontamination of the metal during the melt process. Montana Tech is working on the Induction Slag chemistry. Oregon Graduate Institute is working in Electro-Slag. Other options for melt refining include plasma arc, vacuum arc, and electron beam. These last 3 processes are very sophisticated, but not cost effective for handling INEL's needs. A lot of companies have capabilities in these last 3 technologies. Plasma Arc has been used successfully in hazardous waste destruction. Many companies have approached INEL wanting to demonstrate their existing Plasma Arc equipment's decontamination potential on INEL's RSM.

Cadmium-bearing scrap from nuclear applications, such as neutron shielding and reactor control and safety rods, must usually be handled as mixed waste since it is radioactive and the cadmium in it is both leachable and highly toxic. Removing the cadmium from this scrap, and converting it to a non-leachable and minimally radioactive form, would greatly simplify disposal or recycling. A process now under development at Savannah River will do this by shredding the scrap; leaching it with reagents which selectively dissolve out the cadmium; reprecipitating the cadmium as its highly insoluble sulfide; then fusing the sulfide into a glassy matrix to bring its leachability below EPA limits before disposal. Alternatively, the cadmium may be recovered for reuse. A particular advantage of the process is that all reagents (except the glass frit) can easily be recovered and reused in a nearly closed cycle, minimizing the risk of radioactive release. The process does not harm common metals such aluminum, iron and stainless steel, and is also applicable to non-nuclear cadmium-bearing scrap such as nickel-cadmium batteries.

The other process option is surface decontamination. Surface decontamination can be handled by traditional chemical methods, electro refining, or abrasion. With chemical cleaning, disposal of the waste becomes an issue.

Specific non-destructive decontamination techniques are:
1) Pressurized-water washing;
2) Steam jetting;
3) Vacuuming;
4) Freon decontamination for motors, precision parts, and electronic equipment;
5) Chemical decontamination using gels, foams, pastes and chelates;
6) Chemical depotting;
7) Lasers.

Destructive surface decontamination techniques include:
1) Dry abrasive;
2) High-pressure (35,000 psi) water jet cutting;
3) Flame spalling;
4) Mechanical devices such as jackhammers and scarifiers.

The subject of dust control should be addressed in all surface decontamination techniques.

INEL has been looking at specific products that could be manufactured from their scrap. Two products that are of special interest to INEL are: low-level waste boxes or universal fuel storage canisters. Accumulations of metal waste exhibiting low levels of radioactivity (LLCMW) have become a national burden, both financially and environmentally. Much of this metal could be considered as a resource. The Department of Energy was assigned the task of inventorying and classifying LLCMW, identifying potential applications, and applying and/or developing the technology necessary to enable recycling. One application for recycled LLCMW is high-quality canisters for permanent repository storage of high-level waste (HLW). As many as 80,000 canisters will be needed by 2035.

9.6 INORGANIC DEBRIS TREATMENT

Inorganic debris treatment includes a group of several distinct technologies. Each specific technology will be applicable to specific mixed wastes depending on the available waste characteristics data. The following technologies are included in this group:

1) Incineration may be applicable in cases where the mixed wastes includes a mixture of organic and inorganic materials.
2) Physical separation or decontamination may be applicable in cases where the hazardous or radioactive component of the mixed wastes is present only on the surface of the wastes or can otherwise be readily separated from the waste matrix. In these cases, the radioactive or hazardous waste component can be most effectively removed by technologies such as washing, steam cleaning, abrasive blasting, etching, cutting and disassembly and, for wastes characteristically toxic for mercury, roasting. The technologies in this category only change the form of the wastes but do not eliminate the

hazardous waste component. Therefore, treatment of mixed wastes using these technologies can result in the generation of wastewater or other matrices (e.g., liquid mercury) with hazardous characteristics.

3) Immobilization technologies such as stabilization or microencapsulation may be applicable in cases where the hazardous component cannot be readily separated from the waste matrix.

According to DOE, in determining the needed physical separation treatment capacity, the waste form is a major factor to be evaluated. Many mixed wastes in this treatability group have unusual shapes or configurations whereby waste volume does not provide a representative measurement of the time, equipment, labor, and other resources needed to effectively treat the wastes. A small object with a complex configuration may require a considerably greater treatment "capacity" than a large object with a simple configuration.

9.7 TREATMENT OF TRASH

9.7.1 Sorting

As radioactive trash is generated, it usually receives some form of pretreatment, generally consisting of sorting the material, such as separating combustible from noncombustible material, prior to incineration or separating compactable from non-compactable material prior to compaction. Hand sorting is the most direct method of segregating wastes into constituents that are amenable to treatment by a particular technology, or into radioactive and nonradioactive components.

Pneumatic sorting by an air or inert gas stream can also separate lower density combustible materials, such as paper, plastic, and rags, from higher density noncombustible material such as glass and metal. Manual sorting for radioactivity consists of using a sorting table where bags with low radiation levels are segregated. Radiation readings used for this initial screening have been reported as about 1 mrem/h for typical nuclear reactor facilities. The contents of these bags are opened, and the individual items are scanned and segregated. Automated trash monitors that are more sensitive and reliable for segregating radioactive from nonradioactive waste also are available. DAW volume deductions of 31% through the use of a trash sorting table have been reported.

9.7.2 Compaction

Compaction is one of the simplest and most effective techniques for reducing the volume of dry active waste (DAW). As such, it is particularly suitable for generators of large volumes of lightly contaminated wastes.

Compactors are simple to operate, inexpensive, and available in various designs, forms, and sizes.

Compaction is a process by which material is physically compressed into a smaller volume. Designs of compactors range from the less expensive hand-operated types to fully automated systems using electronically controlled hydraulic systems. Waste can be compacted inside a 55 gallon drum, wooden box, steel box, or other container, depending on the design of the compactor. Three types of compactors are used to reduce low-level waste volumes:

1) Conventional compactors;
2) Box compactors;
3) Supercompactors.

Each of these has its range of operating capabilities. Conventional compactors compact wastes directly into 55 gallon drums, exerting forces from 10 to 30 tons. Box compactors are capable of accepting larger objects and developing compressive forces up to 250 tons. Their rectangular-shaped containers also utilize space more efficiently than conventional compactors. Supercompactors (also called high-force or high-pressure compactors) are the most powerful types available.

As a general rule, supercompactors can exert forces of greater than 1,000 tons. Consequently, they can accept and compact nearly all DAW including steel piping and metal components that fit into the final disposal container. Manufacturers in Belgium, France, and Germany have been leading the development of these supercompactors.

The volume reduction efficiency of a compactor depends on the applied force, the bulk density of the waste material, and the spring-back characteristic of the material when compaction pressure is released.

9.7.3 Other Techniques

Shredders: Shredding of radioactive wastes as a volume reduction technique has been conducted for over 20 years in Europe and for over 10 years in the United States. Shredders are devices that tear, rip, shatter, and/or crush waste materials into smaller pieces. In nonradioactive waste applications, shredders are commonly used in conjunction with either incinerators, compactors, baling, or landfilling. For incineration, shredders are used to reduce particle size for feeding by rams, gravity, or stokers. Shredders are used in conjunction with compactors, balers, or landfilling to reduce void spaces between individual waste objects, thus reducing the volume of the disposed waste. For compaction, size reduction also reduces the amount of spring-back that occurs. A volume reduction of about 3:1 can be obtained.

High speed shredders have several disadvantages in a nuclear environment, and the low-speed, shear-type shredder is more appropriate for low-level waste applications.

Cryofracture: Cryofracture is a size-reducing process whereby objects are frozen to liquid nitrogen temperatures and crushed in a large hydraulic press. Material at the cryogenic temperatures have low ductility and are easily size reduced by fracturing. The main application being investigated for the DOE is for retrieved buried and stored transuranic (TRU) waste.

Cutting: Contaminated plant hardware can be cut into pieces for better packaging and storing. Often the discarded contaminated piping at nuclear reactor stations is sectioned with cutting equipment to fit into (and to reduce the empty volume of) transportation casks or storage containers. For this type of hardware, there is usually no intention of decontaminating and reusing the material. In addition to traditional saw cutting, specialized cutting technologies—such as oxyacetylene cutting and plasma arc torch cutting—can be used, as appropriate.

10

Buried Waste

It was estimated that some 3.1 million cubic meters of buried waste is located throughout the Department of Energy (DOE) complex (as of 1990). This waste is predominately located at DOE facilities at Hanford, the Savannah River Site, Idaho National Engineering Laboratory, Los Alamos National Laboratory, Oak Ridge National Laboratory (X-10), and the Rocky Flats Plant. The wastes at these various sites have been buried or stored in several different types of structures, including trenches, pits, buildings, storage pads, or other specific structures. Waste disposal activities at DOE sites were conducted in accordance with applicable regulations and laws and accepted engineering practices at the time.

Approximately half of all DOE buried waste was disposed of before 1970. Disposal regulations prior to 1970 permitted the commingling of various types of waste (i.e., transuranic, low-level radioactive, hazardous). As a result, much of the buried waste throughout the DOE complex is presently believed to be contaminated with both hazardous and radioactive materials. DOE buried waste typically includes transuranic-contaminated radioactive waste (TRU), low-level radioactive waste (LLW), hazardous waste per 40 CFR 261, greater-than-class-C waste per 10 CFR 61 55 (GTCC), mixed TRU waste, and mixed LLW. Interstitial soils are also believed to be contaminated as a result of these disposal practices, and subsequent releases, thereby significantly increasing the volume of materials requiring remediation.

Typical buried waste includes such items as construction and demolition materials (i.e., lumber, concrete blocks, steel plates, etc.), laboratory equipment (i.e., hoods, desks, tubing, glassware, etc.), process equipment (i.e., heat exchanger, valves, ion exchange resins, HEPA filters, etc.), maintenance equipment (i.e., hand tools, cranes, oils and greases, etc.), and decontamination materials (i.e., paper, rags, plastic bags).

At INEL, the host location for the DOE Buried Waste Integrated Demonstration (BWID), a variety of containers have been used for the disposal of buried waste. Containers used for the shipment of waste at the INEL have included steel drums (30-, 40-, and 55-gallon), cardboard cartons, and wooden boxes (up to 105 in. x 105 in. x 214 in.). Larger individual items were disposed of separately as loose trash. Degradation of the waste containers is believed to have resulted in contamination of the immediately surrounding soil. Estimates of contaminated soils at INEL are on the order of 8 million cubic feet. The specific geology at the INEL in which waste is buried involves approximately 10 to 30 feet of surficial sediments. Underlying the surface sediments is a layer of fractured basalt. There are 2 major thin sedimentary interbeds in the basalt layer, 1 interbed at a depth of approximately 110 ft., and another interbed at 240 ft. Data show that radioactive contaminants have migrated from the buried waste seams to the 110 ft. interbed and possibly to the 240 ft. interbed. Organics have been detected in groundwater beneath INEL (about 600 ft. depth).

INEL buried waste characteristics are generally representative of other DOE buried waste sites. Remediation technology demonstrations performed at the INEL should provide data useful for remedial decision-making at the INEL, as well as at other DOE buried waste sites. Significant remediation challenges are presented by buried waste, particularly the pre-1970 TRU-contaminated waste.

Systematic evaluation of alternatives for buried waste sites resulted in 3 systems approaches: in situ hydrologic isolation or containment, in situ treatment, and retrieval followed by treatment. Each of the 3 systems entails a series of interrelated activities, each of which is essential for the success of the system. The in situ isolation/containment system includes definition of the design basis, which encompasses the design life, health/safety standards, structural criteria, environmental parameters, waste-form requirements, and cost limitations; site characterization; structural stabilization; actual isolation technologies; monitoring; and support activities (transportation, health & safety, decontamination of equipment, etc.). The in situ treatment system includes definition of the design basis; characterization of waste and site, and treatment technologies; verification of treatment effectiveness; site closure and monitoring; and support activities. The retrieval system includes definition of the design basis; waste and site characterization; retrieval; staging/storage; segregation/sorting/packaging; treatment/recycle/recovery/decontamination; disposal or recycle of residue; and support activities.

The majority of the structural stabilization technologies have been demonstrated and are available. Science and technology needs include testing to determine material compatibility for structural caps and solidification. Other needs include field tests with full-scale dynamic compaction equipment.

Needs for isolation/containment technologies include methods to minimize waste generation during application, guidance systems for grout injection/tunneling, and development of waste-resistant materials for capping.

The applicability of thermal destruction technologies for buried wastes and the characteristics of the treated residue need to be determined. Methods for treating off-gas from thermal destruction are also needed.

Science and technology needs for fixation technologies include development of new chemical mixtures and application/injection equipment. This is especially critical to ensure and verify that containers are ruptured and grout/polymer has mixed with contents for in situ technologies. The development of compaction methods is needed to prevent flashing/overpressurization of waste during ISV.

Testing is needed to estimate the effectiveness and identify additives for freeze purification and thermal desorption. Soil simulation methods to promote conductivity for fracted air stripping are needed.

Methods are needed to safely and effectively segregate buried waste during and after retrieval. The development of remotely controlled/robotic retrieval technologies needs to be continued.

Some technologies may plan an important role but may not be able to solve entire problems by themselves. For instance, some technologies can remove or destroy only one contaminant, and other technologies are only useful for retrieval of contaminated soil, wastes, etc. New technologies are likely to be conceived and developed in the future that are not covered.

Buried pipes exist throughout DOE sites, many in the main laboratory area. Some of these pipes have carried radioactive wastes to underground storage tanks or to treatment facilities in the past. Many of the pipes have deteriorated, and most are no longer in use. Most pipes are relatively near the surface, and the levels of radioactivity and other hazards do not appear excessively hazardous. Only 2 technologies were considered for treating the pipes themselves—excavation and fixation (probably with cement-based grout). Where the pipes have leaked considerably, there may be significant amounts of contaminated soil around pipes.

For contaminated soil or buried waste, it is not clear whether removal and destruction of some contaminants on-site (e.g., through incineration) or removal and disposal elsewhere offer greater benefits. Some disadvantages of the removal option include worker health risks and the likelihood of increased air emissions when incineration is used. A possible disadvantage of the in situ approach is partial destruction of the toxic elements (in the case of mixed waste). If the disadvantages of waste removal are fairly significant, the alternative of leaving waste or contamination in place and stabilizing it may be the most prudent approach.

10.1 RETRIEVAL

Previously developed technologies will be integrated with newly identified technologies into a full-scale retrieval system for demonstration at the INEL Cold Test Pit. This full-scale retrieval system demonstration will include site characterization, retrieval, treatment and assaying technology. The retrieval portion will include 2 scenarios currently identified by DOE as potential remedial alternatives: hot spot and full-scale retrieval. Hot spot retrieval consists of retrieving a selected portion of a buried waste site in conjunction with an overall waste containment/stabilization remedy. Full-scale retrieval includes the complete excavation and conveyance of waste from a buried waste site for treatment and disposal. The retrieval system demonstration will be conducted at a simulated buried waste pit at the INEL (Cold Test Pit). Individual technology demonstrations will need to evaluate technology potential to minimize secondary waste generation, costs, and personnel exposure.

The primary retrieval technology needs are presently under development, including remote excavation, contamination control, and waste conveyance systems. Emphasis will be placed on technologies supporting the integrated retrieval system demonstration at the INEL Cold Test Pit. Key technology gaps include movable containment structure and visual simulation, and control systems for an unstructured retrieval environment to oversee the actions of the individual subsystems.

Hot spot retrieval technology needs include:

1) Remote delivery system to move retrieval and support equipment to the hot spot location.
2) Movable containment structures in which operations will be performed.
3) Contamination control.
4) Waste form removal.
5) Pit stabilization.
6) Methods to remotely exhume waste areas of varying size.

Primary full-scale retrieval technology needs include:

1) Innovative remote exhumation techniques to support 80 cubic yards/8 hr. shift retrieval rate minimum.
2) Contamination control techniques to maintain airborne contaminant levels below 10,000 X maximum permissible concentration in air.
3) Dust-free methods for placing exhumed waste into containers for subsequent conveyance (98% dust removal efficiency for dry soil).
4) Conveyance systems to move waste from dig face to processing area.
5) Innovative containment structures in which operations will be performed and that can be mobilized from pit to pit as remediation is completed.

6) Automated data tracking system to record retrieved waste source location and transportation/conveyance status.
7) Visual simulation/sensing capability to monitor and control operations status and equipment location for collision avoidance and enhanced safety and production.

Other full-scale retrieval technology needs include:

1) Systems to size-reduce objects (e.g., tanks, lathes, etc.) encountered at the dig face for subsequent conveyance and processing.
2) System health monitoring techniques to predict component failures prior to occurrence to allow timely and cost-effective maintenance.

10.1.1 Remote Excavation System

The Remote Excavation System (RES) is a military tractor, the Small Emplacement Excavator (SEE), which has been modified for tele-robotic operation by the Oak Ridge National Engineering Laboratory (ORNL). The primary applications for this remote excavation technology are buried waste retrieval for the Department of Energy (DOE) and unexploded ordinance retrieval for the U.S. Army. The hazards of buried waste retrieval are significant if performed by conventional manned operations. The potential hazards include exposure to radiation, pyrophorics (capable of spontaneous ignition when exposed to air), toxic chemicals, and explosives. Consequently, it is highly desirable to excavate and retrieve these wastes by using remotely operated equipment. The RES will be used to excavate and remove buried waste and contaminated soil for ex situ treatment.

The SEE is a multi-use vehicle developed for the U.S. Army that has been configured with a backhoe and a front-end loader. The backhoe is an adaptation of the Case 580E commercial backhoe, and the vehicle is a modified Mercedes Benz Unimog truck. The SEE is not necessarily the excavator of choice for large-scale waste retrieval campaigns; however, the controls technology developed for the RES/SEE shall be readily transferable to other mechanical systems.

The ORNL alterations to the vehicle center around modifying the hydraulic systems for computer control. High-performance servovalve components will be used to greatly improve the dexterity over the existing manual valves. Hydraulic pressure sensors will provide limited indications of force exerted by the backhoe. The backhoe and front-end loader will also be outfitted with position encoders for use in robotic operations. Remote viewing will be provided by 2 color television cameras with pan and tilt mechanisms mounted on the truck body and a camera mounted on the backhoe boom. In the second phase of the project, the vehicle drive system will be modified for remote driving. A hydraulic motor system is being considered for propelling the vehicle during remote driving operations. Options for remote steering are still being investigated.

The RES was demonstrated at the INEL Cold Test Pit and at the U.S. Army Redstone Arsenal to evaluate the feasibility of excavating buried waste and unexploded ordinances with a remotely operated vehicle. The system was developed utilizing 5 national laboratories (INEL, Oak Ridge, Sandia, Lawrence Livermore, and Pacific Northwest). The Department of the Army also provided the platform, which was remotized for use by both DOE and the Army.

10.1.2 Dust Suppression

During retrieval operations there is a need to reduce the risk to personnel and to minimize the spread of contaminated dust. This is particularly true for sites where the risk of contamination is high and the work must be performed remotely.

Handling of TRU-contaminated waste during retrieval operations requires contamination control techniques due to the highly mobile nature of the contaminants. There are mandatory operational and safety limits, which may not be exceeded, and extremely small amounts of plutonium/americium are allowed for personnel uptake.

The DOE has commissioned an easily deployed, mobile unit to efficiently and inexpensively suppress contaminated dust during excavation of buried waste. The Contamination Control Unit (CCU) is a 9 by 26 foot trailer outfitted with a powerful waste vacuuming system and 3 dust suppression systems. The vacuum, a nuclear-grade system, with a High-Efficiency Particulate Air (HEPA) filter and a critically safe 55 gallon waste container, can pull 1 inch debris through 100 feet of hose. The dust suppressing system applies a water mist, a soil fixant and a dust suppressant. The water misting system, Dryfog (manufactured by Sonics, Inc.), uses compressed air to force demineralized water through 4 nozzles. The nozzles discharge a mist curtain that captures and removes aerosolized soil. The concentrated soil fixant, Foamer (manufactured by 3M Inc.), is combined with a stabilizer and water pumped from a trailer-mounted 325 gallon tank. The stabilizer and concentrate are stored in 5 gallon cans near the nozzle and are delivered by suction to the nozzle. The dust suppressant, Flambinder (calcium lignosulfate manufactured by Flambeau Corporation), is commonly used on forest service roads to suppress dust. This product is mixed with water, stored in a 325 gallon tank and pumped to a nozzle via a hose and reel system. Flambinder cures in a few hours, and can withstand traffic of hundreds of vehicles without reapplication.

In a field-deployable setting, the unit is capable of fixing 500 ft^2/minute with a total encapsulated (3M-Foamer), or apply 1,000 ft^2/15 minutes of dust suppressant (flambinder) while simultaneously providing 2 gpm mist from 4 misting nozzles. The vacuum system can be operated continuously during the spray operation. In an emergency response situation, the system is capable of spraying dust suppressant at a rate of 1,000 ft^2/3 minutes.

The CCU provides a reliable, easy to use, easily deployable system to control dust and windblown contamination at DOE and industrial buried waste sites during excavation and retrieval operations. Use of the CCU improves personnel safety and reduces the risk of contamination spread by airborne dust. Furthermore, all the dust control products are non-hazardous,and can be disposed of in sanitary landfills.

The CCU was designed in FY92, and assembled and its performance tested in FY93. The unit is completely developed and ready for Environmental Restoration use throughout the DOE Complex. Currently it is being used by the Environmental Restoration Program at The Hanford Site for a retrieval treatability study. The unit was also used by the INEL Environmental Restoration Program during a capping action at EBR-1 in which Wendon dust suppressant was sprayed on a slightly contaminated soil site of about 1.5 acres. Fernald is evaluating the use of the unit to supply contamination control during actual retrieval actions.

10.2 EX SITU TECHNIQUES

After the buried waste has been retrieved, there are a number of processes available for treatment depending on the composition of the waste.

A DOE effort will perform demonstrations on various interrelated waste pretreatment technologies, primary treatment technologies, and associated secondary systems (i.e., off-gas systems) for the immobilization, detoxification, volume reduction, and/or stabilization of waste components. These demonstrations should provide sufficient information on technologies to support the ER Remedial Investigation/Feasibility Study (RI/FS) process. Three interrelated areas, pretreatment, primary treatment, and secondary treatment, should be addressed from a systems viewpoint for the demonstration and evaluation of technologies for this effort. System recycling should be considered, as well as secondary waste stream treatment.

In FY 95 and out-years, the BWID program will focus ongoing treatment technology development activity, and initiate radioactive testing in those areas. Radioactive testing on bench and pilot scale technology will facilitate ER decision-making through the RI/FS process.

10.2.1 Pretreatment

Pretreatment of various retrieved waste streams may be required (depending on the primary treatment technology) to minimize the amount of waste to be treated or to optimize the primary treatment of the waste. A throughput goal of 1-14 tons/hr is desirable for these technologies. Advanced technologies are needed for the pretreatment of retrieved buried waste and contaminated interstitial soils. These technologies must be capable of:

1) Advanced sorting of retrieved waste into waste types (soils, metals, combustibles, TRU, LLW, hazardous, etc.).
2) Assaying organics, hazardous, elemental, and radioactive constituents using continuous or semi-continuous (batch) processes. This assay will be for large bulk volumes of waste on a real-time or short-time basis.
3) Size reduction/volume reduction (i.e., shred, compact, cryofracture, incineration of combustibles, etc.).
4) Pre-processing retrieved waste to optimize treatments.

10.2.2 Primary Treatment Technologies

Treatment of the waste will include processing solid, mixed, hazardous, and radioactive waste associated with combustibles, metals, soil, and sludges and their associated structures (i.e., drums, boxes, large metal objects, etc.). Treatment may be conducted using thermal, chemical, biological, physical, and/or radiological methods for all or part of the waste streams. A "robust" primary treatment system developed to minimize characterization and pretreatment is desirable. The processing limitations (i.e., reliability in operation, maintenance, non-processable components, system lifetime, feed size, heterogenous feed compositions, etc.) of the technologies should be determined. Emphasis should be given to the volume reduction of the waste and the minimization/treatment of secondary waste streams. This would also include the separation of delisted/recyclable portions of the waste. The development of enhanced waste forms from the waste should be also evaluated. Final waste forms must meet minimum waste acceptance criteria, including TCLP and other criteria that are currently being established.

These technologies must be able to:

1) Treat a significant portion of contaminants retrieved from the waste stream.
2) Minimize the amount of untreated waste.
3) Contain all contamination during treatment.
4) Process any off-gases.
5) Minimize the generation of secondary waste.
6) Be verified.

10.2.3 Secondary Treatment Technologies

Pretreatment and primary treatment technologies may require additional secondary systems. Some of these systems will be integral parts of the technologies, but may by improved by new technologies. Advanced technologies are needed, such as:

1) *Feed systems* - Including the delivery of waste streams in their associated structures to the treatment system. The reliability in remote operation and process control must be optimized.
2) *Off-gas systems* - Various types of off-gas systems needed depend on the type of treatment technology. The primary need lies in the treatment of off-gas for high temperature thermal treatments. The off-gas systems must remove regulated and radioactive components from the off-gas while minimizing secondary waste streams. Closed off-gas systems will be given a high priority.
3) *Secondary waste stream treatment* - The potential for various secondary waste streams exists, which adds to the final disposal cost. Any methods to detoxify, reduce, or treat secondary waste streams may be considered.
4) *Final waste form verification* - As the waste form is produced, the product quality of the waste form must be verified prior to disposal. Extensive sampling is not desirable due to the high analytical cost. NDA/NDE methods to verify final waste form quality on a real-time or short-time basis is needed. PNL is studying the treatment of simulated INEL buried wastes using a graphite electrode DC arc furnace.

10.2.4 Cryogenics

General Atomics, San Diego, California, developed the Cryofracture technology to demilitarize chemical and biological weapons. The process uses liquid nitrogen to freeze materials to -320°F, which are then crushed by a large hydraulic press. The freezing process immobilizes the active components, which are then crushed to fragment the material into treatable pieces. The method has been adapted and demonstrated as a pretreatment size-reduction step in processing TRU contaminated waste stored in drums and boxes.

The system can reduce 55 gallon drums or 4x4x4 foot boxes of waste materials with no probability of fire or explosion. The resultant debris, less than 6" mean size, is much more compatible with most treatment processes.

Potential applications include waste disposal and treatment facilities. The process can replace conventional shredder operations which are challenged by heavy structural materials, cables, cloth, and aerosol containers.

Another process involves an ice electrode. An ice electrode is a conventional electroplating electrode coated with a thin sheath of ice produced by liquid nitrogen-cooled nitrogen gas flowing through an electrode. Bench scale tests indicate that metals that can be electrodeposited on a conventional electrode can also be electrodeposited on an ice electrode. Preliminary work with metals that cannot be recovered by conventional electrodeposition, including oxides of uranium and tungsten, suggests that they can be retrieved and easily recovered as an ice electrode. In addition, the ice electrode minimizes waste generation because the electrode is not destroyed due to the presence of the ice sheath.

10.3 IN SITU TECHNIQUES

In situ treatment of buried waste can serve 2 separate objectives, depending on the selected remediation concept: (1) in situ treatment with long-term monitoring for in-place disposal; and (2) in situ treatment prior to retrieval of the treated waste form, and, ultimately, disposal at an appropriate facility.

The performance characteristics of the treated waste must provide suitable containment or destruction of both hazardous and radiological waste constituents in both the wastes and interstitial soils. This treatment must represent a "permanent" solution, in which long-term contaminant release to groundwater, surface soils, or air is reduced to acceptable levels. Advanced technologies are needed for the in situ treatment of buried waste and contaminated interstitial soils for in-place disposal. These technologies must:

1) Treat all contaminants within the waste stream to provide long-term containment of residual contaminants.
2) Treat to depths from 15 to 25 feet.
3) Minimize the amount of untreated wastes.
4) Contain contaminants during treatment to below regulatory levels.
5) Treat process off-gases to below regulatory levels.
6) Minimize generation of secondary wastes.
7) Be verifiable.

If the material is to be retrieved at a later date, the primary purpose of the in situ treatment is to minimize the potential spread of contamination during the retrieval process. This combination of in situ treatment followed by retrieval must provide a significant advantage in contamination control over the retrieval of untreated wastes. Advanced technologies are needed for the in situ treatment of waste and contaminated interstitial soils. These technologies must:

1) Produce a retrieval product using conventional retrieval technology.
2) Treat to depths from 15 to 25 feet.
3) Minimize the amount of untreated wastes.
4) Contain contaminants during treatment to below regulatory levels.
5) Treat process off-gases to below regulatory levels.
6) Minimize generation of secondary wastes.
7) Be verifiable.

10.3.1 Cryogenic Retrieval

Cryogenic Retrieval (CR) of buried waste is a technology that relies on liquid nitrogen (LN_2) to freeze soil and buried waste in order to immobilize hazardous waste and reduce the spread of contamination while the buried material is retrieved with a series of remotely operated tools. CR is proposed for application to any type of radionuclide or hazardous contaminant that may be contained in transuranic (TRU) buried waste. CR has the potential to be used at any buried waste site within the Department of Energy (DOE) complex.

To freeze the soil for the CR process, a series of carbon or stainless steel small diameter freeze pipes, approximately 10 to 12 ft in length are driven into an area of soil and buried waste to be frozen and removed. LN_2 is delivered into the pipes and small quantities of water are injected to promote cohesion of the soil and waste particles in the frozen area. While the pit to be excavated is frozen, the perimeters of adjacent pits are also frozen. Once the area to be removed is frozen, the center of an access pit in clean soil is excavated.

The access pit serves to create a dig face from which excavation of the waste can proceed. A gantry with remotely operated tools such as a jackhammer, a hydraulic jack, shears, and a grapple is moved over the frozen area to be retrieved. With the gantry tools in place, the tools are remotely operated, and the frozen soil and waste is broken, chipped, cut, and loaded into transport boxes. The jackhammer chisels and breaks up soil and debris that falls into the access pit. The shears are used to cut and size material, while the grapple picks up the debris and loads it into the transport boxes. The hydraulic jack is used to pry or bend the freeze pipes away from the dig face. During the excavation process, a series of air monitors detect the dispersal of rare earth tracers. The output of the process is the excavated soil and waste material. The principal benefit of this technology is contaminant mobility reduction, since by freezing the soil both airborne and liquid contaminants are immobilized.

A weather shield (a large portable cover or tent) is used with this technology to minimize waste distribution by air motion and to permit operation in all weather conditions. On the other hand, if the freeze pipes cannot be extracted and reused they would require treatment as another buried waste type. If intact containers exist in the waste volume, the drilling of injection pipes might rupture the containers, causing an initial spread of contamination. Furthermore, injection of water to help the freeze process may also contribute to some contamination spread if it were to flow out of the retrieval column.

The major technical challenges for this technology are developing a method for placement of freeze pipes in all types of soil and waste; conservation of LN_2; dispersion of water evenly through the soil and waste matrix; reduction of secondary waste created by the freeze pipes; selecting or developing more productive tools for the removal and handling of frozen waste; and improving methods for the measurement of thermal characteristics and for the detection of moisture migration.

A full-scale Cryogenic Retrieval demonstration was completed at the INEL Cold Test Pit in FY92. A series of carbon and stainless steel 2 inch schedule 40 freeze pipes, approximately 10 feet to 12 feet in length, were driven into 3 areas of soil and simulated buried waste. For the field demonstration, about 65 pipes were driven into each of 3 areas that measured 9 feet x 9 feet x

10 feet (810 ft^3). Liquid nitrogen (LN_2) was circulated through the pipes, and small quantities of water are injected to promote cohesion of the soil and simulated waste particles. Besides freezing the test pit, the perimeter of an adjacent access pit was also frozen. Once the area to be removed was frozen, the center of the access pit was excavated (the access pit was in clean soil). The access pit served to create a dig area from which excavation of the waste could proceed. A gantry with a remotely operated jackhammer, a hydraulic jack, shears, and a grapple was moved over the frozen area to be retrieved.

With the gantry and tools in place, the tools were remotely operated and the frozen soil and simulated waste were excavated and loaded into transport boxes. The jackhammer was used to break up the soil and debris. The shears cut and sized the material, and the grapple picked up the debris and loaded it into the transport boxes.

10.3.2 Barriers (Hydraulic and Diffusion)

Sites with components that are not controllable with contaminant-specific sorbents will require barriers that prevent the movement of contaminated groundwater. Systems that use direct injection of barrier reagents as liquids, and subsequent in situ formation of the barrier, are preferred over those involving excavation.

Currently approved clay-based barriers may not be as effective as some synthetic polymers and resins. These agents may cement and seal sediments to form impermeable, chemically-resistant barriers to water movement. In some situations where in situ formation is not possible, and excavation costs are not prohibitive, barriers constructed with synthetic binders and an inert matrix provide an alternative.

Brookhaven will develop and test new barrier materials specifically for buried waste control. Tests will determine the long-term durability of the material, permeability to groundwater, ionic diffusivity, response to wet/dry cycling, and chemical resistance to acid, base, and organic solvent conditions that might occur at waste sites.

This study will also examine the effects of aggregate type and quantity on barrier performance. Inert aggregate substances such as clean sand and recycled glass used to produce the test specimens will be optimized to binder, geological, environmental and waste conditions.

10.3.3 Containment/Stabilization/Long-term Monitoring

Technologies are needed to stabilize and contain radionuclide and hazardous buried waste contaminants. The objective of these technologies is to reduce and/or eliminate the potential for contaminants to migrate from the buried waste matrix. At the INEL Subsurface Disposal Area, the primary release mechanism for buried waste contaminants is leaching from the waste matrix through soils and fractured basalt to groundwater. Secondary release mechanisms include airborne dispersal at the ground surface and vegetative uptakes.

Buried waste technology needs focus on containing and stabilizing contaminants within the buried waste streams.

Stabilization and containment technologies must:

1) Demonstrate an effective reduction or elimination of contaminant releases through surface dispersal and/or leaching.
2) Reliably deliver, mix and disperse stabilization agents (grouts, etc.), including delivery of grout, into the waste matrix with sufficient energy to rupture containers and mix with waste forms.
3) Serve as a barrier for inhibiting contaminant migration and be resistant to waste contaminants and potential subsidence.
4) Not preclude subsequent treatment or waste retrieval.
5) Be verifiable to ensure the buried waste of concern is adequately stabilized and contained.
6) Minimize generation of secondary waste (e.g., during capping, jet grouting, and well plugging).

In order for containment and stabilization technologies to be considered a long-term remedy for buried waste, adequate capability to predict and monitor performance of the contained buried waste is necessary. Therefore, technology is needed to:

1) Provide long-term monitoring of contained and/or stabilized buried waste contaminants through innovative intrusive and/or non-intrusive techniques.
2) Verify that the buried waste source term of concern is adequately stabilized and contained.
3) Improve predictive capabilities to evaluate long-term performance of various containment and stabilization technologies for multiple waste types; includes increased understanding of the grout/waste chemistry, long-term grout stability in the presence of wastes and surface and groundwater, sensitivity of grout performance, and development of new grouting agents.

10.3.4 In Situ Vitrification (ISV)

Research and development data indicate several potential benefits of ISV for remediation of buried transuranic waste. Major benefits of the technology include the following:

1) **Incorporation of inorganics into a durable product:** Test results to date indicate that the durabilities of the ISV products greatly exceed the durability of a typical high-level waste form. Both radionuclides and hazardous metals are dissolved into the molten mass and incorporated into the waste form. (Volatile materials are captured in the off-gas processing system.) Data indicate that the durability of the ISV product is comparable to obsidian or granite, natural analogs which have been observed to be durable for geologic time

periods. Although some work remains for assessing durabilities of specific product phases resulting from processing of buried waste with high metal content, the current data indicate that the ISV product has low enough leachability characteristics to support leaving the processed product in place. However, the option to remove the vitrified product for final disposal elsewhere is viable. If removal is required, the removal of a vitrified product will likely be significantly safer than removal of unprocessed waste.

2) **Destruction of organics by high temperature:** The high temperatures obtained during processing (1500-2000°C) are sufficient to destroy volatile organic materials. This eliminates concerns associated with these particular components of the waste and can result in a change in classification of the waste from mixed to low-level waste (LLW) or TRU. As indicated below, additional research is necessary to fully evaluate the transport of volatile organics in the vicinity of a melt.

3) **Volume reduction of waste:** Vitrification of soils results in a volume reduction of approximately 30-40% due to the densification of the vitrified product relative to normal soil. For cases of buried waste, with more void space in the waste region, the expected volume reduction is greater. Test data indicate volume reductions up to approximately 75%.

4) **Ability to process heterogeneous wastes:** Test results to date are encouraging in showing the ability of ISV to handle a mixture of waste types. Although the exact limits of processing ability are not yet defined, the feasibility field testing for buried waste processing shows successful ability to process high combustible (2.5 wt%) and high metal (11 wt%) content waste in both randomly disposed and stacked configurations. The ability to handle heterogeneous waste reduces the complications and costs associated with waste stream separation into fractions for specific treatments. Such separations may be necessary for ex situ treatment alternatives.

5) **Life cycle cost savings:** In 1991 a preliminary system study was performed to evaluate various options for the treatment of buried waste at the INEL. This evaluation included implementability, effectiveness, and cost. The cost data indicate that significant life cycle cost savings may result from remediation using either ISV or ISV/retrieve options.

The ISV technology has been under development since 1980. Initial development efforts at Pacific Northwest Laboratory (PNL) were directed at contaminated soil applications. The technology is ready for deployment on contaminated soil applications, and it has been transferred to a licensee (Geosafe) for commercial applications. It is discussed further in chapter 6.

Two in situ vitrification (ISV) field tests were conducted at the Idaho National Engineering Laboratory (INEL) during the summer of 1990 to assess ISV suitability for long-term stabilization of buried waste that contains transuranic and other radionuclide contaminants. The ISV process uses electrical resistance heating to melt buried waste and soil in place, which upon cool down and resolidification, fixes the waste into a vitrified (glass-like) form. Methods of controlling off-gassing during ISV would be expected to improve the overall retention of such heavy oxide contaminants during melting/vitrification of buried waste.

10.4 DISPOSAL AND/OR INTERIM STORAGE

Following retrieval and treatment of buried wastes, the resulting waste forms will be packaged for either interim storage followed by disposal, or immediate disposal. Interim storage durations may range from several days to years, pending the availability of a disposal facility. The waste form packaging (drums, large boxes, etc.) and type (vitreous, grouted, untreated) may also vary, depending on the selected retrieval and treatment technologies. Prior to interim storage or disposal, waste forms must be characterized to ensure that disposal and storage regulations and acceptance criteria are met.

Currently, no disposal site exists for buried transuranic-contaminated wastes. The DOE Waste Isolation Pilot Plant (WIPP) is presently undergoing testing and evaluation. The DOE seeks innovative disposal alternative for transuranic-contaminated retrieved buried waste. Primary technology needs include:

1) Investigative innovative disposal concepts for transuranic-contaminated buried wastes, including reclassified TRU waste (e.g., TRU transmutation to short-lived or stable isotopes).
2) Develop proposed specific waste acceptance criteria to define buried waste treatment and packaging requirements.
3) Develop technologies that minimize the volume of waste requiring disposal (waste segregation, treatment, packaging, certification).

11

Immobilization

The generally accepted strategy for disposal of nuclear wastes requires that they be incorporated into a stable solid host material of high integrity for retaining the radionuclides under the conditions of disposal. Such host materials or matrices are designated "nuclear waste forms." Generic examples of waste forms are glasses, crystalline ceramics, or cementitious materials.

The inherent differences between high-level waste (HLW) and low-level waste (LLW) also dictate differences in the suitability of waste forms for final disposal. HLW is characterized by relatively large amounts of radioactivity in a relatively small volume; suitable waste forms must have superior stability and integrity, and are associated with high cost of manufacture. On the other hand, LLW contains small amounts of radioactivity in a large volume; waste forms for LLW require less stability and integrity than those for HLW, and can be produced at lower cost. Transuranic (TRU) waste is a specialized form of LLW that contains significant quantities of long-lived actinides and, thus, may require a specialized waste form for final disposal.

Cementation by grout is the current baseline treatment technology for low-level waste. Vitrification into a glass form is the baseline for stabilization of high-level, concentrated waste streams.

DOE defense HLW is generated and stored at 3 DOE sites: Hanford, Savannah River, and Idaho. Both the Hanford and Savannah River sites have selected borosilicate glass as the final disposal form for HLW generated at those sites. A Defense Waste Processing Facility has been constructed at the Savannah River site for vitrification of HLW. A similar facility, the Hanford Waste Vitrification Plant (HWVP), is being designed for eventual construction at the Hanford site.

Defense HLW from the Idaho Chemical Processing Plant (ICPP) is routinely calcined as it is generated; the dry powdered calcine is stored pending final disposal. The final selection of waste form type for ICPP calcine has not been made. Glass, ceramic, and glass-ceramic are all reasonable candidates.

TRU wastes are not typically incorporated into specific waste form matrices. Generally, residues remaining after volume-reduction operations such as compaction, shredding, or incineration of defense TRU wastes will be stored in 208-L (55 gallon) drums for eventual transportation to the Waste Isolation Pilot Plant, a salt geologic repository near Carlsbad, New Mexico.

Future pretreatment of DOE wastes stored at the Hanford, Savannah River, and Idaho sites may generate concentrated TRU element fractions. New forms may be required for final disposal of such concentrated TRU wastes.

Both low- and high-level wastes can be treated to immobilize or solidify the radioactive materials for subsequent storage and final disposal. Low-level waste, either "as generated" or reduced in volume, can be solidified in a variety of ways: mixing it with asphalt, bitumen, concrete, polymers, or dry salt; incorporating it into a ceramic; or vitrifying it into glass.

High-level waste from past commercial reprocessing operations, defense programs, and any future reprocessing must be solidified before it can be transported to a geologic repository. The final waste form must meet a number of different requirements at various stages of the waste disposal process, including processing that is safe and practical at acceptable cost and unaffected by small variations in waste composition and process conditions; a final form that can withstand handling, short-term corrosion, and coolant loss or sabotage without dispersing its contents; and a final form that can resist transportation accidents, such as impacts and fires. In addition, the final form must meet requirements for emplacement in a repository; the requirements include structural integrity, resistance to surface contamination and fire, dimensions, weight, retrievability, low leachability under both static and flowing water conditions, compatibility with the host rock, and resistance to dispersal after accidents or deliberate intrusion.

To date, borosilicate glass has been the most studied waste form; alternative forms are also being evaluated. Waste can be fired to form a mixture of oxides (calcine) at 300-700°C. Waste can be solidified by mixing it with clay to absorb water; the mixture can also be fired to form a ceramic. Waste can be mixed with concrete; the mixture can be hot-pressed to eliminate excess water. Calcine can be agglomerated with additives to reduce water solubility, or treated to form supercalcines, which are highly stable, leach-resistant, silicate minerals. Titanate and zirconate minerals similar to natural minerals are known to have been stable in a wide range of geologic and geochemical environments for billions of years. Vitrified wastes can be converted to a more stable crystalline form (partial devitrification); high-temperature glasses are also being studied. Pellets of glass, supercalcine, or other waste forms can be incorporated into a metal binder (matrix); a similar alterative is to form small waste particles in situ in the metal matrix (this is known as cermet). Waste can also be coated with carbon, aluminum oxide, or other impervious materials before encapsulation in metal to form multiple barriers.

Although many of the final forms for immobilizing high-level wastes may prove suitable for TRU wastes as well, the large volume of these wastes and their present containment status will probably necessitate further treatment. The technology used will vary with the type of waste. Low-density materials can be compacted, and large metal objects can usually be cut into pieces. Better volume reduction for many types of wastes may be obtained with some form of incineration. Sophisticated incineration processes usually produce an ash, fine powder, or sludge. Decontamination of metallic surfaces can also reduce waste quantities. Immobilizing ashes, residues, and sludges would be the final treatment step.

Borosilicate glasses have received more research attention than other HLW form types. In the United States, the DOE office of Civilian Radioactive Waste Management is conducting research related to the repository-site-specific behavior of this waste form. Nevertheless, there is still a need to conduct basic research related to: (1) the fundamental mechanisms for aqueous corrosion of glass; (2) the effects and mechanisms of vapor-phase (steam) alteration of glass; (3) the relationships between glass composition, solution composition, and aqueous dissolution kinetics for borosilicate waste glasses; (4) the identification of suitable glass waste forms for the calcined HLW at the ICPP; (5) thermodynamic data on solid and aqueous species produced by the aqueous corrosion of borosilicate glasses.

As an example of difficulties associated with immobilization techniques was the clean up of solar ponds used to store and evaporate low-level radioactive and hazardous liquid waste at Rocky Flats. The contractor improperly mixed the pond sludge with cement in making large concrete blocks for disposal, causing thousands of blocks to subsequently crumble and crack. In addition, the contractor packaged the blocks in fiberwall boxes that were unsuitable for long-term storage when exposed to the weather and deteriorated. These problems have contributed to a more than $100 million cost growth and a more than 1 year schedule delay.

11.1 CEMENT SOLIDIFICATION

Cement is the generic term used for inorganic materials that are used to bind together sand, stones, or other materials in order to make an artificial rock-like material (free-standing monolith). Concrete consists of larger aggregates with or without fine materials, bound together by cement. This section discusses the use of cement for low-level wastes.

Cement solidifies liquid radioactive waste by both chemical reaction (hydration) and physical encapsulation of the waste. It is the hydration reaction that causes cement to harden into a free-standing monolith. As cement cures, free water in the cement mixture is chemically bound until essentially all the water is incorporated into the hardened matrix. Three general types of cement can be used to solidify LLW: Portland, gypsum, and masonry cements.

Cement is an alkaline medium and is highly sensitive to the pH of the final mixture. Cement mixtures will not cure if the pH is too low. Although cement itself is quite effective in raising the pH of most wastes, its capability do to do so is limited, particularly with highly acidic wastes. Additives, such as lime, are often used to raise the pH of the waste prior to mixing with cement. Typical power plant radioactive waste streams with low pH include boric acid wastes (PWRs) and carbonic wastes. Untreated detergent wastes, oils, and other organic liquid can also be difficult to solidify with cement because they tend to coat the cement particles and prevent them from interacting with water required for the hydration process. They can be solidified, however, with a gypsum cement and emulsifier.

Cement has been successfully used to solidify most of the waste streams generated from nuclear power plants, and can be used to solidify most of the liquid wastes generated by industry and institutions. There are numerous commercial cement solidification systems available on the market. Some of these systems have been designed to be permanently installed as part of the radioactive waste processing system at nuclear power plants, while others are mobile systems that provide services on a contract basis. These mobile systems are either skid mounted or truck mounted for transport to designated locations.

In general, there are 2 types of mixing processes: in-container mixing and in-line mixing. In-container mixing processes involve mixing the wastes and solidification agent inside the disposable containers. In-line mixing involves mixing of the solidification agent and wastes before transferring the mixture into individual containers for disposal. Specialty equipment is also available that can process wet solid wastes to remove any moisture from the waste stream prior to mixing with cement for solidification.

Apart from solidifying low-level waste streams, cement also has been a potential candidate agent for the solidification of mixed wastes (LLW with a hazardous waste component) and hazardous wastes. However, application to mixed waste types must be evaluated on a case by case basis, depending on specific waste characteristics.

Additionally, while cement is suitable for immobilizing contaminated metal scraps and certain hazardous compounds, it is incompatible with a number of metallic salts and organic materials. These areas of incompatibility can be improved through further laboratory research and development efforts.

Generally, the waste is incorporated into a matrix from which the leaching of radionuclides can be expected to be negligible (under natural conditions) from either storage or disposal operations. Cementation processes have been widely used in the U.S. and abroad. Portland cements are the most commonly used matrix, but the use of high alumina cements, as well as pozzolanic cements is also becoming more widespread. Cements can also be used for immobilizing sludge and miscellaneous solid waste, and for embedding spent ion-exchange resins, decladding hulls, and contaminated hardware.

Data suggest that silicates used with lime, cement, or other settling agents can stabilize a wider range of materials than cement-based technologies, including oily sludges and sludges and soils contaminated with solvents. Several vendors use organophilic proprietary compounds as additives to bind organics to the solid matrix. Both the cement-based and pozzolanic-based methods have been applied to radioactive wastes as well. The presence of solid organics such as plastics, resins, and tars often increases the durability of the solid end product.

An example is the Cement Solidification System (CSS) at the West Valley Demonstration Project (WVDP), West Valley, New York. The CSS, designed to operate within an existing process cell, automatically and remotely solidifies low-level nuclear waste by mixing it with Portland Type 1 cement. The qualified waste form mixture is placed into square, 270 liter (71 gallon) metal drums. The drums have an integral polyethylene liner to protect the carbon-steel material from potential corrosion. The CSS produces drums at a continuous operation rate of 4 drums per hour. All system processing data is monitored by a computerized Data Acquisition System (DAS).

Grouting is a commonly used process for stabilization of waste. Another example is the Hanford Grout Disposal Program. Radioactive liquid wastes generated at Hanford are stored in underground, double-shell tanks (DST). The waste currently stored in the DST has been generated by a variety of operations. Some of the wastes will undergo pretreatment to remove transuranics and key fission products, such as Cs-137 and Sr-90. The resulting non-transuranic, low-activity fraction is processed into grout. The liquid waste is mixed with a preblend of dry materials to produce a grout slurry. The dry materials consist of Portland cement, blast furnace slag, fly ash, and clays.

The slurry then is pumped to the near-surface vaults where it solidifies into a solid grout.

The U.S. Army Engineer Waterways Experiment Station (WES) developed a grout based on portland cement, Class F fly ash, and bentonite clay for the Hanford Grout Vault Program. The purpose of this grout was to fill the void between a waste form containing 106-AN waste and the vault cover blocks. Following a successful grout development program, heat output, volume change, and compressive strength were monitored with time in simulated repository conditions and in full-depth physical models. This research indicated that the cold-cap grout could achieve and maintain adequate volume stability and other required physical properties in the internal environment of a sealed vault.

DOE is facing technical uncertainties with the grout process. When radioactive materials are grouted, heat is produced, and generally speaking, the amount of heat rises with the level of radioactivity. If the temperature rises above 90 degrees centigrade, the grout may not effectively immobilize liquid wastes.

Test results revealed that the estimated temperature of the grout would likely exceed 90 degrees centigrade due to the heat generated from the solidification of grout and the decay of radioactive waste components in the grout. As a result, DOE may have to change the grouting process or process the low-level waste in another pretreatment sequence to remove more radionuclides.

Even if the process works from a technical standpoint, the contents of the low-level waste have raised questions about the appropriateness of using grout vaults as a disposal method. The low-level waste designated for disposal in grout vaults will contain materials that have a high level of radioactivity. These materials include cesium 137, strontium 90, technetium 99, iodine 129, and transuranic waste. On the basis of an October 1990 Westinghouse analysis of the radionuclide content of double-shell tank waste, the grout in each vault could contain about as much radioactivity as would be contained in 8 canisters produced by the high-level waste vitrification plant. Under the current program about 240 grout vaults will be needed. Compared to the total amount of grout in a vault, however, DOE Richland anticipates that the amount of high-activity materials will be small enough that the grout will meet the Nuclear Regulatory Commission's criteria for shallow-ground disposal. In contrast, DOE headquarters told GAO that Hanford's low-level waste may not meet the Nuclear Regulatory Commission's criteria for shallow-ground disposal.

Washington, Oregon, and the Yakima Indian Nation have challenged the adequacy of DOE's disposal plans for this waste. Among their concerns is that the waste DOE considers low-level waste may be high-level waste. Some of the highly mobile radioactive material in this waste will retain its radioactivity past a grout vault's ability to keep the material in place. Technetium 99 and iodine 129 require 230,000 years and 16 million years, respectively, before half of their radioactivity has decayed. These materials are also highly mobile if they enter the ground. The manager of Hanford's grout facility acknowledged to GAO that these radioactive materials will eventually leak into the ground, but he stated that they represent a small fraction of the total radioactive content of the grout vaults. DOE claims the grout vaults will retain the waste for up to 10,000 years, but acknowledges that this is an assumption not based on empirical evidence.

At Savannah River the waste inventory has been concentrated via evaporation to about 120,000 m^3 and is stored in 51 large, underground carbon-steel tanks. This concentrate contains about a billion curies of radioactivity; the principal isotopes being Sr-90 and Cs-137. The concentrate is treated in the waste tanks to remove the soluble cesium and strontium. The resulting decontaminated salt solution is blended with residues from the F-Area/H-Area

Effluent Treatment Facility (ETF) prior to transfer to the Saltstone Processing Area (SPA). The sludge and precipitates will be vitrified into borosilicate glass in the Defense Waste Processing Facility.

In the Saltstone formulation, the salt solution is mixed with a preblend of dry materials in the saltstone mixing unit. The preblended dry materials consist of ground blast furnace slag (grade 120), Class F fly ash, and cement or calcium hydroxide to promote setting of the slag mixture. The resulting mixture is gravity fed into an agitated holding tank. A centrifugal pump then moves the grout slurry to an aboveground disposal vault. The disposal unit is a surface vault with 4 cells constructed of reinforced concrete containing blast furnace slag. After filling with saltstone, the vaults will be decommissioned by mounding with earth and installing a clay cap.

The western New York Nuclear Services Center near West Valley, New York, was the only commercial nuclear fuel reprocessing facility in the United States and operated from 1966 to 1972. In 1980, the U.S. Congress authorized the Department of Energy to implement a nuclear waste management program at the facility. Approximately 2,400 m^3 (650,000 gallons) of high-level waste is stored in an underground, carbon-steel tank, of which about 90% is a supernatant liquid containing 7.4 million curies of predominantly Cs-137.

Liquid from the waste storage tank is passed through zeolite ion exchange columns to remove Cs-137. The decontaminated supernatant is evaporated to raise the concentration of the remaining sodium nitrate/nitrite salts to 37-41% by weight.

On a batch basis, precise amounts of liquid waste, cement, and chemical additives are fed into a high shear mixer. After the mixing cycle, the cement/waste slurry is transferred to square, 269 L (71 gallon) drums. Filled drums are moved to a crimping station where the lids are remotely attached, and the drums are tested for external contamination and surveyed for dose rate. The drums are stored on site, pending the completion of an Environmental Impact Statement.

Brookhaven is developing technology that aims to develop, demonstrate, and implement advanced grouting materials for in situ stabilization of contaminated soils and the placement of impermeable, highly durable subsurface barriers. The developmental effort focuses on cementitious and soil cement mixtures compatible with commercially available placement techniques.

The developed superplasticized grouts and soil cements have significantly superior mechanical, physical and durability properties than those of conventional formulations. The permeabilities are 2 to 5 orders of magnitude less than for other materials frequently used as caps and barriers such as clay, soil-bentonite and cement bentonite slurries. Therefore, the dimensions of the barriers can be reduced significantly.

Fernald is investigating the possibility that 2 types of waste found at the site can be effectively mixed together to form cement, instead of using cement as an additive to stabilize waste. The principal ingredients of cement are ash and lime, both of which are considered waste requiring treatment and safe disposal. A solid can be formed from these 2 types of waste by adding a small amount of cement. Thus, treatment can be achieved using available resources without purchasing additives.

11.2 BITUMEN SOLIDIFICATION

Bitumen (asphalt) has been used in Europe and Canada as a solidification agent for LLW, but only recently it has been used in the United States for solidification and stabilization of radioactive waste. Bitumen systems are considered to be both waste stabilization and volume reduction technologies, as the heat that is required to melt the bitumen assists in evaporating the liquid waste.

Bitumen does not react chemically with the majority of materials comprising low-level radioactive waste. Bitumen solidifies waste materials by entrapment within its structure, isolating the wastes from contact with water and providing structural stability.

The main advantages of stabilization using bitumen are its leach resistant characteristics, low operating cost, and handling ease. On the other hand, bitumen has several disadvantages. One disadvantage is that it does not perform well with certain dehydrated salts, such as sodium sulfate, sodium nitrate, magnesium chloride, and aluminum sulfate. When a dehydrated waste containing these salts is exposed to water, rehydration occurs, which could cause the solidified monolith to deteriorate. Another advantage of bitumen solidification is its high carbon content which may limit its resistance to biodegradation. The issue of biodegradation is still undergoing extensive laboratory testing.

Other operational difficulties with bitumen include the solidification of organic resins. These difficulties can be overcome by clay additives. These additives also adsorb waste oils and organics that would otherwise prevent bitumen from hardening at room temperature. Additionally, clay helps retard flammability of bitumenized wastes. Clay sometimes is also used as an additive to further immobilize radionuclides such as Sr-90 and Cs-137 because of clay's adsorptive properties for these elements. Lastly, any bitumen-processing system used to evaporate liquids must not overlook the potential for generation of volatile organics that may be included in distillates. Potential for generation of volatile organics is minimized with the use of harder forms of bitumen (e.g., oxidized bitumen).

During the solidification process, heat is required to melt bitumen into a viscous form to mix with the waste materials. The potential for fire resulting from vaporization of volatile organics caused by heating during the mixing process has been a major criticism of the use of bitumen as a solidification agent. Additionally, bitumen itself can have a low ignition temperature. Some types of bitumen can be ignited at temperatures as low as 315°C.

There are 5 basic methods for solidifying waste with bitumen. Of the 5 methods, only 2 are used for commercial application in the U.S. Currently, oxidized bitumen is used with the screw-extruder process, and the direct-distilled (non-oxidized) bitumen is used with the thin-film evaporator process. The other methods—the stirred bitumen process, the temporary emulsion process, and the sedimentation process—are either available only in Europe or are in the experimental stage.

11.3 POLYMER ENCAPSULATION

Incorporation of radioactive waste into plastics is a relatively newer technology. The main plastic materials used for this purpose are polyethylene, ureaformaldehyde, polyester, and polystyrene. For dry solid waste (generally the structural parts), a polymer—impregnated cement matrix has also been used as the embedding matrix.

Polymer encapsulation of mixed wastes encloses waste products in thermoplastic or thermosetting materials using commercially-available processing technologies. Two primary polymer processes are being tested.

In 1 process, thermoplastic polymers, such as polyethylene (a commonly-used plastic that is resistant to chemicals and moisture), are combined with dried waste in a commercially-available extruder, which melts the polyethylene and mixes it with the waste. The waste encapsulated in polyethylene is extruded into a drum, where it solidifies upon cooling. The process operates at a low temperature, requires no off-gas treatment, and generates no secondary waste. Since high loadings of waste may be incorporated into the polymer, a substantial reduction in volume may be possible relative to cementation, which has been used to immobilize wastes at the RFP.

A second process, in which bulk materials (i.e., "debris") are suspended in a drum and encapsulated with molten or liquid plastic, is also being investigated. The solidified polymer surrounds the waste and immobilizes hazardous contaminants. The use of recycled polyethylene is being investigated for this application. Thermosetting plastics (resins combined with hardeners, similar to epoxy) have also been evaluated for encapsulating wastes.

Chemical processes associated with polymeric solidification are somewhat complicated. In general, the processes involve mixing liquid monomeric chemicals that react with a catalyst and linking individual molecules to form long-chain hydrocarbon molecules. This process is called polymerization. Sometimes, a promoter is also added to the process, causing the catalyst to decompose and accelerating polymerization. These processes are usually carried out at room temperature and require no additional heat. The wastes themselves do not participate in the chemical reaction of polymerization.

Polymers solidify the liquid waste by entrapping waste elements among complex linkages of the long-chain molecules.

Of the many commercially available solidification agents, polymeric systems were found to achieve among the highest waste loading factors. Vendor test data have shown samples with waste loading factors as high as 60 to 67% by weight that are still able to meet all 6 BTP stability criteria. In contrast to cement, polymeric systems do not require water to solidify. Consequently, these systems often result in significant volume reduction. Polymer-solidified wastes possess compressive strengths of 1,500 to 9,000 psi and exhibit good leach resistant characteristics. Test results presented to the NRC in vendors' topical reports demonstrate that polymer-solidified wastes remain impervious to the effects of radiation, temperature fluctuation, water immersion, and microbial attack.

While polymeric systems possess all the favorable characteristics required to meet the NRC's stability requirements, they are slightly more expensive than other stabilization systems. Preparation procedures also require precise measurement, handling, and mixing of chemical ingredients. Lastly, the potential for explosions, fires, and releases of toxic fumes caused by some of the chemicals used in the process requires serious consideration in the design and operational procedures of a system.

The advantages of using a solidification and encapsulation process involving a thermoplastic material rather than a hydraulic cement derive primarily from the processes by which the 2 binder materials solidify. Thermoplastic materials solidify as they cool, usually in a matter of hours. Furthermore, thermoplastic materials are inert, so they cannot react with waste of any kind. By contrast, hydraulic cement takes days to cure and solidify through a series of hydration and chemical reactions. These reactions increase the chance of chemical interaction between the waste and the cement, which limits the amount of types of waste that can be solidified and can compromise the integrity of the final waste form.

The Brookhaven National Laboratory is very active in polymer encapsulation research and development, and has developed the PERM process. Polyethylene Encapsulation of Radionuclides and Heavy Metals (PERM) is a waste treatment and stabilization technology for high-level mixed waste. Specific targeted contaminants include radionuclides (e.g., cesium, strontium, cobalt), and toxic metals (e.g., chromium, lead, cadmium). A polyethylene encapsulation process was developed several years ago at Brookhaven National Laboratory (BNL) for solidification of low-level radioactive waste (LLW) such as evaporator concentrate salts and ion-exchange resins. Recently, it has been successfully applied for treatment of hazardous and mixed waste streams such as sodium nitrate salts and sludges.

Brookhaven National Laboratory (BNL) has also identified non-thermal stabilization mechanisms for waste streams that cannot be stabilized by vitrification. Polyethylene encapsulation and modified sulfur concrete encapsulation are thermoplastic methods that can stabilize mercury and chloride salts. Polymer impregnated concrete can stabilize tritiated aqueous wastes.

Liquid mercury, chloride salts and tritiated aqueous waste streams are not easily treated by incineration or vitrification methods. DOE facilities at Oak Ridge, Savannah River, Fernald, Hanford and Rocky Flats need improved technologies for these areas.

Research efforts at Rocky Flats Plant (RFP) focus primarily on the development of a polyethylene extrusion process to stabilize low-level nitrate salt waste, which is one of the largest mixed waste streams at RFP. Pilot-scale studies and an integrated system demonstration are planned to obtain the operational data and design criteria necessary to implement a polymer solidification system for this waste stream.

Polyethylene waste forms are superior to grout waste forms. Polyethylene is less dependent on the waste chemistry, accepts a wider range of waste types, and increases waste loading. Finally, polyethylene is easier to process under heterogeneous waste conditions. It can also handle high nitrate loadings, as well as liquid mercury, chloride salts, and tritiated aqueous waste streams.

Dow Chemical has been marketing a vinyl ester-styrene process. General Electric has developed a process based on vinyl toluene and a polyester polymer (AZTECH system). Polyethylene epoxies are also being evaluated.

11.4 SULFUR CEMENT ENCAPSULATION

The advantages of sulfur cement encapsulation over hydraulic cement encapsulation are similar to those of PE encapsulation. Like PE, sulfur cement does not require a chemical reaction to set and attains full strength within hours rather than days. In general, sulfur cement waste forms have much higher waste loadings than those of hydraulic cement waste forms, although these loadings vary with the type of waste being encapsulated. Sulfur cement waste forms have greater compressive and tensile strengths and are highly resistant to corrosion by acids and salt.

An additional advantage of sulfur cement encapsulation is that waste sulfur is in abundant supply from the desulfurization of incinerator flue gas and the clean up of petroleum products. Currently, most of this supply, which is expected to increase to 30 million tons per year by 2000, is disposed of as waste. Therefore, sulfur cement encapsulation essentially used one type of waste to encapsulate another.

Sulfur Polymer Cement (SPC) is an encapsulating waste-immobilization material. The wastes are encapsulated in the sulfur matrix with the exception of a few sulfide-forming metals. SPC has a high mechanical strength in a short period of time, high resistance to many corrosive environments, and low porosity. One restriction of SPC is that the prospective waste must contain less than 1% water. The promising characteristics of strength and resistance to corrosion, along with ability of the material to meet the criteria for radiation stability, compressive strength, and the EPA leachability tests, make this a promising final waste form.

A draft study states that an improved approach to grout would be to vitrify Hanford low-level waste, mix the vitrified product with sulfur polymer cement, and pump the mixture into large near-surface vaults like the 5 DOE has already constructed. The vitrified waste would be more effectively immobilized for an indefinite period of time and would be retrievable for processing sometime in the future if a better technology were developed.

11.5 POZZOLANIC MATERIALS

In this solidification and stabilization process by Advanced Remediation Mixing, Inc., pozzolanic materials react with polyvalent metal ions and other waste components to produce a chemically and physically stable solid material. Optional accelerators and precipitators may include soluble silicates, carbonates, phosphates, and borates. The end product may by similar to a clay-like soil, depending upon the characteristics of the raw waste and the properties desired in the end product. When combined with specialized binders and additives, this process can stabilize low-level nuclear wastes.

Typically, the waste is first blended in a reaction vessel with pozzolanic materials that contain calcium hydroxide. This blend is then dispersed throughout an aqueous phase. The reagents react with one another and with toxic metal ions, forming both anionic and cationic metal complexes. Pozzolanic accelerators and metal precipitating agents can be added before or after the dry binder is initially mixed with the waste. When a water soluble silicate reacts with the waste and the pozzolanic binder system, colloidal silicate gel strengths are increased within the binder-waste matrix helping polyvalent metal cations.

A large percentage of the heavy metals become part of the calcium silicate and aluminate colloidal structures formed by the pozzolans and calcium hydroxide. Some of the metals, such as lead, adsorb to the surface of the pozzolanic structures. The entire pozzolanic matrix, when physically cured, decreases toxic metal mobility by reducing the incursion of leaching liquids into and out of the stabilized matrices. With modifications, the system may be applied to wastes containing between 10 to 100 percent solids.

11.6 VITRIFICATION

This section discusses ex situ vitrification in general. Specific examples of both ex situ and in situ vitrification are discussed throughout this book.

Vitrification, the process of converting materials into a glass or glass-like substance, is increasingly being considered for treating various wastes. Vitrification is conceptually attractive because of the potential durability of the product and the flexibility of the process in treating a wide variety of waste streams and contaminants. These characteristics make vitrification the focal point of treatment systems for high-level radioactive waste (HLW) around the world.

Vitrification is the process of converting materials into a glass or glassy substance, typically through a thermal process. Although heat is not necessarily required for vitrification (for example, vapor deposition, solution hydrolysis, and gel formation can also form glassy materials), this section considers only vitrification processes which use heat.

When accomplished through a thermal process, vitrification may also destroy hazardous organic contaminants.

Vitrification was selected as the immobilization technology best suited to the majority of DOE high-level waste: the process equipment performs well in remote operation and the borosilicate glass product tolerates considerable variation in waste composition. Vitrification has also been approved by EPA as the best demonstrated available technology for disposal of this waste under RCRA.

Currently the preferred waste form for permanent disposal of HLW is vitrified borosilicate glass inside a stainless steel canister. Approximately 25 to 30 volume percent of the vitrified glass is HLW and the balance is glass frit, unless the HLW contains aluminum, phosphorous, or chromium.

Vitrification plants are planned, or under construction, at the Savannah River Site and Hanford Site. The Defense Waste Processing Facility (DWPF) at Savannah is completed, but not yet operational because of extensive retrofitting required.

The West Valley Demonstration Project (WVDP), a joint program by DOE and the New York State Energy Research and Development Authority, will vitrify the high-level waste now stored at that site. Studies to identify a suitable treatment process and waste form for calcine and liquid wastes at INEL continue. Cumulative production will be about 21,000 canisters, each containing on average approximately 2 metric tons of waste form (volume of approximately 1 cubic meter).

The vitrification process selected for the DOE high-level waste treatment facilities at the Hanford Site, (HWVP), Savannah River Site (DWPF), and West Valley Demonstration Project (WVDP) incorporates high-level waste into a borosilicate glass matrix, thus reducing the mobility of radioactive and other hazardous constituents. Waste and borosilicate glass-forming materials will be fed continuously as a slurry into a glass melter and heated to temperatures above 1,000°C. After becoming molten and homogeneous, the melt will be poured into stainless steel canisters. Sealed canisters will be cleaned and stored at each site pending transfer to a Federal repository for disposal.

The treatment schemes at these HLW treatment facilities vary somewhat because of the difference in waste composition and the quantities of HLW to be treated. At all 3 facilities, HLW will be pretreated and partitioned into low- and high-level radioactive fractions to reduce the amount of waste requiring immobilization in glass.

In the DWPF scheme, the current inventory of HLW, 34 million gallons of sludge and salt solution/cake, is pretreated by in-tank processing. The salt portion is decontaminated by precipitation and sorption for disposal as low-level waste (LLW) in concrete. The sludge is washed with water/caustic to remove soluble non-radioactive components. The salt precipitates, sorption products, and washed sludge are then mixed with glass frit and other additives and subsequently delivered to the melter. They will be solidified in stainless steel canisters. Plans are to accumulate previtrified waste in 2,500 m^3 batches, each of which is expected to supply DWPF for 2 to 3 years of operation.

Two types of HLW, approximately 660,000 gallons in total, were initially stored at WVDP. Supernatant from one of the waste types is presently being treated by ion exchange. The sludge, as in the DWPF scenario, will be washed to remove the non-radioactive soluble components. Both the decontaminated supernatant and the sludge wash waters will subsequently be disposed of as LLW. The washed sludge, spent ion exchange media, and other waste type will be combined into 1 batch, mixed with glass-forming chemicals, and subsequently fed to the melter.

Four types of HLW stored in the double-shell tanks at Hanford, approximately 6.7 million gallons in total, will be treated at HWVP. Some or all of the waste in the single-shell tanks may also be treated at HWVP. Depending on their characteristics, these wastes will undergo pretreatment by sludge washing, cesium separation, solids dissolution, transuranic separation, and/or organic destruction. Once pretreated, these wastes will be blended with frit and additives and subsequently transferred to the melter. The number of volumes of batches to be processed at HWVP are as yet undetermined.

Vitrification operations at WVDP are projected to last 30 months and produce approximately 300 canisters of vitrified HLW. HWVP will operate for about 10 years and produce an estimated 1,780 canisters. If single-shell tank waste is treated, it will operate longer. These facilities will be subsequently decontaminated and decommissioned. The current inventory of HLW at DWPF will produce about 5,600 canisters in operations lasting 17 years. DWPF will remain on-line to treat any newly generated HLW.

The borosilicate glass form was selected as the end product at the 3 facilities because of its ability to accommodate variations in waste composition while maintaining acceptable processability. Borosilicate glass was one of the many candidate waste forms considered by DOE for immobilizing radioactive constituents in HLW. The waste form ranked highest in several studies when collectively considering parameters such as its ability to handle fluctuations in waste composition, leach resistance, waste loading, mechanical strength, radiation stability, thermal stability, and overall processability. Remote operation of process equipment to produce borosilicate glass is a well developed technology.

Waste streams must be treated prior to vitrification to ensure efficient downstream processing. Primary separations are (1) removal of suspended and dissolved solids from aqueous organic streams; (2) separation of water from organic liquids; (3) treatment of wet and dry solids; (4) mercury removal and control; (5) decontamination of waste classified as debris. Potential problem areas include processing chlorides, nitrates, high sulfur, phosphorus, and chromium-bearing salts.

Unfortunately some anionic species which are present in the nuclear waste streams have only limited solubility in oxide glasses. This can result in either vitrification concerns or it can affect the integrity of the final vitrified waste form. The presence of immiscible salts dan also corrode metals and refractories in the vitrification unit as well as degrade components in the off-gas system. The presence of a molten salt layer on the melt may alter the batch melting rate and increase operational safety concerns. These safety concerns relate to the interaction of the molten salt and the melter cooling fluids.

To date, borosilicate glass has been the most studied waste form; alternative forms are also being evaluated. Waste can be fired to form a mixture of oxides (calcine) at 300-700°C. Waste can be solidified by mixing it with clay to absorb water; the mixture can also be fired to form a ceramic. Waste can be mixed with concrete; the mixture can be hot-pressed to eliminate excess water. Calcine can be agglomerated with additives to reduce water solubility, or treated to form supercalcines, which are highly stable, leach-resistant, silicate minerals. Titanate and zirconate minerals similar to natural minerals are known to have been stable in a wide range of geologic and geochemical environments for billions of years. Vitrified wastes can be converted to a more stable crystalline form (partial devitrification); high-temperature glasses are also being studied. Pellets of glass, supercalcine, or other waste forms can be incorporated into a metal binder (matrix); a similar alternative is to form small waste particles in situ in the metal matrix (this is known as cermet). Waste can also be coated with carbon, aluminum oxide, or other impervious materials before encapsulation in metal to form multiple barriers.

It may become necessary to collect and dispose of krypton-85 and iodine-129. Various techniques are being investigated for their collection, immobilization, and disposal. Krypton-85, with a half-life of less than 5 h, can be stored until its radioactivity has decayed.

Although many of the final forms for immobilizing high-level wastes may prove suitable for TRU wastes as well, the large volume of these wastes and their present containment status will probably necessitate further treatment. The technology used will vary with the type of waste. Low-density materials can be compacted, and large metal objects can usually be cut into pieces. Better volume reduction for many types of wastes may be obtained with some form of incineration. Sophisticated incineration processes usually produce an ash, fine powder, or sludge. Decontamination of metallic surfaces can also reduce waste quantities. Immobilizing ashes, residues, and sludges would be the final treatment step.

Although vitrification has been considered the "Best Developed Available Technology" for high-level waste, it is also being seriously considered for low-level waste and mixed waste.

It is important that the host medium, glass, can incorporate significant quantities of waste and possess a structure with the ability to immobilize not only the 40 or more different elements that characterize waste chemistry, but also variations in composition. Glass possess a random network structure which is relatively open and "forgiving" of elements to be incorporated. All components, including added glass formers as well as waste constituents already present, can play 1 or 3 basic roles in the glass structure; network formers, intermediates, or modifiers. Glass forming components are added to the system to optimize durability or processing conditions. A ratio of about 70% glass formers to 30% waste components forms a very good waste glass product. An important point to note is that when solidified, the various elements of the waste become an integral part of the glass structure via primary and/or secondary forces.

If mixed waste or some portion of it is destined for HLW disposal, the presence of certain species in the waste stream may inhibit the formation of a durable glass or may generate a toxic, difficult to handle, off-gas stream. Potential problem species include:

1) *Sulfates and Halogens:* Substantial amounts of sulfates can result in liquid phase segregation in the melter, leading to increased volatility of alkali, alkali earth, and radionuclide elements. The molten sulfate salt liquor is corrosive, and can lead to rapid and severe attack of stainless steel and high nickel alloys. Halogens, and in particular chlorides, are volatile, tend to be corrosive, and can produce volatile metal from being contained in the glass. There is a need to remove sulfur and halogens from the incoming waste stream.
2) *Volatile Metals:* R&D on high-temperature metal processing in glass melters indicates that glass may not be able to capture a high percentage of the metal contaminants during waste form processing. The percentage of metal that remains in the glass is dependent on factors such the melter temperature, chloride content, carbon content, metal volatility, presence of a cold

cap, and excess air. However, most of the heavy or toxic metals exited with the off-gas or remained in the melter head space deposits. There is a need to remove readily volatized metals, both radioactive and RCRA regulated, and metals that form volatile oxides from the waste stream. These include Cs, Pb, Hg, Rb, Bi, Cd, Ag, and As.

Separation of radioactive materials from the toxic materials could allow each portion of the waste to be handled under only one set of regulations, thus reducing handling and disposal costs. For example, removal of regulated organics prior to a thermal treatment system may allow the mixed waste stream to be treated as a low-level waste if the organics were the only RCRA regulated materials.

The majority of stored mixed wastes at DOE facilities are solid, but substantial volumes of liquid mixed wastes also exist. For some solids, innovative methods such as selective leaching may be considered for certain wastes. It is important to minimize the secondary wastes produced, to recycle as much of the process reagents as possible, and to concentrate the recovered contaminants for disposal or reuse. More standard adsorption and extraction methods may be adapted for separating components dissolved in waste solutions.

The multibarrier isolation system in the underground repository is designed to prevent groundwater from contacting waste forms for hundreds to thousands of years. During the first 300 years, most of the activity and hazard is due to radionuclides such as Sr-90 and Cs-137, each with relatively short 30 year) half lives. For times greater than 300 years, the activity of the waste is dominated by several long-lived fission products, actinides and daughter products of actinides. As a result of radioactive decay, the activity of SRP waste glass will decrease by a factor of about 10,000 in 1,000 years, and for very long time periods, the total radioactivity of the HLW will actually be less than the naturally occurring uranium from which the waste was derived. Even under the premise that groundwater breaches all containment and leaches the waste glass, glass leaching is very slow and the transport of radionuclides in a repository is solubility limited. Most of the radionuclides, especially the long-lived constituents, form insoluble compounds subject to sorption and precipitation on waste package components and surrounding geology so even if leached, they cannot easily be transported.

11.6.1 Minimum Additive Waste Stabilization (MAWS)

The minimum additive wastes stabilization (MAWS) technology is being demonstrated at the Department of Energy (DOE) Fernald Environmental Management Project (FEMP) in Fernald, Ohio. The MAWS system integrates into 1 single process 3 primary treatment technologies that are usually employed separately for site remediation: vitrification, soil washing, and ion-exchange wastewater treatment. The MAWS system is centered on vitrification and incorporates all primary and secondary waste streams into a final, stabilized glass waste form. The integrated system is innovative because the waste streams are viewed as additive resources to vitrification, and a "portfolio" approach is adopted to maximize the economic benefit.

Vitrification has been used in the DOE Complex for treatment of low-volume, high-level radioactive wastes with low waste loading and high treatment cost. However, the economic attractiveness for treating large volumes of low-level waste/mixed waste (by maximizing waste loading) and the feasibility of production-scale processing have yet to be demonstrated. This program will demonstrate both the economics of total life cycle cost savings, through increased waste loading and final waste volume reduction, and the production-scale feasibility of vitrification of large volumes of low-level waste/mixed waste through a synergistic approach. In addition, it will demonstrate the capability of producing a leach-resistant (long-term) waste form, clean soil and water for placement back into the environment, and an off-gas effluent that meets regulatory requirements.

Individual component technologies may include:

1) Vitrification;
2) Thermal destruction;
3) Soil washing;
4) Gas scrubbing/filtration; and
5) Ion-exchange wastewater treatment.

The glassy slags are composed of various metal oxide crystalline phases embedded in an alumino-silicate glass phase. The slags are appropriate final waste forms for waste streams that contain large amounts of scrap metals and elements with low solubilities in glass, and that have low-flux contents. Homogeneous glass waste forms are appropriate for wastes with sufficient fluxes and low metal contents. Therefore, utilization of both glass and glassy slag waste forms will make vitrification technology applicable to the treatment of a much larger range of radioactive and mixed wastes. The MAWS approach was applied to glassy slags by blending multiple waste streams to produce the final waste form, minimizing overall waste form volume and reducing costs. The crystalline oxide phases formed in the glassy slags can be specially formulated so that they are very durable and contain hazardous and radioactive elements in their lattice structures.

11.6.2 Heating/Melting

In **joule heating**, an electric current flows through the material. As the material internally resists the current, the current loses power and transfers heat energy to the material. The dissipated power is predicted by Joule's Law. Thus, with increased electrical resistance, if current can be maintained, additional power is dissipated and the material heats more rapidly. However, unless the voltage is increased, an increase in resistance will also decrease current.

Joule process heating furnaces for the treatment of hazardous wastes evolved directly from glass melters in the glass industry. The electric furnace/melter category includes processes that use a ceramic-lined, steel-shelled melter to contain the molten glass and waste materials to be melted.

Some melters are much like electric glass furnaces used to manufacture glass products (e.g., bottles, plates). Such melters receive waste materials and glass batch chemicals directly on the surface of a molten glass bath. Most melting occurs at the waste/molten glass interface as heat is transferred from the molten glass. As waste is heated, volatiles may be released and organics are either pyrolyzed (in an oxygen-poor environment), or oxidized (in an oxygen-rick environment). Off-gas treatment is required to minimize air emissions.

The joule heated continuous ceramic melter features continuous vitrification of mixture of glass frit and HLWW concentrate or calcine in refractory-walled glass melter at 1,100/1,200°C. HLW glass is drained into storage container.

Advantages include: efficient heating technique; high production rate; flexibility in HLLW composition; low off-gas effluent; and a relatively simple process. Disadvantages include: refractory sludge buildup; relatively complex equipment; and corrosion of containment refractories.

A related process, glass marbles with metal coatings features production of HLLW glass in Joule heated glass melter, marble formation in marble-making device from molten glass, and coating of marbles by plasma spraying of selected metals (e.g., lead, aluminum, etc.).

Advantages include more homogeneous and leach-resistant than monolithic HLW glass form, and highly impact resistant. Disadvantages include complex process, requiring a larger number of canisters than that of monolithic process.

A suitable treatment technology for Mixed Low Level Waste (MLLW) does not currently exist, although vitrification to glass form appears promising. Converting waste to glass has been successfully applied to high-level radioactive wastes, and has been utilized since the late 1960s. The joule heater will be tested on Rocky Flats mixed waste.

Process variables such as joule heater temperatures, glass compositions, glass rheology, and feed rates will also be examined after glass compositions are made from Rocky Flats Sludge. An analysis of the testing should yield process data on melting rates, organics destruction efficiencies, combustion rates, and off-gas aerosol emissions. The overall culmination of this effort is to provide the necessary data for design and permitting of a full-scale MLLW treatment facility.

Joint participants in this vitrification effort to treat Rocky Flats waste include PNL, and the Savannah River Site, where glass development will take place.

Multi-fuel Glass Melter: Vortec has developed a multi-fuel glass melter with application to hazardous wastes. The Cyclone Melting System (CMS) is composed of 3 primary components: a multi-fuel-capable batch preheater, a cyclone melter, and a glass melter reservoir. Preheated combustion air, pulverized coal, and glass-forming ingredients enter the preheater from the top. The batch rapidly preheats in suspension by radiative and convective heat transfer. The preheater is designed to burn pulverized coal or a variety of gaseous, liquid, and coal-slurry fuels. The preheated batch ingredients are separated against the walls of the cyclone melter by centrifugal forces. The liquid phase reactions occur along the walls, and the melted glass and combustion gases exit the melter to the melt reservoir. The melt reservoir gives material more time to form a glass, and is designed to hold an adequate supply of glass for level control or temperature conditioning. The melted glass may then be delivered to a glass forming process, or other glass conditioning device, for integration with a glass manufacturing process. The combustion gases exit the melt reservoir to a high-temperature recuperator where waste heat is recovered and recycled to the preheater. Off-gas contaminants may also be recycled to the preheater to increase process destruction efficiencies (DEs).

Another project, the Vitrification Technology Plan for DOE waste was completed. Preliminary glasses for Rocky Flats process sludge, Los Alamos National Laboratory (LANL) process sludge, LANL ash, Oak Ridge Y-12 sludge, and TSCA bottom ash, and small-scale melters tests on surrogate Rocky Flats process sludge and LANL sludge were completed. A 28 day MCC-1 leaching of Rocky Flats waste glass was completed at MIT. Melter testing at Clemson University will be performed on Oak Ridge Y-12 glass compositions. Joint participants include Clemson University and Massachusetts Institute of Technology.

Another project is a collaboration between Clemson University, the Westinghouse Savannah River Company, and private companies that are providing equipment and technical support. EN VitCo, Inc. has a high-waste loading, transportable melter, and StirMelter, Inc. has a high-rate, low-cost melter. Rust Remedial Services (Chem Waste Management, Inc.) is providing chemical analysis services, waste form characterization, and engineering support. The Savannah River Site is contributing its expertise on how the glass should be made. This project is being supported by MWIP, Savannah River Site's High-Level Waste Program, WeDID, and the Waste Management and Environmental Restoration International Programs Office. Bench scale vitrification of low-level waste has been conducted at Fernald.

Another project at Savannah involves vitrification of waste sludge into glass. The M-Area operations at the Savannah River Site (SRS) in Aiken, South

Carolina, produced reactor components for nuclear weapons materials for the U.S. Department of Energy. The resulting waste is currently being stored in 9 tanks. The total volume in storage was initially (approximately) 1,200,000 gallons of which (approximately) 1/3 is a gelatinous hydroxide sludge. Vitrification of the sludge into glass is an attractive option because it reduces the waste volume by (approximately) 85% and reduces final disposal volume by 96% compared to alternative stabilization technologies.

A new process is called the **Hybrid Treatment Process (HTP)**, so named because it is built on the 20 years of experience with vitrification of wastes in melters, and the 12 years of experience with treatment of wastes by the in situ vitrification (ISV) process. The process will be tested at INEL.

The process as currently envisioned begins at the disposal site, where a hole is excavated. The containment vessel for the process, a mild steel metal tank, is brought onto the site and assembled so that air flows around all its sides to keep it from melting during the high-temperature processing period. This batch-type processing is similar to that used with the in-can melting technology.

The major advantages of the HTP concept include very low relative costs, short development and deployment schedule, flexibility in waste acceptance, quality of the final waste product, reduced worker exposure to hazardous chemicals and radiation, treatment of reclassified TRU without plutonium concentration, and minor generation of secondary waste.

11.6.3 Thermal Process Heating

Thermal process heating differs from electric process heating in that the heat for melting is produced by the burning of the waste and/or fuel. The melting most commonly occurs in a rotary kiln operated in a slagging mode to produce a glass product, but other incinerators are also used to vitrify wastes. Fossil-fuel-fired glass furnaces have been used in the glass industry and may also be applicable to waste vitrification.

Rotary Kiln Incineration: A rotary kiln is a cylindrical, refractory-lined shell mounted at an incline from a horizontal plane. This cylinder is rotated to facilitate mixing of wastes under incineration with combustion air, as well as to promote transfer of wastes through the reactor. Constant rotation of the kiln also provides continuous exposure of fresh surfaces to oxidation to promote destruction. A rotary kiln system includes the waste feed system, rotary kiln incinerator, auxiliary fuel feel system, afterburner, and air pollution control systems.

Rotary kiln incinerators operated in the slagging mode may produce a vitrified product. At high enough temperatures, the material in the kiln will deform, producing an amorphous state in that material. This molten slag can then be tapped and may harden into a glass or glass-like product upon cooling, based on material composition.

The vitrification system operates continuously. Material is pumped from the holding tank into the kiln at a controlled rate. Kiln temperature is varied, based on the composition of the feed. A pool of molten material forms in the kiln and rises to an overflow level. When it reaches the overflow level, the molten material flows out of the kiln and into the exit system.

11.6.4 Induction Heating

Currently, induction heating application to hazardous and radioactive wastes is represented by the French AVM process (Atelier de Vitrification Marcoule) and its descendants. However, because induction heating is also used in commercial glass manufacturing, it is potentially applicable to hazardous and radioactive wastes.

Induction heating is accomplished by inducing currents in the material to be heated. For example, a solenoid can be used to create a variable magnetic field inside the coil and around it. If an electrically conductive body is placed inside the magnetic field, the variation in the magnetic field causes a variation in the magnetic flux passing through the material and induces an electromotive force (EMF) current. The EMF current causes eddy currents, and these are converted into heat due to the Joule effect. Induction heating can also be created using highly varied induction configurations (flat inductors, linear inductors, tunnel inductors, etc.) and a wide range of relative part/inductors.

11.6.5 Resistance Heating

Initial large-scale testing of vitrification for HLW was done in crucibles heated by external resistance heaters. Their design represented a direct increase in scale from glass development crucible tests. Crucible heating was discarded as a treatment option for HLW because of low melt rates caused by slow heat transfer and lack of agitation and because temperature non-uniformities made it difficult to homogenize the glass.

11.6.6 Cyclone Furnace

The Babcock & Wilcox Co. (Babcock & Wilcox) cyclone furnace is designed to combust high organic (high-ash) coal. Through cofiring, the cyclone furnace can also accommodate highly contaminated wastes containing heavy metals and organics in soil or sludge. High heat-release rates of 45,000 British thermal units (Btu) per foot of coal ensure the high temperatures required to melt the high-ash fuels. The inert ash exits the cyclone furnace as a vitrified slag.

The cyclone vitrification technology is applicable to highly contaminated inorganic hazardous wastes, sludges, and soils that contain heavy metals and organic constituents. The wastes may be solid, a soil slurry (wet soil), or liquid. To be treated in the cyclone furnace, the ash or solid matrix must melt (with or without additives) and flow at cyclone furnace temperatures (2,400 to 3,000°F). Because of the technology's ability to capture heavy metals in the slag and render them nonleachable, the technology is an important treatment application for soils that contain lower-volatility radionuclides such as strontium and transuranics.

11.6.7 Plasma Heating

Plasma heating is an electrical heating process which relies on the conversion of a gas into a plasma through the application of energy by an electric arc. Plasma heating offers high operating temperatures and high power densities. Unlike joule heating vitrification, which grew out of the glass-making industry, plasma heating vitrification has grown out of the specialty metals industry.

Plasma Arc Vitrification occurs in a plasma centrifugal furnace by a thermal treatment process where heat from a transferred plasma arc torch creates a molten bath that detoxifies the feed material. Solids melt and are vitrified in the molten bath at 2,800°F to 3,000°F. Metals are retained in this phase which, when cooled, forms a non-leachable, glassy residue which meets TCLP criteria.

A plasma is an ionized gas. At high enough temperatures (e.g., 20,000°K for argon), electrons are stripped from their nuclei and the matter exists as a mixture of negative lectrons, positive nuclei, and atoms. The ionized particles make plasma an excellent electrical conductor.

Plasma heating equipment must perform 2 basic functions: creating the plasma and effectively heating the product.

Plasma is commonly created by passing a gas through an electrical arc. The arc can be generated by direct current (DC) or alternating current (AC). With a DC arc the cathode generally consists of tungsten and the anode generally consists of copper. The anode also typically functions as a nozzle directing the plasma. In contrast, in a single phase AC arc plasma generator, the electrodes act as the cathode and anode alternately, and must therefore be made from the same material.

Gases used in generating a plasma arc include nitrogen, oxygen, noble gases, air, and mixtures of these gases. Electrode life is a major concern and is influenced by electrode material, the gas used, and electrical current levels. Electrode structure, gas injection method, and nozzle design help shape the plasma and determine heating efficiencies.

The product is heated in 1 of 2 ways: by a non-transferred arc or by a transferred arc. A non-transferred arc uses 2 internal electrodes. A small column of injected gas is heated by the electric arc, creating a plasma flow that extends beyond the tip. Non-transferred arcs heat only via conduction and produce a dispersed heat that is needed for tasks such as air and gas heating and drying. Non-transferred arcs have been applied to hospital wastes.

Plasma arc torches are being used for the vitrification of low-level mixed waste. One drawback of existing plasma arc torches is that their anodes are short-lived. For more effective use of this technology for management of radioactive and mixed waste, the efficiency of the torches need to be improved. One way of doing this is by prolonging the life of the anodes of existing torches. A longer lasting torch would also be beneficial in other ways, for example, by minimizing the intrusion need by the workers into a radioactive environment and allowing for fewer electrodes to be discarded as low-level waste.

The Mixed Waste Integrated Program (MWIP) is adapting the plasma torch, developed for use in metals processing, for the treatment of mixed low-level wastes. This robust technology is advantageous due to the possible acceptance of a wide range of heterogeneous waste streams with minimal prior characterization. The plasma process is a fixed-hearth process in which whole drums are fed into the stationary hearth. This high-temperature process destroys organics and stabilizes the residuals into a non-leaching, vitrified waste form. Off-gas systems insure complete destruction of organics and removal of particulates before atmospheric discharge.

The system consists of a material-handling system for moving wastes into and out of the hearth, a primary combustion chamber, a secondary combustion chamber, and an off-gas treatment system. The volatilization of organics occurs in the primary chamber along with combustion of inorganic material; the products of incomplete combustion will be fed into the secondary combustion chamber operating with excess air above stoichiometric levels and a natural-gas burner to maintain temperatures above 980°C. The off-gas treatment system for the proof-of-principle unit consists of an air-atomized water quench, a high-temperature pulse-jet baghouse filter, and an induced air draft fan to maintain a slightly negative air pressure in the system.

Plasma refers to a highly energized gas. In the plasma heath system, the plasma is contained within a dc torch with power levels up to 1.2 MW and nitrogen as the primary plasma gas. The torch uses the flowing gas to stabilize an electrical discharge between 2 electrodes. One of the electrodes is contained within the torch, and the other electrode is solid material being treated.

The ability of the system to accept poorly characterized wastes (including full 55 gallon drums), the high-efficiency destruction of organics, the resulting volume reduction, and the high integrity of final waste forms make this technology very promising in the treatment of many heterogenous waste streams. The current pilot-scale efforts are intended to provide design data for future upgrades to the hearth and off-gas system, baseline the process for comparison to future system upgrades, provide information to other key activities such as cost/risk/performance analysis, and provide overall direction for the development of the plasma hearth process.

At INEL, proof of concept tests for fixed hearth plasma thermal treatment unit were completed on surrogate compacted waste drums and buried waste.

MSE Butte, MT will demonstrate the applicability of the Plasma Centrifugal Furnace (PCF) for the treatment of mixed (hazardous and radioactive) wastes and contaminated soil into an extremely durable waste form for disposal. The PCF is a rotating hearth plasma torch technology used for the refining of titanium and is now being developed for treatment of hazardous and mixed wastes and contaminated soils.

The Plasma Hearth Process (PHP) evaluation is being conducted with Science Applications International Corporation (SAIC) and Retech, Inc. of Ukiah, CA. SAIC is planning and conducting the testing at the Retech facility in Ukiah, CA.

The plasma arc process can accept a wide variety of waste types including paper, cloth, plastics, metals, glass, soil, and sludges. The ongoing projects are directed to demonstrate the application of the plasma arc process to representative surrogate waste streams. This project is a collaboration between Idaho National Engineering Laboratory (INEL), Oak Ridge National Laboratory (ORNL), MSE, Science Applications International Corporation (SAIC) and Retech.

The PHP system is potentially applicable to industrial process and buried radioactive wastes. It could be used as a component system along with criticality considerations to decommission spent nuclear fuel rods. The engineering design and development may have application for the advancement of vacuum metallurgy.

Another project is to design, fabricate, demonstrate, and evaluate a graphite-electrode DC arc furnace for its effectiveness in treating hazardous, radioactive (both low-level and TRU), and mixed wastes and soil, both buried and stored. An engineering-scale furnace, the Mark I, was tested in FY92 at MIT's Plasma Fusion Center to gain preliminary information on the treatability of buried waste surrogates using DC arc technology.

The Mark II furnace is a refractory lined carbon steel vessel 23 feet (7 m) high and 7 feet (2.1 m) in diameter with 4 soft patch panels around the circumference to provide access for waste feed, glass discharge, and diagnostic equipment. The off-gas system incorporates components that will probably be used in the full-scale production furnace.

This technology has been developed by MIT and EPI, Inc., and presently a contractual agreement is being negotiated with a large company for licensing of the furnace and diagnostic technology.

A new calcination/dissolution pretreatment process combines calcination and dissolution in a new and innovative manner in order to thermally destroy organic and inorganic constituents. A large scale plasma arc heats the tank waste in excess of 800°C to destroy organic compounds, ferrocyanide, nitrate, and nitrite and to separate the strontium and transuranic waste into a small volume via water dissolution.

Target process flow is about 20-80 gpm for a typical operating unit. Previous attempts at calcining high sodium waste material using conventional calcining concepts such as rotary kiln, spray tower, and fluidized bed have encountered problems due to plugging of the reactors by molten material.

The calcined product is expected to be further treated by dissolution, which produces high-level and low-level waste streams. The high-level stream will be vitrified, and the low-level stream is slated for near surface disposal.

Large scale pilot plant tests to determine the feasibility for plasma arc calcination will be conducted using Hanford waste stimulants. Major test objectives include equipment performance, such as continuous operation without plugging, and process performance, such as organics, ferrocyanide and nitrate destruction efficiency.

A large-scale plasma arc demonstration at the Westinghouse Science and Technology Center successfully calcined 3,000 pounds of simulated Hanford tank waste continuously without plugging. This demonstrates that large scale, high calcination of high sodium wastes is possible. Future plans include a second demonstration of plasma arc calcination of a 101-SY tank simulated waste. This longer test will couple calcination and dissolution operations with enhanced sample analysis.

11.6.8 Electric Arc Furnaces

Electric arc furnaces also are being applied to vitrification; they heat by creating current flow between 2 electrodes in an ionized gas environment. They differ from plasma furnaces in that plasma is not created and therefore not part of the heat transfer mechanism. The electric arc furnace was first developed in the metal industry.

A group from Electro-Pyrolysis, Inc. is working with a group from Massachusetts Institute of Technology to develop an innovative vitrification process. In this process, a DC electric arc is used in connection with a plasma heating arc to pyrolyze solid hazardous materials. The electric arc provides the primary energy for the heating and melting of the target material. This occurs in a sealed unit, thus reducing overall the amount of gases produced during pyrolysis and allowing the gas to be removed from the system in a non-oxidizing atmosphere. Furthermore, because the chamber is sealed, generated gases are forced to exit upward through the hollow arc-generating electrode and must pass through the electric arc. In addition, a plasma-heated zone created by electron-beam ionization and microwave heating is located at the tip of the electrode; gases must also pass through this. Thus, the plasma functions as a scrubber for off-gases generated by the electric arc. The electric arc provides target material heating and also off-gas treatment.

An electric arc is also being used in the vitrification tests in Albany, Oregon of MSW bottom ash and fly ash and the ash from sludge incineration. The Bureau of Mines and the American Society of Mechanical Engineers are the primary sponsors of these tests. The Japanese are also working on electric arc vitrification.

11.6.9 Microwave Heating

In microwave heating, a form of dielectric heating, the body to be heated absorbs electromagnetic radiation. More specifically, a dielectric is a material which is an electrical insulator. A dielectric becomes polarized when it is placed in an electric field. If the electric field is alternating, successive distortion of the molecules causes heating. Ceramic-like wastes such as incinerator ash, thermal insulators, concrete, soil, and sand are mostly composed of dielectric material and can be directly melted by microwave radiation.

Dielectric heating is usually classified into 2 sub-categories on the basis of frequency ranges used: radio frequency heating using frequencies between 10 and 300 MHz, and microwave heating using frequencies between 3,000 and 30,000 MHz. Of these 2 forms of dielectric heating, only microwave heating has been used to vitrify hazardous wastes.

A microwave installation consists of a microwave generator, a waveguide, an applicator, and ancillary monitoring, handling, and safety devices. The microwave generator produces the microwaves that dielectrically heat the load material. The waveguide directs the microwaves from the generator to the load material by reflecting the microwaves from its metal walls; it also keeps radiowaves from propagating in all directions. Applicators define the way in which the microwaves are applied to the load material. There are many types of microwave applicators. These applicators vary depending on the type of process, continuous or batch, and the nature and shape of the load material. Ancillary monitoring, handling, and safety devices work much as those used in other types of treatment processes.

The main advantage of microwave heating is that the heat is produced directly and solely in the mass of the material to be heated. Another advantage is high power density.

The main disadvantage is relatively high energy consumption and corresponding costs. Arcing resulting from induced currents in metallic components of waste may damage the microwave generator unless special provisions are made.

Microwave Solidification utilizes microwave energy to glassify waste solids to a stable final waste form. In contrast to cementation, the microwave approach avoids a large increase in the volume of waste for disposal. Another benefit of this approach is the ability to process waste directly in a drum. Work on microwave solidification has proceeded through a pilot-scale demonstration using surrogate waste and bench-scale using actual wastes. The program has prepared a preliminary definition of an off-gas system. The current emphasis is aimed at developing a complete system for use with the microwave and on evaluating the acceptability of the final waste form.

The Microwave Fluidized Bed at Rocky Flats is a thermal treatment process that uses microwave energy to heat a bed of silicon carbide particles (SiC) that is being fluidized by a water vapor stream containing hazardous organic liquids. The fluidized bed breaks down the liquid into less hazardous constituents. The precise mechanism is unknown, and trial and error approaches will be required to optimize the process.

This technology is being evaluated as a potential method to treat several mixed wastes, including process sludges, incinerator ash, and miscellaneous wastes, such as crucibles and foundry materials. In contrast to the use of cementation to process these wastes, the microwave approach results in a reduced volume of waste for disposal. In addition, the waste is processed "in-drum," which reduces material handling and generation of another waste stream.

Development of the microwave system at Rocky Flats has been achieved in collaboration with Microdry, Inc. and Rocky Flats Technologies. Investigators at the Oak Ridge National Laboratory and Los Alamos National Laboratory, have also provided input to the microwave project.

11.7 CERAMIC FORMS

Tailored ceramics or glass-ceramics may prove to be forms of choice for disposal of Idaho site HLW. A full suite of fundamental research studies to identify specific waste forms and to establish their properties is needed. DOE is currently planning to construct the Idaho Waste Immobilization Facility (IWIF) at INEL.

DOE states that the final treatment for the high-level liquid waste and the high-level waste calcine will be glass-ceramic type vitrification. This process will be implemented in the IWIF. DOE's supporting studies have concluded that the glass-ceramic process is the preferred method for INEL high-level waste immobilization. Based on technical evaluations, and laboratory and pilot plant mockup tests, EPA believes the glass-ceramic process is more efficient than the glass process for calcine waste forms.

DOE has previously identified the glass-ceramic waste form as similar to the Savannah River glass vitrification waste form. DOE states that the glass-ceramic waste form of vitrification will "provide treatment of the high-level mixed waste" pursuant to applicable nuclear waste acceptance specifications.

The glass-ceramic waste form contains a glass phase that encapsulates a crystalline ceramic structure (containing the radioactivity) on a microscopic scale.

The LLW form development will focus on testing 2 alternatives to the current disposal form for low-level waste (grout): nitrate to ammonia and ceramic (NAC) and polyethylene. The NAC process destroys nitrates and produces a ceramic LLW form in 1 process. The resulting ceramic can be sintered, which would destroy all organics by the high heat added during the final phase.

Ceramic formations such as titanate-based dense ceramics, and titania-silica tailored polyphase ceramics have also been investigated, however, they suffer from problems of high cost and complexity. Chemically bonded ceramic phosphates can decrease costs incurred from secondary waste due to solidification and stabilization at lower temperatures.

Both the crystalline and tailored ceramic formations feature low leachability, high waste loading, and high temperature stability. The disadvantages are high cost and complexity.

The Synroc process developed in Australia is an example of a synthetic ceramic material that can be utilized to encase high-level radioactive wastes.

The coated sol-gel particles process features mixing of additives as sol with HLLW (sludge or liquid), formation of spheres and gelation, washing, drying, and sintering, followed by fluidized-bed coating with 3 SiC layers and a pyrocarbon layer. Advantages include: high retention of radionuclides; inherent ease of quality assurance sampling; high-temperature stability. However, it is a costly and complex process.

Crystalline phosphates are potentially well suited as an advanced ceramic form for immobilization of TRU elements. Basic research is needed to achieve optimum synthetic routes for crystalline phosphates via low-temperature thermal processes and to study the bonding, stability, and electronic properties of such phosphate ceramics. Particular attention should be given to the utility of TUCS (Thermally Unstable Complexants) compounds (i.e., diphosphonic acids) for low-temperature synthesis of phosphate-based ceramics containing high TRU element concentrations. The complexes formed by TUCS compounds with actinides break simply upon mild ($\leq 100°C$) heating of HNO_3 solutions to produce solid residues that have been identified, at least in one case, as crystalline orthophosphates (e.g., $NdPO_4$) with the monazite structure.

Oxides with the fluorite structure, such as CeO_2, ThO_2, and PuO_2, are extremely inert. Fluorite-based matrices could be based upon simple oxides such as ZrO_2 or upon complex oxides such as zirconolite, $CaZrTi_2O_7$, or pyrochlore, which are fluorite-related structures. The principal obstacle to development of oxide-based matrices for TRU wastes is the development of new and facile routes to synthesize these hosts and to dope them uniformly with TRU ions.

Fundamental research studies need to be scoped and executed to identify and develop optimum glass and ceramic forms suitable for incorporation and immobilization of separated Sr-90 and Cs-137 fractions. A large inventory of separated Sr-90 (as SrF_2) and Cs-137 (as CsCl) already exists at the Hanford site. Additional amounts of separated Sr-90 and Cs-137 may be produced in future waste management activities at DOE sites.

Mixed waste streams, which contain both chemical and radioactive wastes, are one of the important categories of DOE waste streams needing stabilization for final disposal. Recent studies have shown that chemically bonded phosphate ceramics may have the potential for stabilizing these waste streams, particularly those containing volatiles and pyrophorics. Such waste streams cannot be stabilized by conventional thermal treatment methods such as vitrification. Phosphate ceramics may be fabricated at room temperature into durable, hard and dense materials. For this reason, room-temperature-setting phosphate ceramic waste forms are being developed to stabilize these "problem waste streams."

Another process, the Quantum-Catalytic Extraction process has been developed from the Catalytic Extraction Process (CEP), by Molten Metals Technology, Inc. It allows processing of mixed waste streams, and preparation for final form disposal.

Quantum-CEP allows both destruction of hazardous components and controlled partitioning of radionuclides. This leads to decontamination and recycling of a large portion of the waste components to commercial products as well as volume reduction and concentration of radionuclides for final disposal.

At the core of both CEP and Quantum-CEP is a molten metal bath which acts as a catalyst and solvent in the dissociation of the feed, the synthesis of products and/or the concentration of radionuclides in the desired phase. Upon introduction to the bath, feeds dissociate into their constituent elements and go into a metal solution. Once in this dissolved state, addition of co-reactants enables reformation and partitioning of desired products. The partitioning control afforded by co-reactant addition is a distinguishing feature of CEP and Quantum-CEP.

A typical CEP system consists of the Catalytic Processing Unit (CPU), the reactor holding the liquid metal catalyst and solvent, followed by a hermetically sealed gas handling train. The reactor can handle feeds of most physical forms: gases, fine solids, pumpable liquids and slurries can be fed through tuyeres at the bottom of the reactor. Larger solids (e.g., bulk) can be added through top-mounted, submerged lances.

Previous results reported in the literature using melt refining techniques with contaminated metals demonstrated that uranium and plutonium can be selectively removed from the metal phase and concentrated in a separable, vitreous oxide phase.

11.8 METALLIC ALLOYS

Advanced waste pretreatment and partitioning schemes may produce concentrated Tc-99 and I-129 fractions. Because these radionuclides tend to form soluble, highly mobile chemical species when contacted with groundwater, special forms are needed to satisfactorily immobilize wastes containing very high concentrations of Tc-99 and I-129. In addition to the conventional glass and ceramic waste forms, metallic alloys may prove to be highly attractive forms for disposal of separated Tc-99. Basic research that will lead to methods of preparing alloys of Tc-99 should be performed. Research that will lead to non-leachable forms for isolation of concentrated I-129 wastes, such as insoluble, reducing salts based upon CuI, also needs to be carried out.

11.9 CALCINATION

HLW at the Idaho National Engineering Laboratory (INEL) is now being pretreated by a fluidized bed calcination process to achieve a more environmentally stable, non-corrosive, dry, granular solid consisting of metal oxide and metal salts. Operation of the New Waste Calcining Facility (NWCF) results in a generation of 465 m^3/yr of calcine solids.

The NWCF treats liquid mixed wastes by calcination, resulting in conversion of the liquid to a solid granular form. Calcination of the liquid waste was accomplished by blending aluminum- and zirconium-bearing waste (created from spent fuel reprocessing) with sodium-bearing waste generated from decontamination of plant equipment. The blending of a sodium waste with either aluminum or zirconium raffinates has been essential for the calcination process.

Currently, the facility is operating under RCRA interim status and has a design liquid input of 5,200 gal/day. However, current research on different waste treatment techniques is not complete; therefore, an operating capacity feed rate to the NWCF will be expressed as cubic meters of calcine produced. Approximately 465 m^3/yr of calcine is produced during a typical campaign.

The WIF (formerly referred to as the Idaho Waste Immobilization Facility (IWIF)) is a proposed facility for the processing of mixed wastes at the ICPP. The objective is to convert the ICPP wastes into a single HLW form and a single LLW form. Candidate technologies and waste form options have been identified based on waste characteristics, plant compatibility, process feasibility, and cost.

The planned, unapproved IWIF process was identified for treatment of ICPP calcine. This facility would have converted the calcined waste into a vitrified glass-ceramic waste form in a hot isostatic press process. However, since the waste would not be separated into HLW and LLW fractions, all the vitrified glass-ceramic waste would have been disposed of in a geological repository. The selected WIF processes will separate the wastes into HLW and LLW fractions. After the separation processes, the HLW fraction will be relatively small in volume compared to the LLW fraction. Reducing the amount of HLW is desirable since any HLW will have to be sent to a geological repository, and therefore disposal will be expensive. Options include direct immobilization as a glass-ceramic and volume reduction pretreatment to separate actinides and fission products followed by immobilization as a glass or glass-ceramic.

The HEPA Filter Leach System at INEL is used to remove radionuclides and other hazardous constituents from used HEPA filters by chemical extraction. The resulting leachate from the chemical extraction may be solidified using the fluidized bed calciner.

11.10 HIGH-INTEGRITY CONTAINERS (HIC)

The low-level radioactive waste (LLW) form stability requirements of 10 CFR 61.56(b) require that structural stability of the waste be achieved either by the waste form itself, by processing the waste to a stable form, or by placing the waste in a disposable container or structure that provides stability after disposal. A container that provides stability to the waste after disposal is called a high-integrity container (HIC).

Use of a HIC can provide a convenient and economical means for handling, transporting, and disposing of low-level waste. HICs are most frequently used in conjunction with dewatering or drying systems for wet solids such as ion-exchange resins and filter sludges. Since the HIC eliminates the need to solidify wastes to achieve a stable waste form, the use of a HIC can reduce the total volume of waste disposed. Considering this advantage, the HIC may be considered a volume reduction as well as a stabilization technology.

HICs are used primarily for the disposal of Class B and C wastes and those Class A wastes which are required by Washington and South Carolina to be stabilized (wastes with half-lives greater than 5 years with concentrations in excess of 1 uCi/cm^3). Due to their cost, HICs are rarely used to stabilize other Class A wastes.

HIC materials:

1) Steel fiber reinforced polymer impregnated concrete.
2) Stainless steel/polyethylene.
3) Ferralium/family.
4) Reinforced concrete for EPICOR-II liners.
5) Polymer encapsulated carbon steel.
6) Fiberglass/polyethylene.
7) Polyethylene.

11.11 MIXED WASTE

Waste contaminated with chemically hazardous and radioactive species is defined as mixed waste. Significant technology development has been conducted for separate treatment of hazardous and radioactive waste, but technology development addressing mixed-waste treatment has been limited. In response to the need for a comprehensive and consistent approach to mixed-waste technology development, the Office of Technology Development of the U.S. Department of Energy (DOE) has established the Mixed Waste Integrated Program. The program is identifying and evaluating treatment technologies to treat present and estimated future mixed wastes at DOE sites.

The Final Waste Form Technical Area encompasses the development of technologies that are suitable as a final waste form for storage and/or disposal. Performance standards and evaluation criteria for the final waste form will be developed. Emphasis is placed on the ability of the final waste forms to exhibit high waste loading, significant volume reduction, low-leachability, and high-durability. The DOE/Industrial Center for Vitrification Research was established at Clemson University.

The development of stabilization technologies including vitrification and polymer encapsulation will be continued. Molten metal processing will also be emphasized. Data requirements for the assessment/characterization of final waste forms to meet regulatory requirements will be developed. A major effort in the Final Waste Forms Technical Area will be an expedited vitrification demonstration on actual radioactive waste in the DOE inventory. Research in support of this demonstration is being conducted at Savannah River and Pacific Northwest Laboratory (PNL), and the demonstration will take place at 1 or both or these 2 sites.

The Plasma Heath Process (PHP) demonstration project is one of the key technology projects in the Department of Energy (DOE) Office of Technology Development (OTD) Mixed Waste Integrated Program (MWIP). Testing to date has yielded encouraging results in displaying potential applications for the PHP technology. Early tests have shown that a wide range of waste materials can be readily processed in the PHP and converted to a vitreous product. Waste materials can be treated in their original container as received at the treatment facility, without pretreatment. The vitreous product, when cooled, exhibits excellent performance in leach resistance, consistently exceeding the Environmental Protection Agency (EPA) Toxicity Characteristic Leaching Procedure (TCLP) requirements.

At Hanford a planned DOE facility, the Waste Receiving and Processing (WRAP) Module 2A, Building 2337-W, is scoped to provide required treatment for containerized contact-handled (CH), mixed low-level waste (MLLW). The core processes in WRAP Module 2A include cement stabilization of particulate waste, polyethylene encapsulation (via extrusion) of particulate waste, and cement encapsulation (via vibratory infilling) of hard and soft debris. The WRAP Module 2A will begin construction in 1996 after a detailed design effort and pilot testing activities.

12

Mill Tailings

Mill tailings are the primary residue of uranium milling operations that separate uranium from uranium ore. At a few very old sites, the tailings are the result of radium mining operations in the early 1900s. Mill tailings contain daughter products of uranium—particularly radium-226 with its 1,600 year half-life. The tailings may also contain hazardous contaminants from milling operations; much of the residue is sand. Mill tailings are of concern due to possible water contamination by tailings pore fluid contaminants and air exposure by radon.

In 1978, public concern regarding potential human health and environmental effects from uranium mill tailings led Congress to pass the Uranium Mill Tailings Radiation Control Act (UMTRCA) (42 U.S.C §7901 et seq.). In the UMTRCA, Congress acknowledged the potentially harmful health effects associated with uranium mill tailings and designated 24 inactive uranium processing sites for remediation. Pursuant to the UMTRCA, the U.S. Environmental Protection Agency (EPA) developed standards, which include exposure limits from surface contamination and concentration limits for groundwater, to protect the public and the environment from potential radiological and non-radiological hazards from the abandoned processing sites. In 1987, the EPA proposed new regulations to replace some sections and change others.

The U.S. Department of Energy (DOE) is responsible for performing remedial action to bring surface and groundwater contaminant levels at the 24 sites into compliance with the EPA standards. The DOE is accomplishing this through the Uranium Mill Tailings Remedial Action (UMTRA) Surface and Groundwater Projects. All remedial action must be performed with the concurrence of the U.S. Nuclear Regulatory Commission (NRC) and the affected states and Indian tribes. A listing of the 24 sites is given in chapter 3.

Former uranium processing activities at most of the 24 inactive mill sites resulted in the contamination of groundwater beneath and, in some cases, downgradient of the sites. This contaminated groundwater often has elevated levels of hazardous constituents such as uranium and nitrate. The purpose of the UMTRA Groundwater Project is to protect human health and the environment by meeting the EPA standards in areas where groundwater has been contaminated.

Uranium production in the United States has declined dramatically from a peak of 43.7 million pounds U_3O_8 (16.8 thousand metric tons uranium (U) in 1980 to 3.1 million pounds U_3O_8 (1.2 thousand metric tons U) in 1993. This decline is attributed to the world uranium market experiencing oversupply and intense competition. Large inventories of uranium accumulated when optimistic forecasts for growth in nuclear power generation were not realized. The other factor which is affecting U.S. uranium production is that some other countries, notably Australia and Canada, possess higher quality uranium reserves that can be mined at lower costs than those of the United States. Realizing its competitive advantage, Canada was the world's largest producer in 1993 with an output of 23.9 million pounds U_3O_8 (9.2 thousand metric tons U).

The U.S. uranium industry, responding to over a decade of declining market prices, has downsized and adopted less costly and more efficient production methods. The main result has been a suspension of production from conventional mines and mills. Since mid-1992, only nonconventional production facilities, chiefly in situ leach (ISL) mining and by-product recovery, have operated in the United States. In contrast, nonconventional sources provided only 13 percent of the uranium produced in 1980.

12.1 HISTORY

In the United States, the history of the uranium production industry can be divided into 3 periods. Prior to 1940, uranium was produced as a minor commercial commodity. Almost overnight during World War II, its military importance made uranium exploration and production literally explode. During the early sixties, uranium production for peacetime nuclear-powered electrical generation again brought a surge in exploration and production capacity.

The forties military demand was met from known sources of supply with most uranium coming from Belgian Congo pitchblende and the Great Bear Lake deposit in Canada. These sources were supplemented by production from treatment of previous tailings left over from earlier rare metals extraction facilities and several small new mines in the Colorado Plateau area. The extraction processes used were essentially those developed at the turn of the century with recovery relatively low compared to modern methods.

The United States enacted the first legislation for the control of the uranium production industry with passage of the Atomic Energy Act (AEA) of

1946. This precipitated research efforts to improve extraction processes and eventually led to the use of lower-grade sources than considered practical before. The 1946 AEA (and its amendment in 1954) did not include provisions for environmental restoration of production facilities. In the United States, uranium production was encouraged by incentives such as government provided ore stations, access roads, hauling subsidies, bonuses and a guaranteed fixed price for U.S. Government purchases. This resulted in a surge of exploration and production leading to the peak years of 1960-1963, with an annual production of about 15,000 metric tons of uranium concentrate extracted from about 7 million metric tons of ore per year in this 3 year period.

By the end of the fifties, most large ore reserves had been identified, and in 1958, the U.S. Government canceled its agreement to purchase future uranium from ore reserves as yet unidentified. This removed the incentive for further exploration and eventually led to many mill shutdowns as their contracts with the government expired. By the early seventies, mining and milling was conducted by private industry, with free market economics dictating the supply and demand of uranium concentrate. Most of the facilities which had produced uranium concentrate for the U.S. Government has been shut down and abandoned. Facilities still commercially producing uranium under contracts with the nuclear power plant industry continued operation well into the seventies.

Through 1979, the total uranium production from sources in the U.S. is estimated at about 280,000 metric tons, extracted from about 150 million metric tons of ore. During the seventies and eighties, the U.S. uranium industry steadily declined. In 1991, only 2 conventional mills operated, producing less than 1,100 metric tons of uranium and 550,000 metric tons of tailings. This can be attributed in large part to nuclear power plant cancellations or deferments, leading to lower uranium demand and lower prices. At the same time, environmental restoration costs have continued to climb leading to increased reliance on uranium imports.

The uranium in the U.S. was found in a variety of geological settings and the uranium extraction industry developed a wide diversity of both mining and milling facilities. Mills tended to be located in river valleys near small to medium-sized communities because of the availability of water and a stable work force. The mines providing ore to these mills tended to be numerous and in most cases, substantially removed from the mills. In addition, the mine operators generally were not captive to the mill operators. Free market economics were allowed to dictate the best market for the ore provided by most mines. Therefore, the tailings left behind at most mill sites contain residual materials from processing ore from many sources.

12.2 ACTIVE MILLS

Although there are no conventional mills now operating in the United States, a discussion of uranium ore milling is included, in order to develop a background for an appropriate presentation of remediation techniques.

Uranium ore milling, like any commercial ore milling, requires environmental controls for dusts and emissions of various toxic chemicals. However, in addition to dust and toxic chemicals, uranium milling produces radioactive emissions, both during the active milling phase and the inactive phase after tailings have been disposed of. Operations at active uranium mills, including management of the tailings, are regulated by NRC. The 2 major concerns with respect to radioactive emissions are atmospheric emissions during active milling operations and disposal and isolation of tailings piles.

Radioactive releases from existing mills constitute the largest routine releases from the nuclear fuel cycle. Four major milling activities produce radioactive particles and gases: ore stockpiling, ore crushing and grinding, yellow-cake drying and packaging, and tailings disposal. The primary mill product, called yellow cake (U_3O_8), is highly refined uranium ore that is ready for conversion and enrichment facilities. It presents substantial potential for occupational exposure to radiation.

Uranium mill tailings, which are estimated to total 130 million m^3, or over 150 million tons, make up the single greatest volume of nuclear waste. The tailings piles are located in several western states at more than 20 sites, which are generally remote from population centers. Emissions of radon-222 gas appear to be the most likely source of radiation exposure from the piles, although other radionuclides are present. Health hazards of equal or greater importance may also exist from toxic non-radioactive elements—such as arsenic, selenium, molybdenum, cadmium, and lead—that are commonly found in tailings.

Piles at active mills are kept wet to reduce radon release. Greater risk comes from inactive mills where the piles have dried out and may be dispersed by wind. Local groundwater may also be affected by rain water leaching radioactive and toxic materials. The radioactivity and toxicity of the tailings are no greater than those of the original ore, but since the ore was brought to the surface of the earth and processed into more dispersable forms, it now poses some hazard.

Processing mined uranium ore at a mill can produce radioactive atmospheric emissions and large quantities of sand and slimes (fine-grained material), known as tailings. The tailings contain low concentrations of the uranium-238 decay chain: uranium-238, uranium-234, thorium-230, radium-226, radon-222, and the radioactive decay products of radon-222. They retain about 85% of the total radioactivity of the uranium ore from which they were produced. Radium concentrations range from 26 to 600 pCi/g and thorium concentrations range from 70 to 600 pCi/g, depending on the ore, the process used, and the distribution of tailings particle sizes considered.

Many simple and well-developed technologies are suitable for controlling atmospheric emissions from uranium milling activities. The natural moisture content of uranium ore usually keeps dust emissions to a minimum, but dust problems arise as the ore dries. The major controls for dust from ore piles are windbreaks designed to minimize ore drying, wetting the ore with water, and wetting the ore with chemical agents, or covering the tailings with clean soil. Fugitive dust control methods are estimated to reduce emissions by 20-90%.

During ore crushing, 97-100% of emissions can be eliminated by using stack controls, such as wet impingement scrubbers and bag filters, and modifying processes, such as replacing dry crushing with wet semiautogenous grinding. Emissions from yellow cake can be virtually eliminated by using wet scrubbers on the stacks; impingement or venturi scrubbers and demisters are currently used. Wet scrubbers remove particles by spraying them with water droplets.

The tailings are pumped as slurries into earthen impoundments, which represent the single largest source of radioactive emissions at a mill. While the disposal area is being used, emissions can be reduced by keeping the piles wet with tailings solution or water. This control method can be supplemented with chemical stabilizers, such as resinous adhesives, lignosulfates, electrometric polymers, milk of lime, wax, tar, pitch, asphalt, potassium and sodium silicates,and neoprene emulsions. Annual or more frequent maintenance is generally required because considerable crust breakage and erosion occurs over time. Tailings-pile control methods can reduce emissions by 38-90%.

Progressive reclamation of active tailings involves dividing a large tailings area into smaller cells with sequential construction, filling, and reclamation. Thus the surface area of exposed tailings is substantially reduced, resulting in up to 85% lower emissions.

In one approach, an initial basin would be formed by building low earthen embankments on the 4 sides of a square. Mill tailings would be slurried into the basin; as the basin filled, coarse fractions of the tailings (sands) would be used to raise and broaden the embankments. The embankments would be compacted on the outer side for strength.

12.3 IN SITU LEACH MINING

Although the in situ leach (ISL) mining of uranium in the United States started in the 1960s, the real expansion of this form of mining, which is also called solution mining, took place in the early to mid 1970s in Texas. Some of the early test work used an acid lixiviant, but was soon recognized that, because of environmental considerations, the use of alkaline lixiviants would be preferable to the regulatory agencies and the public. In the past, the 2 types of alkaline based lixiviants used at U.S. ISL mines were ammonia bicarbonate and sodium bicarbonate. A few ISL mines get by with just adding carbon dioxide to the well field solution. Ammonia bicarbonate is no longer being used today in the U.S. because of the difficulties and expense of restoring the water quality to acceptable standards following mining.

The growth in solution mining in Texas in the 1970s was followed closely by the start-up of ISL mines in Wyoming and some pilot testing in New Mexico. In some states, the expansion of ISL mining was slowed by the regulatory process which required a project to first go through a time consuming and expensive R&D pilot test phase before starting commercial licensing and production. From the standpoint of the regulators, the pilot test had to demonstrate that the groundwater contaminated by solution mining could be restored to its original quality following mining. This clean up of the contaminated groundwater at ISL mines is called aquifer restoration. With the drop in yellow cake prices starting in the early 1980s, the number of operating ISL uranium mines steadily declined from a peak of about 12 to the present level of 2 mines in Wyoming, 1 mine in Nebraska, and 1 mine in Texas.

Uranium in situ leach mining is the process of recovering uranium from a water saturated ore body in a manner which leaves the overlying rock strata and the land surface undisturbed. The process involves the installation of a series of water wells through which a non-toxic chemical solution (lixiviant) and an oxidant are injected into the uranium bearing sandstone formation. The solution passes through the formation, dissolves the uranium, and is pumped back to the surface via recovery wells. The water solution with the dissolved uranium is piped to a surface plant where a series of conventional hydrometallurgical processes,, including ion exchange, extract the uranium. The resulting solution, now barren of dissolved uranium, is refortified with leach chemicals, piped to the well field and injected back into the mineralized formation. This continuous loop process takes place in a well field until the dissolved uranium concentration (head grade) in the recovered solution drops to a level where mining is no longer economical.

Once the uranium has been removed from the ore body, the groundwater affected by the leaching solutions must be restored, or cleaned up, to a condition that is acceptable to the regulatory agencies. Generally, the regulatory agencies require that the groundwater affected by the mining be returned to (or very near) the pre-mining water quality.

ISL mining has much less impact on the environment than conventional operations. Since no mill tailings or rock waste are produced, ISL operations incur lower reclamation and environmental monitoring costs. ISL mining also poses less radiation and dust exposure to employees. If enacted, proposed reductions in the limits of radiation exposure would significantly add to the cost of underground mining. Because of its reduced environmental impact, ISL uranium mining is expected to be more widely accepted by the public than conventional mining and milling.

12.3.1 ISL Aquifer Restoration

Step One - Groundwater Sweep: Following a planned and controlled sequence of events, mining in an exhausted well field is gradually terminated and aquifer restoration activities commence. The first step in groundwater restoration is normally referred to as the groundwater sweep phase. During this phase injection of lixiviant into the ore body is stopped, but fluid continues to be pumped from the recovery wells. This action removes contaminated water from the formation and brings surrounding uncontaminated groundwater into the ore body aquifer. The contaminated water removed in this process is either injected into a new well field or disposed of on the surface in lined evaporation ponds or treated and surface discharged. The groundwater sweep step was, at one time, thought to be sufficient to completely restore a contaminated aquifer but the large volume of water that has to be removed and disposed of, coupled with the fact that the dilution efficiency of the process decreases as the water quality approaches baseline conditions, makes additional restoration steps necessary. The groundwater sweep step is most efficient at the early stage of restoration when the differential between the ISL affected groundwater quality and the natural groundwater quality is the greatest. Typically, 1 to 3 pore volumes of formation water are removed during this step. A pore volume is defined as the volume of water in the ore body portion of the aquifer.

Step Two - Water Treatment: The second step in the aquifer restoration process is normally referred to as the water treatment and re-injection phase. In this phase the contaminated water in the formation, remaining after step 1, is pumped to the surface and fed to a water treatment unit in a continuous process. The clean water (permeate) from the water treatment unit is sent back to the well field and injected back into the ore body aquifer. The brine, or reject water from the treatment unit, is normally sent to an evaporation pond or a deep disposal well. The 2 most common types of water treatment units are reverse osmosis (RO) units and electrodialysis (ED) units. The water treatment step takes place until the groundwater quality in the old body aquifer is back to pre-mining values which normally takes about 2 to 6 pore volumes of treated water circulated through the ore body aquifer. At some mines the completion of this second step is sufficient to return the groundwater quality to acceptable conditions. However, at the majority of mines it is necessary to go to the third step which involves the addition of chemicals to the water being re-injected.

Step Three - Reductant Addition: During the ISL mining process, the addition of chemicals and an oxidant into the ore body aquifer creates an imbalance in the natural geochemical environment. The formation is taken from a reduced state to an oxidized state. This oxidized condition remains, to some degree, during aquifer restoration and as a result uranium and other heavy metals continue to solubilize making it difficult to return these chemical species to the baseline levels. To counter this situation step 3 is initiated which involves the continuation of step 2 plus the addition of a chemical reductant, such as hydrogen sulfide, to the solution being injected into the ore body aquifer. The addition of a reductant extinguishes the chemical reactions by returning the formation to a reduced state thus restoring the pre-mining geochemical environment. This action usually allows the dissolved concentration of uranium and other heavy metals to stabilize at acceptable levels.

Step Four - Water Recirculation: A final step in the restoration process at some ISL mines is to just circulate the formation water through the ore body aquifer for 1 or 2 pore volumes following step 3 to make the water quality homogeneous throughout the well field. This action tends to eliminate spacial and temporal variations in water quality during the stability monitoring period.

12.4 RECLAMATION

Reclamation efforts follow the termination of milling operations. The major objectives of controlling tailings piles are to provide effective long-term stabilization and isolation from the biosphere, control random radon and gamma emissions from the tailings, and protect local water quality. Active and passive control technologies are available to meet these objectives. Active controls require that some institution, usually a government agency, continually monitor and maintain the stabilized and isolated pile. However, it is unlikely that active control methods could be maintained for longer than a few hundred years. Passive controls are designed to maintain tailings pile integrity over long periods with little or no active involvement by human agencies.

Current and potential control technologies that represent a complete tailings disposal and isolation program fall into 4 groups: uranium ore process alternatives, disposal area location, tailings area preparation, and tailings stabilization and covering.

Basic uranium ore processing consists of grinding raw ore and mixing it with sulfuric acid to dissolve the uranium. Uranium in the form of U_3O_8 is then separated out with solvent extraction techniques. Alternative processes can remove substantial fractions of the radium and thorium during processing, thereby reducing the radiological hazard of the tailings. The alternatives include nitric acid leaching; separating the slimes, which contain the bulk of the radioactivity, from the sands in the tailings; neutralizing the tailings with lime to precipitate radioactive and toxic elements; treating the tailings with barium chloride; and using ion-exchange treatment for slurries. The alternative methods all cost substantially more than current processes and are not generally developed to a commercial level. Combining the tailings with asphalt or cement at the conclusion of ore processing has been suggested to decrease leachability and reduce diffusion of radon. Another suggested alternative involves dewatering tailings piles by appropriate drainage construction to prevent seepage of drain water into the groundwater.

The conventional location for a disposal area has been above grade near the mill; the area is surrounded by 10- to 30 m embankments. However, tailings could also be disposed of below grade in open mine pits or special excavations. Similarly, tailings could be moved to abandoned deep mines. Both methods substantially reduce erosion and emissions but entail greater cost. Groundwater contamination can occur in any below-grade disposal.

If above-grade or near-surface disposal is selected, existing techniques can prepare disposal areas to minimize seepage into groundwater. The techniques include compacting the soil to reduce permeability and using clay or synthetic liners as sealants.

Remedial actions at tailings sites consist of stabilizing and covering the tailings to separate them from the environment. A cover of soil or combined soil and clay can protect piles from erosion and reduce radon emissions, depending on the moisture content of the cover. Gravel, rock, and riprap can also provide substantial resistance to wind erosion and water infiltration, thus minimizing radon diffusion.

On average, tailings reclamation activities account for approximately 54 percent of the decommissioning costs for conventional uranium mills. Average decommissioning costs for a conventional production facility are $14.1 million: $7.7 million for tailings reclamation, $2.3 million for groundwater restoration, and $0.9 million for mill dismantling and $3.2 million for indirect costs.

Because nonconventional (in situ leaching) operations do not require removal of ore to the surface, there are no mill tailings and surface disturbance is kept to a minimum. Groundwater restoration accounts for the largest share of decommissioning costs (40 percent of the total) at nonconventional operations. Of the estimated $7 million average decommissioning cost for nonconventional sites, groundwater restoration accounted for $2.8 million. The wellfield reclamation costs were $0.9 million and the plant dismantling costs came to $0.6 million. Other costs (such as evaporation ponds, disposal wells, and radiological surveys) averaged $1.2 million. The indirect costs averaged $1.4 million.

12.5 PILE LOCATION AND CONFIGURATION

Contaminated materials can be stabilized either on site, in place, or at an alternate site. The final location and configuration of a pile affects every other design consideration and is influenced in turn by any or all of the considerations. Each of the 3 types of stabilization can vary from above-grade disposal to different degrees of below-grade disposal. The ultimate goal of remedial action is to assess technically acceptable alternatives in determining the most cost-effective option.

The level of effort expended to protect against settlement and ensure slope stability at mill tailings disposal sites depends greatly upon the site-specific conditions and the design concept used. Both surrounding natural slopes and embankment slopes are analyzed. Because of the flatter slopes required to promote surface water runoff and because of erosion considerations, slope stability of the embankment will only be critical for sites exhibiting high seismicity or soft soil zones within or below the embankment. If critical, slope stability considerations could lead to designing flatter or buttressed slopes.

Overall magnitude of settlement is not a critical design control; however, differential settlement, which tends to increase along with increasing total settlement, can cause cracking of the radon cover or drainage flow concentrations. Either condition, if not adequately designed for, could lead to failure of the pile cover system.

To demonstrate the ways in which the proposed design minimizes the need for active maintenance, emphasis should be placed on (1) the use of natural, durable materials; (2) shaping the pile to resist natural forces such as water and wind erosion; and (3) minimizing infiltration by design features of the cover. Specifically, the pile components will be designed such that:

1) The radon barrier of compacted soil is protected by the overlying filter, riprap, or soil and vegetation layers, which resist erosion, frost, and biointrusion.
2) The sand filter, drain, and bedding layer is designed to be clean and durable. It is designed to avoid plugging by piping of soil particles, and is protected from erosion by the overlying rock layer.
3) Rock erosion protection is constructed of durable rock, sized to resist runoff resulting from a PMP or PMF. Soil/vegetated covers are also composed of natural, durable materials with sufficient thickness and material properties so as to withstand long-term and maximum precipitation events.

It should also be emphasized that the piles will be designed with accepted factors of safety to prevent slope, settlement, and deformation failures. Therefore, maintenance requirements will be minimal. The remedial action designs also incorporate seismic, liquefaction potential, and geomorphological considerations.

12.5.1 Stabilization In Place (SIP)

An adequate assessment of SIP includes the following design considerations.

1) Reconfigure the pile to have stable slopes with a minimum movement of contaminated materials.

2) Buttress the sideslopes with clean material to form stable slopes as necessary if exposure of slime pockets will produce unstable conditions.
3) Identify the modifications that exist (if any) to avoid the impact of upland drainage areas or nearby large streams.
4) Minimize the final pile area that will remain restricted. This is in conflict with minimum material movement unless the depth of the materials is shallow enough that it is more cost-effective to consolidate the material and reduce the final area requiring cover.
5) Achieve a balance between minimizing slopes for erosion protection and maximizing slopes for a greater volume-to-pile-area ratio.
6) Identify any modifications that exist to avoid differential settlement due to slimes concentrations. The design should evaluate the potential effects on the radon cover and the effect on erosion protection due to drainage flow concentrations.
7) Increase radiological characterization as necessary to properly design the radon barrier.
8) Reduce the necessary radon barrier through effective use of windblown material spread evenly over the top of the tailings. However, if the quantity to spread is not substantial, it may be more cost-effective to place the windblown materials randomly and provide a thicker radon barrier.
9) Ensure that the contaminated materials are well above the groundwater table.
10) Identify the risks associated with location in a floodplain (not automatically a reason for relocation). However, it may not be possible to protect the pile from floods depending on the proximity to the stream and the constricted nature of the floodplain.

12.5.2 Stabilization On Site (SOS)

An adequate assessment of SOS includes the following design considerations.

1) Evaluate the necessity for the movement of all, or the major portion, of the contaminated materials in order to protect them from 1 or more conditions (e.g., high groundwater, proximity to large streams, slope instability, potential differential settlement, configuration effect upon flood conditions, or geomorphic instability).
2) Assess the modifications available to avoid any existing hydrologic impacts. Greater control of hydraulic conditions is possible with SOS than with SIP, thereby reducing erosion protection requirements.
3) Minimize the overall pile area to effectively reduce overall cover requirements without adversely affecting pile stability.

4) Avoid large slime pockets by more complete mixing of the materials to reduce the potential for differential settlement.
5) Base the final location of the pile upon construction requirements that avoid excessive double handling of materials.
6) Improve stability conditions with more complete compaction of the embankment.
7) Reduce radiological characterization data needs and simplify the radon barrier design process since mixing the materials averages the emanation rates. Ensure sufficient data are available to determine the averages adequately.
8) Evaluate the necessity for substantial erosion protection requirements due to flooding, runoff from upland drainage basins, and flow in stream channels.

12.5.3 Relocation to an Alternate Site

Factors included in the selection of an alternate site are as follows:
1) Exercise maximum flexibility in selecting a site and choosing a configuration that minimizes hydrologic impacts.
2) Improve conditions involving slope stability and differential settlement due to the mixing and recompaction of the entire pile.
3) Simplify radiological characterization and the radon barrier design process by mixing the materials; this averages the emanation rates. Ensure sufficient data are available to determine the averages adequately.
4) Select the location for an alternate site by evaluating locations near the top of drainage areas with stable existing slopes, seismically stable locations, and locations on government-owned land.
5) Avoid sites with shallow groundwater. Shallow groundwater will affect the pile configuration by limiting the degree of below-grade disposal.
6) Evaluate partial to complete below-grade disposal in order to obtain cover materials from the disposal site using cut and fill procedures.
7) Develop an economic comparison of the possible designs. Different designs may vary from a thick pile above-grade that minimizes the final pile area and cover requirements to a completely below-grade disposal that uses shallow slopes and additional soil below cover depth to eliminate expensive imported rock for erosion protection.

12.5.4 Hydrologic Impacts

The type of stabilization used on mill tailings disposal sites is greatly influenced by the expected magnitude of hydrologic impacts. The level of effort expended for erosion protection depends upon the site-specific conditions and the final pile location and configuration.

Three primary design situations affect the stabilized tailings:

1) On-pile surface runoff.
2) Upland watershed runoff.
3) Flooding associated with nearby large streams or large watersheds.

Hydrologic impacts from upland watershed runoff and flooding from nearby streams can prove to be the most disruptive and can necessitate relocation of the pile on the site or to an alternate site. Under SIP or SOS, there are greater restrictions with regard to improving surface water drainage conditions. Upland watershed runoff is generally allowed to drain around a pile in direct contact with protected sideslopes, or is controlled with on-site drainage diversion channels. In either case, adequate sideslopes and toe protection must be provided to assure the longevity of the containment system. If materials are to be stabilized in a major floodplain, the design becomes increasingly difficult due to the magnitude of design flow depths and velocities. At this point, the design must also account for the effect that geomorphic changes can have upon the hydraulic conditions at a site. The design normally requires increased sideslope and toe protection to prevent flooding impacts. Geomorphic considerations (e.g., channel migration and undermining) require additional toe protection in the form of thickened perimeter rock aprons or buried riprap walls. Under severe situations where the longevity of the design is questionable or the design is economically impracticable, relocation to an alternate site becomes necessary.

Under the relocation option, the primary goal is to locate an alternate site that is (1) outside of any floodplain, and (2) at or near the head of any drainage areas. Under any of the options, the pile configuration is very important in affecting the hydrologic impacts. Consideration must be given to dividing the topslope drainage to avoid concentration of large drainage areas down 1 sideslope. The topslope can also be pitched away from 1 side to avoid drainage down a longer sideslope. To protect against off-site flooding impacts, it may be necessary to design the overall shape of the stabilized pile based on hydraulic conditions at the site. Within a floodplain, a pile may have to be narrowed substantially to a hydraulically smoother structure so as not to block flood flows. Below an upland watershed, the shape of the pile can be designed to divert drainage so that it affects only a small portion of the pile.

12.5.5 Disposal Cell Design

A disposal cell consists of 2 parts: the perimeter dike or embankment, and the top cover. Its design accordingly involves (1) examining the various possible alternative perimeter dike and cover details; (2) selecting appropriate perimeter and top cover details; and (3) combining appropriate details to constitute a complete disposal.

In design, the selection of the appropriate disposal cell details should lead to a balance of cost effectiveness and control of seepage to the extent necessary to meet applicable site groundwater standards. Furthermore, a balance is to be sought between providing for the physical stability and resistance to erosion of the cell, and the need to control and limit infiltration to the wastes.

While steady state seepage and transient drainage affect concentrations of hazardous constituents in groundwater, other groundwater factors must be considered in the design as they influence the longevity requirement of the disposal cell and protection of human health and the environment or other applicable state standards.

The depth to groundwater is important in that maximum fluctuations of the water table should not extend into the foundation of the disposal cell. If the water table inundates the contaminated materials, hazardous constituents may be leached at a rate that is substantially greater than the long-term seepage rate from the pile. Furthermore, the inundated materials may create stability problems or a liquefaction potential.

The hydraulic conductivity of the foundation materials should be adequate to accept seepage from the disposal cell as unsaturated flow. This will prevent the perching of seepage at the contact between foundation materials and the pile and, thus, the creation of a surface water seep. Similarly, the redistribution of moisture within the materials should be predicted to make certain that surface water seeps do not develop from the creation of a phreatic surface within the pile. In any case, the cover design must ensure long-term compliance with the EPA groundwater protection standards.

Under some circumstances, a geochemical underliner may be necessary to allow compliance with the groundwater standards. The geochemical liner could either be constructed as a layer or a geochemical sump into which seepage drains. Hazardous constituents with the potential to exceed concentration limits in groundwater could either be adsorbed onto the geochemical liner matrix, be chemically reduced, or be precipitated as a result of neutralization. The addition of geochemical layers must not adversely affect the geotechnical stability of the disposal cell. UMTRA never considers liners to be completely impermeable, otherwise they would suffer from the "bath tub effect," which would require leachate collection and treatment systems.

Another disposal cell design concern is the placement of the cell foundation, i.e., above or below grade. The depth to bedrock, depth to groundwater, proximity to an existing pit, availability of cover soil, possible use of excavated soils as restoration material, geomorphic stability, and availability of rock for erosion protection should be considered when locating the disposal cell. The relative importance of these considerations is site specific and should be compared for the most cost-effective design.

12.6 RADON BARRIERS

Radon emanation from unstabilized and uncontrolled tailings was 1 of the main driving forces behind the establishment of the UMTRCA. As a result, the regulations contain a radon design standard (40 CFR 192.02 (b)), which states that remedial actions should be designed to provide reasonable assurance that releases of radon-222 from residual radioactive materials to the atmosphere will not:

1) Exceed an average release rate of 20 picocuries per square meter per second, or
2) Increase the average annual average concentration of radon-222 in air at or above any location outside the disposal site by more than one-half picocurie per liter.

The radon barrier may serve a dual purpose in the design: (1) to reduce the radon emissions (flux) from the contaminated materials, and (2) to reduce the infiltration of water into the contaminated materials, especially in rock-covered piles.

The pile of configuration and type of stabilization influence the required amount of radon barrier. Once a pile exceeds a thickness of 7 to 10 feet, increased thickness does not affect the required radon barrier. Thus, minimizing the areal extent of a pile may, by increasing thickness, reduce the total radon barrier material needed. Contaminated materials can be layered with the less-contaminated material on top, resulting in less radon barrier material being required.

Stabilization in place provides the least amount of control over the location and layering of contaminated materials, and more radiological characterization data must be obtained in order to design the radon barrier properly. Average values of the parameters for the material in each distinct layer should be used to model the radon emissions because the EPA standards reference the site-averaged radon flux. For SIP, effective use of low radioactivity materials (e.g., windblown contaminated soils) as layers on top of the higher radioactivity materials may reduce the necessary amount of radon barrier materials.

Options requiring excavation and handling of materials (e.g., SOS or relocation to an alternate site) result in better mixing of the materials. This simplifies the design and radiological characterization data needs since the average radioactivity and other parameters for the mixed material may be used; however, sufficient characterization is necessary to ensure that the average values are adequately known. It is beneficial to retain low radioactivity materials separately for placement as a layer over the top of the other materials, as this lowers the radon flux and thus decreases the amount of radon barrier material needed.

If appropriate, acceptable mixing ratios of sandy tailings and slime may be established. This will be done only if it is shown by geotechnical analysis that such mixing ratios are required to control differential settlement.

Capping the radon-emitting site does nothing to eliminate the source of radioactivity from the area of concern. It simply impedes release by shielding and trapping. Thus, the cap must remain intact, without penetrations, indefinitely.

12.6.1 Rock Covers

On the UMTRA Project, rock covers have been standard elements in meeting the longevity and performance criteria. Rock covers consist of 3 components: (1) a radon/infiltration barrier of compacted soil; (2) a bedding/filter layer of sand; and (3) a top layer of rock riprap for erosion protection. Rock covers are generally best applied at the more arid sites and can be optimized so that seepage out of the pile is minimized, to promote compliance with the EPA groundwater standards. In addition to erosion and infiltration protection, rock covers must be designed to provide protection against the effects of freeze/thaw and biointrusion.

12.6.2 Vegetated Covers

A vegetated cover may be placed on the topslopes of tailings disposal cells as an alternative to a rock cover. Vegetated covers are generally not recommended for sideslopes because the vegetation may not be able to resist gullies originating on the steeper sideslopes or advancing headward from off the pile.

A vegetated cover consists basically of plants and soil, sometimes with other earthen materials, that have been selected to maximize transpiration and resistance to erosion. The soil and plants in a vegetated cover have specific performance objectives that must be met if the cover is to achieve its intended goal of controlling water balance, resisting erosion, and otherwise contributing to the long-term integrity of the stabilized pile.

The key to vegetated cover design is to use the proper combination of plants and soil to assure that some plants survive (even if dormant) the dry periods so that adequate transpirational capacity will be available after precipitation events to prevent moisture from infiltrating into the contaminated materials. A rock mulch may be required at exceptionally arid sites to resist evaporation (making more soil moisture available to plants during dry periods) and to supplement the plants' ability to resist erosion to the cover.

12.7 GROUNDWATER CONTAMINATION PREVENTION

The 2 most important techniques for prevention of groundwater contamination are:

1) Properly design the cover to prevent precipitation from entering the pile (section 12.6).
2) Properly design the pile to eliminate hydrologic impacts from upland watershed runoff and flooding from nearby streams (section 12.5).

Natural processes that influence contaminant migration through soils and aquifer materials include advection, mechanical dispersion, dilution, filtration of suspended or colloidal solids, biological decomposition of organic and inorganic compounds, ion exchange, specific adsorption, neutralization, physical adsorption, neutralization, precipitation of dissolved chemicals, ion sieving by dense clay layers (ultrafiltration), and decay of radioactive elements. Geochemical interactions between the soil or aquifer matrix and the contaminants may attenuate concentrations of the contaminants along the flow path. Attenuation mechanisms may include neutralization of seepage from the tailings, precipitation of solid compounds, sorption of trace metals and toxic non-metals, and oxidation/reduction followed by mineral precipitation.

The neutralization capacity of the soil and aquifer material beneath an acidic mill tailings pile is the single most important chemical factor in determining the ability of geologic material to chemically attenuate the movement of contaminants. This neutralization is the main driving mechanism for mineral precipitation and for adsorption in these systems. The increase in solution pH that characterizes neutralization in this environment produces a condition in which the solubility of iron and aluminum oxyhydroxides decreases. These solids precipitate and scavenge other contaminant trace elements (uranium, manganese, arsenic, selenium, and molybdenum) from solution.

Various types of barriers may be useful in the protection of groundwater resources at some uranium mill tailings disposal sites. Barriers can be classified as physical flow barriers or geochemical barriers. Physical barriers act to impede the flow of seepage or groundwater through reduction in permeability along the flow path or an induced change in hydraulic gradient. These barriers include liners placed below the contaminated materials, slurry walls, grout curtains, and regions of artificially induced high hydraulic head. Geochemical barriers consist of material placed in the flow path of contaminants that will react with and immobilize hazardous and radioactive constituents.

Physical barriers such as grout curtains and slurry walls can usually only delay the transport of constituents from the disposal facility. They can force the contaminants to follow a longer flow path, which might provide greater potential for geochemical attenuation. However, because the point of compliance is at the down-gradient edge of the area where the waste is placed, these types of barriers are usually not useful for meeting groundwater protection standards. Barriers consisting of areas of high hydraulic head created through injection of water are not acceptable as a long-term remedial action because they rely on maintenance.

Vertical barrier walls may be installed around the contaminated zone to help confine the material and any contaminated groundwater that might otherwise flow from the site. The barrier walls, which might be in the form of slurry walls or grout curtains, would have to reach down to an impermeable natural horizontal barrier, such as a clay zone, in order to be effective in impeding groundwater flow. A barrier wall in combination with a surface cap could produce an essentially complete containment structure surrounding the waste mass.

Slurry walls are constructed by excavating a trench under a slurry. The slurry could be bentonite and water or it could be Portland cement, bentonite and water. In cases where strength is required of a vertical barrier, diaphragm walls are constructed with pre-cast or cast-in-place concrete panels.

Grout curtains are constructed by pressure-injecting grout directly into the soil at closely spaced intervals around the waste site. The spacing is selected so that each "pillar" of grout intersects the next, thus forming a continuous wall or curtain. Various kinds of grout can be used, such as Portland cement, alkali silicate grouts, and organic polymers.

Sheet piling involves driving lengths of interconnected steel into the ground to form a thick, impermeable, permanent barrier to groundwater flow. Sheet piling is generally used with drains or pumping shallow wells that lower the potentiometric surface on the inside of the barrier and allow contaminated groundwater to be contained and extracted. The hydraulic barrier created by sheet piling can prevent lowering of the water table on the other side of the barrier so that existing wells outside the contaminated area are not affected.

A geochemical barrier placed beneath the contaminated materials in a disposal facility is a possible method for immobilizing hazardous and radioactive constituents before they reach the uppermost aquifer. The geochemical barrier would contain materials that adsorb or react with hazardous and radioactive constituents. For example, a site lacking enough natural neutralization potential might use a geochemical barrier of some material with limestone to neutralize acidic seepage from the tailings and react with constituents to form insoluble solids. Limestone ($CaCO_3$), and hydrated lime ($Ca(OH)_2$) have been studied as neutralizing agents for uranium tailings solution.

Vertical barriers could be considered for use to prevent or delay escape of liquids and perhaps gases (if installed in combination with a cap), until a more desirable permanent remediation technology is adopted.

Barrier walls could be considered only for large discrete masses of waste materials or around several smaller masses close together. Barrier walls are not totally impermeable to water.

Liners: Residual radioactive materials at Title I processing sites generally do not contain water above the specific retention of the materials. Therefore, even if the tailings are relocated, a liner is generally not required. If water is added for dust control or other purposes when relocating waste, however, it must be demonstrated that the as-built moisture content is less than the specific retention. Otherwise a liner or "equivalent" is required. The equivalent of a liner is not specified. Presumably, this could be a natural low-permeability soil or rock underlying the disposal site. The material would have to limit flow from the Title I disposal facility to the degree needed to ensure long-term compliance with the EPA groundwater protection standards. The performance assessment would have to show that drainage of residual moisture through the liner or equivalent would not cause regulated constituents in groundwater at the point of compliance to exceed concentration limits during the specified control period.

For Title II sites, the liner system must prevent migration of waste into underlying soil during the designed control period, or it must be demonstrated that an alternate design will prevent the migration of any hazardous constituents into groundwater or surface water at any future time, or it must be demonstrated that an alternate system and operating practices will provide protection of groundwater and surface water that is at least as effective as a liner and leachate collection system. These exceptions to the liner requirement provide flexibility in designing a site-specific disposal facility. Natural conditions and operating practices may be relied on in some cases to provide containment. Such a design would have to be supported by an accurate and defensible analysis of site conditions. Factors to be considered in deciding if an alternate design is acceptable include the nature and quantity of wastes, alternate design, hydrologic setting, attenuative capacity of subsoils between the impoundment and the uppermost aquifer, and all other factors that would influence the migration of hazardous constituents from the impoundment.

12.8 GROUNDWATER RESTORATION

Active restoration methods may be identified for some sites and generally fit in 1 of 2 categories: (1) above-ground treatment methods, wherein the contaminated water is removed from the aquifer, treated, and either disposed of, used, or reinjected into the aquifer; and (2) in situ methods, such as the addition of chemical or biological agents to fix contamination in place. An aquifer restoration program at a site may involve 1 or more restoration methods.

12.8.1 Natural Flushing

With natural flushing, contaminants in groundwater are dispersed or removed by the natural flow of groundwater. Institutional controls would be used along with natural flushing; these controls must effectively protect human health and the environment and, where practicable, satisfy beneficial uses of groundwater during natural flushing.

Natural flushing could be employed as the sole method for groundwater remediation, or it could be used in conjunction with active methods. For example, natural flushing could be used in conjunction with gradient manipulation to control the migration of groundwater contaminants and to ensure that the proposed EPA groundwater compliance standards are achieved within the 100-year period. Natural flushing could be a useful method of groundwater remediation at sites with high groundwater velocities, locations near points of groundwater discharge into surface bodies, or aquifer properties that disperse and/or absorb contaminants.

Generally, aquifers with high groundwater velocities and dispersivities have the capacity to decrease contaminant concentrations by the processes of dilution and dispersion. When a groundwater contaminant plume discharges into a river, groundwater contaminants may be diluted because the groundwater discharge volume is very small compared to the volume of river flow.

12.8.2 Gradient Manipulation

Gradient manipulation uses either wells or trenches to add water to an aquifer to increase groundwater velocity in a specific direction. Gradient manipulation could be used to deflect a contaminant plume into a surface water body that is large enough or flows quickly enough to dilute the groundwater contaminants to concentrations within regulatory limits. Conversely, gradient manipulation could be used to prevent discharge of a contaminant plume into surface water bodies by creating a hydraulic diversion to contaminated groundwater flow. Gradient manipulation could be used in conjunction with natural flushing to decrease concentrations at a faster rate and to prevent the migration of contaminants into areas where groundwater was not previously contaminated or where institutional controls cannot be effectively applied.

12.8.3 Contaminant Isolation

Groundwater contamination sources could be in the form of highly contaminated water or adsorbed hazardous constituents in the unsaturated zone above the water table. Zones of highly contaminated groundwater below a processing site could also be considered a contamination source. These types of groundwater contamination sources could be mitigated or eliminated through engineered measures to control or contain their hazardous constituents.

Surface water control, and groundwater barriers could be used to decrease the rate of contamination or keep a contaminant source from entering the groundwater. These technologies could prevent hazardous constituents remaining in soils at processing sites from migrating into the groundwater or prevent highly contaminated leachate in saturated materials from migrating into an aquifer. For areas of highly contaminated groundwater under a former tailings pile, techniques could be employed to isolate this groundwater for more efficient extraction. Because of the expense involved, these applications would be limited to small areas of highly contaminated material or groundwater.

12.8.4 Groundwater Extraction

Groundwater extraction controls movement of contaminated groundwater and removes it from the aquifer. In many cases, it would be necessary to extract groundwater only from the most highly contaminated zones. Groundwater flow information and groundwater hydraulic parameters would be used to design the number, depths, spacing, and pumping or injection rates of wells. With the aid of groundwater flow models, the time required for the remedial actions could be estimated.

Well systems could be used to extract contaminated groundwater for treatment or to create hydraulic barriers to groundwater flow and increase the efficiency of extraction. These wells would then be pumped at specified rates to control the movement of contaminated groundwater. In some cases, it could be necessary to combine periods of well pumping with periods of no pumping. When pumping has stopped, contaminants can diffuse out of less permeable zones or desorb from the aquifer matrix until equilibrium concentrations are reestablished in the groundwater. Subsequent pumping can then remove the minimum volume of contaminated groundwater at the maximum possible concentration.

Well point systems consist of closely spaced, shallow wells connected to a pipe with a centrally located suction lift pump. These systems can create an effective hydraulic barrier by capturing contaminated groundwater. Well point systems would be used mainly for shallow water table aquifers, because the maximum drawdown obtainable by suction lift is limited to approximately 25 feet (ft) (8 meters (m)). Because well points are smaller in diameter and shallower than monitor wells, they are simpler and cheaper to install.

12.8.5 Contaminated Groundwater Treatment

Once contaminated groundwater is extracted from an aquifer, it may be necessary to treat it to protect human health and the environment. The need for treating extracted contaminated groundwater before it is discharged depends on the concentrations of contaminants in the extracted groundwater and the regulations regarding discharge of effluent to the surface and groundwater. Once treated, groundwater could be discharged to surface water bodies, recharged back into a shallow aquifer, or used as irrigation water for agricultural purposes.

Treatment processes:

a) Precipitation
b) Coagulation/Flocculation
c) Oxidation/Reduction
d) Ion Exchange
e) Membrane Separation
f) Adsorption
g) Biological Systems
h) Sedimentation
i) Filtration

12.8.6 In Situ Groundwater Treatment

In situ treatment entails chemical, physical, or biological agents in the affected ground or groundwater that degrade, remove, or immobilize the contaminants. It also includes methods for delivering solutions to the subsurface and methods for controlling the spread of contaminants and treatment reagents beyond the treatment zone. In situ treatment of groundwater is for the most part an innovative technology, with some of the in situ processes not yet proven beyond the test phase.

In situ treatment processes are generally divided into 3 categories: biological, chemical, and physical. In situ biodegradation, commonly called bioremediation, is based on acceleration, or enhancing the rate of bioflora to metabolize the organic contaminants. At mill tailings sites, bioremediation would be used to solubilize or immobilize inorganic contaminants in the water or soil. In situ chemical treatment involves the injection of a specific chemical or chemicals into the ground or groundwater to degrade, immobilize, or release contaminants that are in the groundwater or attached to the soil particles. Physical in situ methods involve the physical change of the soil or groundwater using heat, electric energy, or other means to immobilize or to expedite the release or movement of contaminants from the soil or water. In some instances, a combination of in situ and aboveground treatment would be required to achieve the most cost-effective treatment.

In situ treatment of contaminated groundwater would require extensive site characterization to determine the contamination level and areal extent of contamination in the soils, the aquifer matrix, and the groundwater. A successful in situ treatment system that utilizes injection of materials for chemical or biological treatment must provide the following:

1) Adequate contact between treatment agents and contaminated solids or groundwater.
2) Hydrologic control of treatment agents and contaminants to prevent their migration beyond the treatment area.

3) Complete recovery of spent treatment solutions and/or contaminants when necessary.

In situ treatment is applied either by gravity flow through infiltration galleries or drains or by pressure through injection and extraction wells. Where it is desirable to treat contamination in the unsaturated zone and the soils are relatively permeable, or the groundwater contamination is relatively shallow and under water table conditions, a gravity flow system could be used. However, if the depth to groundwater is more than 20 feet (6 meters) and the contaminants are distributed over a fairly deep profile within the aquifer, injection and extraction wells would be required. This involves installing a bank of injection wells along the upgradient edge of the groundwater contamination and within the contaminant plume. A treatment agent would be pumped into the aquifer through the injection wells. Extraction wells could be required to capture treatment agents and provide potentiometric control of the system or to allow for additional abovegrade treatment of contaminated groundwater before discharge or reinjection.

Physical in situ treatment systems would generally be useful for a limited area, because the costs of physically manipulating the ground and/or groundwater to immobilize or detoxify contaminants would be high.

Specific treatment agents could be used to mitigate different types of contamination. The 2 major types of in situ treatment hazardous constituents are microbiological and chemical processes.

Microbiological treatment has been applied widely in situations where organic materials, including hydrocarbon fuels, solvents, and pesticides, have been released into the environment through spills, leaking transfer systems, and storage tanks. It also may have some efficacy in the treatment of nitrates in groundwater. Bioremediation could also be used to produce biomass within an aquifer for metal sorption. Under certain conditions, microorganisms could solubilize heavy metals, which would aid in removing them from the aquifer. Environmental factors that influence the effectiveness of in situ biodegradation are the dissolved oxygen level, pH, temperature, predators and competition (including the presence of toxins and growth inhibitors), oxidation/reduction potential, availability of nutrients, salinity, and concentration of compounds that need to be biodegraded.

In situ chemical treatment can be used to immobilize, mobilize (for extraction), or detoxify inorganic and organic hazardous constituents. Technologies for immobilization include precipitation, chelation, and polymerization. Treatments to mobilize contaminants for extraction include the addition of lixiviants (flushing agents), dilute acids or bases, and water. Detoxification includes oxidation, reduction, neutralization, and hydrolysis. to some extent, each of these treatments may serve 1 or more purposes.

Two types of in situ chemical treatment could be considered for the removal of hazardous constituents from groundwater at mill tailings sites. The first is to add chemical lixiviants to the injection solutions to enhance the mobility of hazardous constituents in groundwater during either natural flushing or extraction for treatment. The second is to construct permeable treatment beds that function as geochemical barriers to remove and immobilize hazardous constituents in groundwater.

12.9 INNOVATIVE TREATMENT PROCESSES

Two innovative treatment processes, 1 for soil clean up, and the other for wastewater treatment are described below.

12.9.1 Heap Leaching

Heap leaching for soil clean up is an adaptation of a proven mining method for removing precious and semiprecious metals from low-grade ore. In the mining industry, thousands of tons of ore are processed daily for less than \$100 yd^3. Metals remaining in the depleted ore are measured in parts per billion. The benefits of this form of soil remediation, when proven at field scale, would include:

1) On-site clean up;
2) Low-cost processing;
3) Conservation of expensive repository space; and
4) Elimination/reduction of long-term costs for monitoring, isolation, and habitat protection of disposal areas.

Additionally, the method would provide the ability to closely examine, sort, treat, and verify the cleanliness of the soil. This methodology will provide a permanent solution to the soil contamination problem, not just a relocation of the soil and cost for future care.

Aqueous solutions of carbonate/bicarbonate of either potassium or sodium are being tested because they are the reagent of choice in the uranium mining and processing industry. Future reagents, however, may depart from carbonate/bicarbonate. During leaching tests, reagent is pumped from a dewar to the columns. After being introduced at the top of the column, the reagent flows downward by gravity, solubilizing and mobilizing the uranium from the soil.

Pilot-scale tests are envisioned where contaminated soil will be excavated and placed (heaped) on an impermeable pad on the surface of the ground. The pad will be sloped toward a sump at the bottom edge of the heap.

Selected leaching reagent(s) will be pumped to and distributed on top of the heap with a drip irrigation system or aerial sprayers. Reagent will travel down through soil, solubilizing and mobilizing the contaminant(s). The leachate will then be collected from the sump and pumped to a leachate treatment and regeneration system. This system will remove the contaminant(s) from the leachate and regenerate the leaching reagent for return to the top of the heap.

The process will be continued nonstop until the contaminant(s) in the soil have been reduced to EPA standards.

Participants in this study include: New Mexico State University, University of New Mexico, Rust Geotech, Inc., Hazen Research, Inc., and DOE Los Alamos.

12.9.2 Biological Techniques

A task at Oak Ridge will develop a biological metal binding technology for the removal of uranium and other metals from uranium mill tailings wastewater.

The project will focus on biosorbent material capable of reducing dissolved uranium levels within a simulated wastewater from part-per-million to part-per-billion levels. This biosorbent material will be immobilized and encapsulated in a gel-based matrix that also possess biosorptive capacity, resulting in a combined biosorbent material suitable for deployment within flow-through reactors.

Experimentally, as the uranium-laden water flows through the columns, the bacteria absorb the heavy metal. Immobilization of the biosorbent will permit continuous process operation without generation of unacceptable organic residues within the effluent, in both respects reducing the overall treatment cost. Use of inexpensive sorbent materials may further reduce costs significantly relative to conventional processes.

Pseudomonas aeruginosa was selected as the target organism and customized for use in remediation experiments.

Another project is being studied at McMaster University, Hamilton, Ontario, Canada. The technology's core consists of certain fungi and bacteria that have been killed and immobilized in special polymers. For reasons that aren't fully understood, the microbes can selectively snag atoms of uranium and other radioactive elements. The captured elements then can be separated from the dead microbes and processed into fuel for nuclear reactors or otherwise disposed of.

13

High-Level Waste Treatment

Across the DOE complex, 332 Underground Storage Tanks (USTs) have been used to capture, process, and store radioactive and hazardous wastes generated from the production and processing of nuclear weapons materials since the 1940s.

The majority of the waste is stored in USTs of various design types ranging in capacity from 5,000 gallons to 1,300,000 gallons at 5 DOE sites: the Hanford site in Richland, Washington; the Fernald site in Fernald, Ohio; the Idaho National Engineering Laboratory (INEL) near Idaho Falls, Idaho; the Oak Ridge Reservation in Oak Ridge, Tennessee; and the Savannah River Site in Savannah River, South Carolina.

A small amount of high-level waste is generated in the commercial activities of the New York State Energy Research and Development Authority, and stored at West Valley, NY.

The total DOE high-level waste volume of about 381,000 cubic meters contains some 1.1 billion curies of radionuclides. At this volume Hanford accounts for about 63%, Savannah River (SRP) for about 32%, and INEL about 3%.

While most of the defense waste by volume is contained in waste tanks at Hanford, most of the waste in terms of radioactivity is contained in waste tanks at SRP, with a total radioactivity of the SRP waste of more than 700 million curies.

Two waste storage tank design types are prevalent across the DOE complex: single-shell wall and double-shell wall designs. They are made of stainless steel, concrete, and concrete with carbon steel liners, and their capacities vary from 5,000 gallons to 10^6 gallons. The tanks have an overburden layer of soil ranging from a few feet to tens of feet.

Tank waste consists of several physical forms: sludge, supernatant, and salt cake. Most of the waste is alkaline, and contains the following chemical constituents: nitrate and nitrite salts (approximately half of the total waste), hydrated metal oxides, phosphate precipitates, and ferro-cyanides. The 640 MCi of radionuclides are distributed primarily among the transuranic (TRU) elements and fission products, primarily Sr-90 and Cs-137. In-tank atmospheric conditions vary in severity from near ambient to temperatures over 93°C; tank void-space radiation fields can be as high as 10,000 rad/h.

13.1 HANFORD HIGH-LEVEL WASTE

In 1943, the U.S. Army Corps of Engineers selected the Hanford Site near Richland, Washington to build the first plutonium production reactors and processing facilities as part of the Manhattan Project. Eight plutonium production reactors and 1 combination plutonium-steam production reactor were built and operated between 1944 and 1987. Chemical reprocessing plants were constructed and operated to recover plutonium and uranium from the irradiated fuel produced in these reactors. This reprocessing resulted in the accumulation of a wide variety of radioactive (i.e., transuranic, high-level, and low-level waste (LLW) and chemical waste). High-level and chemical liquid wastes were stored in single-shell tanks (SSTs) and in double-shell tanks (DSTs) and remain there today.

The Hanford Site has been dedicated to cleaning up contaminated wastes in preparation for final decommissioning and closure, which will occur during the next 40 years. Waste will be retrieved from SSTs beginning in the year 2003 and ending in 2018. Wastes in 1 tank will be retrieved beginning in 1997 to resolve a high-heat safety issue.

Hanford Site tanks have been used to store radioactive reprocessing waste since the 1940s. Until the early 1970s, most reprocessing waste was stored in underground, reinforced concrete, carbon-steel-lined SSTs (Single Shell Tanks). A total of 149 SSTs, having capacities from 55,000 gallons to 1 M gallon, have been used. In the 1960s and 1970s, radioactive strontium and cesium were extracted from wastes in some SSTs. Storage of new waste in these SSTs ceased in 1980. The SSTs contain approximately 36.5 M gallons of radioactive waste. The 149 SSTs are grouped into 12 tank farms. Since 1971, underground, reinforced concrete, carbon-steel-lined DSTs (Double Steel Tanks) have been used for storage of liquid waste. By 1981, large quantities of pumpable liquid waste had been removed from SSTs and placed in DSTs.

As a result of using several different plutonium recovery and radioisotope separations processes at the Hanford Site, the chemical and radionuclide compositions of existing individual tank contents vary significantly. Volumes and compositions of the wastes generated were strongly dependent upon the process used. Methods for treating the waste in the tanks also had major impacts on the compositions of tank contents. These treatment methods included the following:

1) In-tank scavenging of strontium and cesium by the precipitation of strontium phosphate and cesium ferrocyanide to reduce the concentration of Sr-90 and Cs-137 in supernatant liquids and disposal of the supernatant liquids as LLW.
2) Removal of Sr-90 and Cs-137 at B Plant to reduce in-tank heat generation and allow concentration of the remaining waste.
3) Concentration of tank contents by evaporation of water to crystallize the waste as a saltcake.

Tank wastes have been released to the ground as a result of leaks from SSTs and associated transfer lines, and other miscellaneous spills. Sixty-seven SSTs are assumed to have leaked a total volume of approximately 600,000 to 900,000 gallons.

In addition to the 67 assumed leaking tanks, at least 100,000 gallons of liquid wastes are estimated to have been released to the ground as a result of unplanned releases and spills. The information available for these releases and spills indicates generally low levels of radioactivity. One significant release of long-lived fission products to the ground occurred in 241-C Tank Farm between 1969 and 1971. A release in 1971 resulted in the addition of an estimated 9.25×10^{14} Bq (25,000 Ci) Cs-137 to the ground.

In addition, perhaps as much as one million gallons of clean cooling water was sprayed into a single-shell tank 241-A-105 in the 1970s to aid in evaporative cooling. It is likely that much of this water (50,000 to 800,000 gallons) did not evaporate and therefore, may have leached some of the sludge and added to the waste released from this tank. Past practice was to exclude the cooling water from the leak volume estimate.

Interim stabilization involves the removal of supernatant and interstitial liquid from the SSTs to the extent technically and economically feasible. Isolation of an SST involves physical modifications to tank structures to preclude the inadvertent addition of liquid to the tank.

The purpose of interim stabilization is to minimize the spread of contamination if the tanks begin to leak. The SSTs containing more than 50,000 gallons of drainable liquid (or more than 5,000 gallons of free-standing supernatant) are pumped. SSTs containing less than this amount are not pumped because attempting to remove the residual liquid would result in no significant decrease in risk to public health and the environment and radiation doses to operating personnel would increase. Approximately 100 SSTs have been interim-stabilized to date. Except for Tank 241-C-106, all interim-stabilization activities will be completed by the year 2000.

The DOE disposition of DST and SST waste, and cesium and strontium capsules at the Hanford Site in the *Final EIS, Disposal of Hanford Defense High-Level, Transuranic and Tank Wastes, Hanford Site, Richland, Washington* (DOE 1987). This document is frequently referred to as the Hanford Defense Waste (HDW) Environmental Impact Statement (EIS) or HDW-EIS. In April 1988, in the Record of Decision (ROD) for the HDW-EIS, DOE decided to proceed with preparing the readily retrievable DST waste for final disposal. Wastes were to be processed in a pretreatment facility (planned to be the Hanford Site B Plant and AR Vault) to separate DST waste into 2 portions. The larger portion would be low activity waste, whereas a much smaller portion would be highly radioactive. The low-activity waste was to be mixed with a cement-like material to form grout. The grout was to be poured into large, lined, concrete, near-surface underground vaults. The high-activity fraction was to be made into a borosilicate glass and poured into stainless-steel canisters at a Hanford Site Waste Vitrification Plant. The canisters were to be stored at the Hanford Site until a geologic repository was ready to receive this waste.

In the HDW-EIS ROD, DOE decided to conduct additional development and evaluation before making decisions on final disposal of SST wastes. This development and evaluation effort was to focus on methods to retrieve and process SST wastes for disposal. The SST waste was to continue to be stored and monitored in the interim. Before a decision on the final disposition on these wastes could be made, the alternatives for the final disposal were to be analyzed in a supplement to the HDW-EIS.

First priority is placed on retrieving waste from tanks on the safety Watch List. Retrieval will then proceed farm-by-farm, based on available funding and completion of necessary infrastructure upgrades, to provide feed for waste treatment operations.

The DOE must retrieve waste from the SSTs to the extent practicable to meet the requirements of the Resource Conservation and Recovery Act (RCRA) for hazardous waste, and the Nuclear Waste Policy Act for nuclear waste. Radioactive waste has been retrieved successfully from Hanford Site underground tanks in the past using sluicing technology. However, the highly variable waste characteristics, degradation of the tank structures and systems, and changes in regulations make the present task of retrieval a more difficult challenge.

The tank system consists of 177 tanks grouped in 18 tank farms. The SST portion of the system consists of 149 tanks grouped in 12 tank farms. The SSTs have the following 4 capacities:

1) Sixteen tanks have a 55,000 gallon capacity.
2) Sixty tanks have a 533,000 gallon capacity.
3) Forty-eight tanks have a 758,000 gallon capacity.
4) Twenty-five tanks have a 1 M gallon capacity.

There are 133 SSTs classified as 100 series tanks and 16 classified as 200 series tanks. All 100 series tanks are 75 feet in diameter with domed tops. Tank volumes are either 533,000 gallons, 758,000 gallons, or 1 M gallons). The 533,000 gallon and 758,000 gallon tanks originally were arranged in "cascades" of 3, 4 or 6 tanks. These tanks were connected by piping in a manner such that when the first tank in a cascade was filled it overflowed to the second tank, which overflowed to the third tank, and so on.

Access to SSTs is by risers penetrating the dome of the tank. Risers are vertical pipes or ducts ranging from 4 inches to 52 inches in diameter. Both sampling and retrieval efforts have been conducted using risers for access. The number of risers available for sampling varies from tank to tank, depending on the number of risers existing on the tank, location in the tank, and the equipment that may be in or around the riser. Technology is being developed to add risers to some tanks if an evaluation finds that the potential for biased results exists because samples are taken using only existing risers. The waste retrieval program has proposed systems using the existing risers in the tanks and new openings of up to 48 feet in diameter.

The 200 series tanks are similar in construction to the 100 series tanks. They feature a 20 foot diameter, a capacity of 55,000 gallons, and a flat top. These tanks are covered with approximately 12 feet of soil. These 16 tanks are located in the 241-B, 241-C, 241-T, and 241-U Tank Farms in groups of 4.

A total of about 37 M gallons of waste is stored in SSTs. Of this waste, about 600,000 gallons is supernatant, 23 M gallons is saltcake, and 12 M gallons is sludge. The saltcake consists of a solid form of the various salts created by evaporating liquid alkaline waste. The saltcake consists primarily (approximately 93 wt%) of sodium nitrate and sodium nitrite. The sludge consists of the solids (hydrous metal oxides) precipitated in the neutralization of acid waste before it could be transferred to the SSTs. Some of the salt associated with hot slurries from the evaporators precipitated with the sludge. As a result of the precipitating salt, roughly 50 percent of the reported sludge volume is saltcake. The liquid solution in SSTs exists as supernatant and interstitial liquids. An estimated 6 M gallons of drainable interstitial liquid is present in SST saltcake and sludge. The amounts of saltcake, sludge, and liquids vary widely in individual tanks.

The SSTs primarily contain inorganic waste, although relatively small amounts of organic wastes, such as solvents, are present that were entrained with the aqueous waste during fuel reprocessing. Also, water-soluble complexing agents and carboxylic acids added in a fractionation process are in some SST wastes. Chemical reactions (e.g., oxidation-reduction, neutralization, precipitation) and radiolysis have converted many of these chemicals into different compounds.

The estimated chemical compositions of SST wastes are based on process records of fuel elements processed, chemicals used, and a limited number of waste sample analyses. Chemical compositions of the wastes vary widely from tank to tank.

The radioactive components of SST wastes primarily consist of fission product radionuclides, such as Sr-90 and Cs-137, and actinide elements, such as uranium, plutonium, and americium.

The 28 newer type V (DST) tanks are built with secondary carbon steel barriers in what is known as the "tank-within-a-tank" design. The lined cylinders are then capped by concrete roofs and covered under 5.9 to 8.8 feet of soil and gravel.

A method is needed to detect the presence of leaks and locate the leaks to assist in soil clean up. Current methods of detecting leakage are not always satisfactory. Careful measurement of tank inventory is a simple method, however, the leak size is limited by complicating factors like temperature variations and the method can be impractical if the tank contains both solid and liquids. Moisture, chemical or radiological sensors can be placed beneath a tank to detect the effects of contamination in subsurface soils. However, in an in homogeneous soil, plumes can be highly channeled and these sensors can easily miss spills.

DOE's approach to cleaning up the single- and double-shell tanks' high-level waste involves 5 basic steps, which can be summarized as follows:

1) **Characterization** involves the determination of the specific physical, chemical, and radiological components of the wastes in each tank. This step is important because DOE's current information about tank wastes is incomplete. Some of the waste was placed in the tanks with little or no documentation of its makeup, and some tanks contain a complex mixture of unknown waste constituents. Detailed knowledge of tank contents is needed to determine how to resolve tank safety issues as well as how to retrieve, pretreat, and treat the wastes. To characterize the waste, DOE plans to analyze samples drawn from each tank.

   Characterization is a difficult, but important task. Not only do waste compositions vary between the sites, but they vary between tanks at the same site. Waste compositions also vary considerably within the same tank.

   An evaluation of the analytical results from 3 core samples taken from one of the first 2 tanks shows that tank waste concentrations vary significantly not only between different cores (horizontal variation), but also between different segments of the same core (vertical variation). For example, GAO found large vertical variations in concentrations for several constituents. Further, although horizontal variations in concentrations were relatively small for many constituents, for other constituents, the variations were very large. The large vertical and horizontal variations in concentrations of important constituents, at least in this one tank, emphasize the difficulty in characterizing the wastes by analyzing a small number of core samples.

   Westinghouse initially estimated that the cost to analyze each core sample was about $200,000. Subsequent changes to the draft waste characterization plan, in response to comments made by a panel of experts from the National Academy of Sciences and officials from Ecology, raised the estimated cost to approximately $430,000 per core sample. The principal change involved analyzing each 19 inch core sample segment, in addition to analyzing a composite sample as originally planned.

2) **Retrieval** is the removal of the waste from the tanks by pumping or other means and its transfer to treatment facilities. Because the waste exists in liquid, solid, and other forms, certain steps may be needed to turn the waste into a form that will allow the pumping to take place. Solid waste that cannot be pumped from any of the single-shell tanks may have to be extracted with specially designed robotic arms.

3) **Pretreatment** is the separation of the high-level fraction of the waste from the low-level fraction and from other non-radioactive elements, such as aluminum, organic compounds, and salts. This step is desirable because it decreases the amount of high-level waste that must be vitrified. The remaining low-level waste can then be treated and disposed of less expensively.

4) **Treatment** involves the immobilization of the waste. DOE plans to vitrify the high-level fraction of the waste separated during pretreatment by mixing it with a glass-forming material and melting the mixture into glass. As planned, the molten glass will be poured into stainless steel canisters to harden; each container is about 2 feet in diameter and 10 feet tall. DOE plans to immobilize the remaining low-level fraction of the waste by mixing it with cement, flyash, and other materials so that it will harden into a cement-like substance called grout. Vitrification is discussed in chapter 11.

5) **Disposal** involves the final emplacement of the immobilized waste so as to ensure isolation from the surrounding environment until it is no longer dangerously radioactive. DOE plans to temporarily store the canisters containing the high-level fraction of the waste at the Hanford Site until an underground repository is ready to receive them permanently. It will dispose of the low-level fraction of the waste in large, underground concrete vaults at the Hanford Site; each vault will hold about 1.4 million gallons of grout.

Tank waste exists in a variety of forms, such as liquid, sludge, slurry, and solid waste. Removing such diverse waste for treatment requires such steps as dislodging, pulverizing, cutting, and pumping. To deal with this diversity, DOE will need to develop several different retrieval methods. DOE has used various waste retrieval techniques over the years, such as dislodging the sludge with high-volume sprayers and pumping it out of the tanks, but other techniques for retrieving the large volumes in Hanford's tanks are still largely in the conceptual stage of development.

DOE's ability to meet waste processing milestones depends in part on the availability of a steady supply of waste from the tanks. DOE faces potential problems both in readying its retrieval methods for the first wastes to be removed and in developing retrieval methods for the remaining wastes to ensure this steady supply. DOE acknowledged that it faces numerous technical challenges, including acceptable retrieval techniques that balance available technologies, retrieval rates, and potential leaks to the soil during retrieval.

The 7 million gallons of double-shell tank waste that will be pretreated consist of 4 different waste types, which result from different nuclear fuels or reprocessing techniques. The 4 waste types are neutralized current acid waste, neutralized cladding removal waste, plutonium finishing plant waste, and complexant concentrate.

The first type of waste DOE intends to remove is called "neutralized current acid waste" (NCAW). According to DOE officials, this waste was placed in 2 double-shell tanks and is relatively well defined. It constitutes about 6 percent of the waste in double-shell tanks. The second type of waste DOE plans to retrieve is from one of its single-shell tanks; this waste is believed to be similar to NCAW.

In retrieving the NCAW waste from the double-shell tanks, DOE plans to first mix the sludge using mixer pumps installed in the tanks. The pumps, which are to be placed under the surface of the tank liquid, will direct a jet stream of liquids already in the tanks onto the surface of the waste to dislodge or dissolve it. The dislodged or dissolved waste is then pumped out in the resulting liquid to a double-shell tank for subsequent pretreatment.

The effectiveness of sludge mixing—at least in the manner DOE will have to use it—has not been fully demonstrated. DOE has tested sludge mixing at Savannah River and in model studies at Hanford, but only with an approach that uses 2 mixing pumps. According to the Westinghouse retrieval manager, 2 pumps may be sufficient only for the limited number of tanks that contain wastes in liquid and near-liquid form. For thicker wastes, DOE plans to use 4 pumps. The Westinghouse manager stated that DOE has not yet specified the number of pumps that individual tanks will require. DOE has expressed concern that using 4 pumps to mix tank wastes may result in unacceptable stress on tank walls that may permanently damage the tanks and result in radioactive waste leaking into the environment. DOE told GAO it has planned a series of laboratory and scale tests to resolve these uncertainties, and it plans to conduct tank retrieval operations within allowable stress limits so that the tanks will not be damaged.

One waste that DOE considers a potential next choice is contained in one of Hanford's single-shell tanks known as tank 106-C. On the basis of limited characterization and available records, DOE believes the high-heat waste in this tank is similar to the NCAW waste that will be retrieved, and it has stated in formal briefings to DOE headquarters that retrieval of this waste and NCAW could provide up to 8 years' supply for the vitrification plant. However, the sludge mixing method to be used to retrieve the NCAW waste may not be suitable for retrieving tank 106-C waste, because the high-pressure pumps could permanently damage the fragile walls of this single-shell tank, allowing waste to leak out. Consequently, DOE may use a retrieval method called "sluicing" to retrieve wastes by directing water from an external source onto the surface of the waste. The dislodged or dissolved waste is then pumped out in the resulting liquid. However, this method may be unacceptable because it increases the volume of waste. In addition, DOE and Washington State are trying to determine whether this tank is leaking. In the meantime, DOE is continuing to develop and test the sluicing method with the goal of minimizing the amount of water to be used.

Under DOE's plans, neutralized current acid waste, the first waste to be vitrified from the double-shell tanks, requires 2 pretreatment processes—in-tank sludge washing and ion exchange.

In-tank sludge washing consists of washing tank wastes with water using large pumps inserted into the top of the tanks. Sludge washing is not the same as sludge mixing. Unlike sludge mixing, it introduces large volumes of additional outside water into the tanks. This approach will allow solid, high-level waste to settle to the bottom of the tank, while lower-level waste will remain near the top where it will be pumped to holding tanks for further pretreatment. DOE will wash the waste in this manner at least twice to separate the high-level waste solids from lower-level waste.

Ion exchange is the second step in pretreating NCAW. The supernate, or liquid waste that remains near the top of the tank during sludge washing, will contain cesium 137, a highly radioactive material that must be separated through further pretreatment. To separate the cesium 137, DOE plans to use an ion exchange process in which the supernate liquid containing cesium passes through tubes containing resin, a material that adsorbs it. As currently planned, the separated cesium will be sent to the vitrification plant for treatment as high-level waste. DOE has used this process before at Hanford to extract cesium from radioactive waste.

DOE has determined that it will need other pretreatment approaches for remaining types of waste. Other methods will be needed for such problems as dealing with potentially explosive tank wastes and with wastes that have vastly different chemical characteristics from the first type of waste to be treated.

Plans are to pretreat 7 to 8 M gallons of DST waste to separate it into HLW, TRU waste, and LLW fractions. The low volume HLW and TRU waste fractions will be stored in DSTs for future vitrification. The high-volume LLW fraction will be stored in DSTs for future immobilization in grout.

Four waste streams will be pretreated in the 244-AR Vault and in B Plant:

1) High-level NCAW from the reprocessing of spent fuel at the PUREX Plant.
2) NCRW from the fuel decladding process at the PUREX Plant.
3) CC waste resulting from past strontium recovery operations.
4) PFP TRU waste from plutonium reclamation and processing.

The potential pretreatment processes include solid-liquid separation and sludge washing, ion-exchange, TRU solvent extraction, selective leaching, and organic destruction. Solid-liquid separation and sludge washing of NCAW solids will be accomplished in the 244-AR Vault. The remaining pretreatment processes will be performed in B Plant.

Liquid radioactive waste and mixed waste currently undergo evaporation in the Evaporator-Crystallizer Facility. Approximately 5 to 10 M gallons of waste volume reduction are achieved annually during normal operations.

Waste concentration has reduced the storage space requirements for DSTs by more than 100 M gallons. The 242-A Evaporator-Crystallizer is the cornerstone of waste management's treatment facilities in that it maximizes the use of available DST space and minimizes the need to construct additional DSTs.

During the late 1960s and early 1970s, a program was undertaken to remove cesium and strontium from liquid waste in SSTs. The cesium and strontium were placed in double-walled metal cylinders approximately 50 cm (19.7 inches) in length and 6 cm (2.4 inches) in diameter which were then stored in the Waste Encapsulation and Storage Facility (WESF).

The capsules are stored in a series of water-filled pools. Storage of the capsules is a continuing activity that requires cooling water, makeup water, ventilation, and facility maintenance.

Solid wastes can be divided into 4 categories: (1) retrievably stored and newly generated TRU waste, (2) LLW, (3) mixed waste which is both radioactive and hazardous, and (4) alkali metals. The TRU waste is either being retrievably stored for future treatment or generated and certified for shipment to WIPP. The LLW is being disposed of on-site except for small quantities stored for treatment. The mixed waste is being stored for treatment or disposal.

DOE evaluating several high-level waste extraction technologies aimed at reducing the volume of high-level waste to be vitrified. These technologies include dissolution and solvent extraction, use of solid sorbents or precipitation, selected leaching, and calcination and dissolution. These processes are experimental and have not been tested at Hanford. The calcination process has been used at DOE's Idaho National Engineering Laboratory, but it has not been tested using Hanford tank wastes. Transuranic extraction (TRUEX), a solvent extraction process for reducing the volume of waste to be vitrified, may be a promising technology.

A subsurface confinement barrier is needed for current remediation activities, such as waste retrieval. One of the dangers of removing waste from the tanks is the possible rupturing of tank walls by retrieval equipment and processing. Existing leaks may also be aggravated by these activities. The objective of the subsurface barrier is to contain the tanks prior to and during retrieval actions. A subsurface barrier will create a subsurface confinement boundary that contains the contaminated tank area. Among those materials being considered for construction of the subsurface barrier are cement, grout, and metal sheet piling. It is very likely that the subsurface barrier will be some combination of these materials. Other processes that could provide additional protection are being explored and considered for future development.

A permanent isolation surface barrier also isolates the waste. Natural material such as fine soil, sand, asphalt, gravel and clay is placed in layers over a tank area to produce an aboveground protective cap. This engineered barrier will reduce waste contact with the environment. The goal of the permanent isolation barrier is to protect the tank area from erosion, water drainage, and plant, animal, or human intrusion for a minimum of a thousand years.

The high-level and TRU wastes will be immobilized in the vitrification plant, and the low-level waste will be stored in grout vaults.

Before shipment to a geologic repository, wastes will be packaged in accordance with repository waste acceptance criteria. However, the high cost per canister for repository disposal and uncertainty about the acceptability of overpacked capsules by the repository suggest that additional alternative means of disposal be considered. Vitrification of the cesium and strontium salts in the Hanford Waste Vitrification Plant (HWVP) has been identified as a possible alternative to overpacking. Subsequently, Westinghouse Hanford Company's (Westinghouse Hanford) Projects Technical Support Office undertook a feasibility study to determine if any significant technical issues preclude the vitrification of the cesium and strontium salts. Based on the information presented, it is considered technically feasible to blend the cesium chloride and strontium fluoride salts with neutralized current acid waste (NCAW) and/or complexant concentrate (CC) waste feedstreams, or to blend the salts with fresh frit and process the waste through the HWVP.

There are also 50 inactive underground radioactive waste tanks. The tanks were formerly used for the following functions associated with plutonium and uranium separations and waste management activities in the 200 East and 200 West Areas of the Hanford Site: settling solids prior to disposal of supernatant in cribs and a reverse well; neutralizing acidic process wastes prior to crib disposal; receipt and processing of single-shell tank (SST) waste for uranium recovery operations; catch tanks to collect water that intruded into diversion boxes and transfer pipeline encasements and any leakage that occurred during waste transfer operations; and waste handling and process experimentation. Most of these tanks have not been in use for many years. Several projects have been planned and implemented since the 1970s and through 1985 to remove waste and interim isolate or interim stabilize many of the tanks. Some tanks have been filled with grout within the past several years. Responsibility for final closure and/or remediation of these tanks is currently assigned to several programs including Tank Waste Remediation Systems (TWRS), Environmental Restoration and Remedial Action (ERRA), and Decommissioning and Resource Conservation and Recovery Act (RCRA) Closure (D&RCP). Some are under facility landlord responsibilities for maintenance and surveillance (i.e., Plutonium Uranium Extraction (PUREX).

However, most of the tanks are not currently included in any active monitoring or surveillance program.

13.2 SAVANNAH RIVER HIGH-LEVEL WASTE

There are 30 to 35 million gallons of HLW stored at the Savannah River Plant (SRP) in Aiken, South Carolina. SRP began operations in 1953 and is currently the leading supplier of plutonium and only source of tritium in the U.S. The acidic radioactive waste stream at SRP resulting from reprocessing of irradiated fuel is neutralized with sodium hydroxide and stored as a highly basic solution in large underground, carbon steel tanks. There are a total of 51 storage tanks at SRP, including 24 original tanks and 27 of new and improved design. Over the past decade, HLW has been systematically removed from older, aging tanks to newer tanks. The new tanks are constructed of a stress-relieved carbon steel and consist of a tank within a tank, encapsulated in a concrete vault. There are many redundant monitoring and safety features associated with each vessel.

There are 51 large concrete reinforced steel tanks located in 2 separate "tank farms" at the Savannah River Site. The tanks are of 4 different designs. Type I and II tanks are closed steel cylinders. Each tank sits inside a 5 foot high secondary steel "pan" enclosed by a reinforced concrete support structure and topped by a thick concrete roof. There are 12 750,000 gallon type I tanks and 4 1,030,000 gallon type II tanks. Type III tanks are of similar design, with the pan forming a secondary barrier under and around the primary tank at full height. Twenty-seven type III tanks have been built to date, each with a 1.3 million gallon capacity. Types I, II, and III tanks also have waste cooling capacity. Type IV tanks are older, uncooled, single-wall tanks used for storage of waste that does not require auxiliary cooling. There are 12 type IV tanks, each capable of holding 1.3 million gallons.

None of the type III tanks has developed any leaks; to date, 5 type I tanks have leaked detectable amounts of waste into the secondary steel pan; all 4 type II tanks have leaked significant amounts. (Waste from 1 tank (16H) overflowed its secondary pan on 1 occasion.) One type IV tank (20) developed leaks in the steel liner, and the waste has been removed from that tank.

As a result of neutralization, a "sludge" forms and settles to the bottom of the tanks. This is a thick, gelatinous precipitate which comprises about 10% of the total waste volume in the tank and consists primarily of aluminum, iron and manganese oxides and hydroxides. The sludge contains actinides, fission products and Sr-90 along with most of the radioactivity. The remaining 90% of the waste is a liquid called "supernate." This liquid consists mostly of sodium nitrate and sodium nitrite resulting from neutralization, along with the principal radionuclide Cs-137. The final component of the waste arises from waste management practices designed to reduce the total waste volume. The supernate is reduced in volume by evaporation and the concentrate is returned back to cooled tanks where "saltcake" crystallizes out of solution. Driving off the water reduces the waste volume about 70%. There are 3 principal waste components; sludge, supernate, and saltcake. These constituents often contain more than 40 different elements. While storage of HLW in tanks has been an effective temporary means to handle waste, many tanks are now past their projected lifetime and a more permanent, long-term solution is needed.

13.3 IDAHO HIGH-LEVEL WASTE

Most INEL HLW is reprocessed naval reactor fuel. Acidic liquid waste is stored in underground stainless steel tanks that are housed inside concrete vaults. The waste is then converted into a calcine powder and stored retrievably in stainless steel bins inside reinforced concrete vaults. There are 3,500 m^3 of HLW stored as calcine, containing 90 percent of the radioactivity, and 8,500 m^3 of liquid HLW containing 10 percent of the radioactivity. The INEL waste is uniform and well characterized, but will not meet Land Disposal Requirements (LDRs).

Since the INEL HLW is in calcine form, processing the waste with aqueous solvent extraction would require dissolving the calcine in nitric acid, which would generate a large volume of aqueous LLW. As an alternative, ESPIP is exploring a glass-ceramic waste form and the possibility of pyrochemical processing. The glass-ceramic waste form would reduce the number of logs from 9,500 to 3,770. If pyrochemical processing was used with a glass waste form, the number of logs generated would be less than 900.

The HLW raffinates from the combined PUREX/REDOX type uranium recovery process were converted to solid oxides (calcine) in a high temperature fluidized bed. Liquid effluents from the calcination process were combined with liquid sodium bearing waste (SBW) generated primarily in conjunction with decontamination activities. Due to the high sodium content in the SBW, this secondary waste stream is not directly amenable to solidification via calcination. Currently, approximately 1.5 million gallons of liquid SBW are stored at the ICPP in large tanks. Several treatment options for the SBW are currently being considered, including the TRansUranic EXtraction (TRUEX) process developed by Horwitz and co-workers at Argonne National Laboratory (ANL), in preparation for the final disposition of SBW.

High-level liquid waste (HLLW) has been previously generated from the reprocessing of irradiated spent fuel and is currently stored in the Idaho Chemical Processing Plant (ICPP) Tank Farm. Additional ICPP Tank Farm waste results from plant decontamination, laboratory analysis, and experiments and is normally classified as LLW. The waste in the ICPP Tank Farm is highly radioactive, acidic, and contain heavy metals. In addition, decontamination activities at the INEL have resulted in a sodium-bearing liquid waste which is managed in the same manner as HLLW (i.e., stored in the same tank system prior to final treatment).

The sodium-bearing liquid waste is currently calcined with fuel reprocessing raffinates. Because the mission of the ICPP changed in April 1992 to no longer reprocess spent nuclear fuel, the material used to blend the sodium-liquid is not longer available.

Approximately 700 m^3/yr of solution is generated and stored in the ICPP Tank Farm. This figure is based on the contribution of several waste streams. Approximately 189.5 m^3/yr of solution from the evaporator bottoms and 227 m^3/yr of solution from the liquid effluent system are generated. Both of these waste streams are sodium-based waste and are currently managed similar to HLW. In addition, 141.5 m^3/yr of HLW solution is generated from the periodic dissolution of the calcine bed. The remaining volume is generated from D&D and miscellaneous sources.

One option for sodium-bearing waste treatment is to blend the sodium liquid with commercial aluminum nitrate, but this option creates more waste volume due to the large quantity of additives required. Therefore, an aggressive R&D program is in place to investigate alternative technologies that could minimize the HLW volume by different separation methods.

13.4 TANK SAFETY

DOE's top priority is to resolve tank safety issues. Cited as the most critical problems at Hanford were 54 tanks containing potentially explosive mixtures of ferrocyanide, organic-nitrate material, and flammable hydrogen gas. While DOE is taking steps to mitigate these safety problems, the first of these tanks may not be pretreated until after 2000 because DOE must first characterize the wastes and then construct the facilities and technologies to process them. In the interim, DOE's plans call for processing waste that is easier to process to ensure that waste will be available for treatment in the vitrification plant in December 1999.

DOE has stated that it plans to perform an oxidation process to destroy potentially explosive ferrocyanides, organic salts, and hydrogen-forming constituents after they are retrieved from the tanks. When these components are eliminated, the potential for an explosive reaction will be eliminated. DOE planned to have a facility operational by December 1997 that would be capable of resolving these problems. Westinghouse developed preliminary plans to build the facility but has not yet developed a conceptual design for it. Completion has already been delayed to at least December 1999 because of significant technical uncertainties and funding limitations, and DOE and Westinghouse officials indicate that the delay could be longer.

In 1990, DOE identified 2 high-priority safety issues associated with high-level waste tank storage at the Hanford Site: (1) accumulation and periodic release of significant quantities of flammable gases (hydrogen and nitrous oxide) in 23 tanks and (2) the potential for explosion of ferrocyanide (FeCN) compounds in 24 tanks.

Processing and waste management practices over the years have contributed to the complexity and safety issues associated with the waste today, especially the use of ferrocyanide to remover cesium-137 from plant and tank wastes. Operators attempted to make more room in the tanks by precipitating out the principal radioactive material and evaporating or pumping off the liquid, sometimes pumping liquid from one tank to another. These practices have resulted in a complex, multiphase, and highly stratified mixture of so-called saltcake (a mixture of carbonates, phosphates, nitrates, and sulfates), sludge, and supernatant inside the tanks.

Over time this has chemically "aged," undergoing thermal and radiolytic breakdown, in some cases generating hydrogen and other flammable gases, or releasing toxic vapors to the tank farm area. Other tanks present a potential explosion hazard because of oxidizers like nitrate in the presence of organic compounds.

DOE established a special task force in 1990 to identify and resolve safety issues and upgrade safety of all tank storage operations. The High-Level Waste Tank Task Force established a program to achieve these goals. The program, which is continually evolving as information becomes available, consists of several major components, including (1) evaluation and resolution of the flammable gases and FeCN issues, (2) identification and resolution of other safety issues and deficiencies, (3) upgrade of safety documentation and criteria and validation of the safety envelope within which high-level waste tank storage activities must operate, (4) upgrade of the conduct of tank operations, and (5) establishment of a workshop program to facilitate communications among all high-level waste sites, headquarters, national laboratories, and other sources of expertise.

Because of the potential consequences of a hydrogen gas or ferrocyanide accident, DOE has made its highest priority evaluation and resolution of these 2 issues at Hanford (the other 3 high-level waste tank sites do not share these issues). Preliminary analyses of flammable gas reactions, potential accident initiation, and consequences and probabilities of occurrence have indicated that if a gas ignition were to occur, it might damage the tank, but containment would likely be maintained.

To control tank corrosion, workers added sodium hydroxide to neutralize the acidic wastes, producing sodium nitrate. And to reduce tank waste volumes, DOE from 1954 to 1957 added sodium and potassium ferrocyanide and nickel sulfate to precipitate out the heat-producing, hazardous, and relatively long-lived radioactive isotope cesium-137.

Thus, large amounts of ferrocyanides were deposited in some tanks, including heat-producing cesium nickel ferrocyanide. Under the right conditions and high enough temperatures, ferrocyanides and nitrates react to release large amounts of heat, potentially explosively.

Neutralized current acid waste (NCAW) from the PUREX Plant is capable of self-boiling because of the heat content. The NCAW can be stored in any of 4 DSTs that are specially designed to contain this waste. These 4 DSTs are called "aging-waste tanks" and are located in the 200 East Area (2 in the 241-AY Tank Farm, and 2 in the 241-AZ Tank Farm). Only the 2 241-AZ aging waste tanks currently contain NCAW.

A unique feature of aging waste tanks is the incorporation of air-lift circulators to control the heat distribution in the waste resulting from radiolytic decay. Circulators are necessary to prevent pressure surges, minimize entrainment of radionuclides in the gaseous effluent caused by uneven boiling, and prevent overheating of tanks from sludge hot spots.

Tank 101-SY, has been periodically releasing high concentrations of a hydrogen/nitrous oxide/nitrogen/ammonia gas mixture into the tank dome vapor space. There are concerns that under certain conditions a detonation of the flammable gas mixture may occur. There are 2 ways that a detonation can occur during a release of waste gases into the dome vapor splice: (1) direct initiation of detonation by a powerful ignition source, and (2) deflagration to detonation transition (DDT).

The first case involves a strong ignition source of high energy, high power, or of large size (roughly 1 g of high explosive (4.6 kg) for a stoichiometric hydrogen-air mixture to directly initiate a detonation by "shock" initiation. The second process involves igniting the released waste gases, which results in a subsonic flame (deflagration) propagating into the unburned combustible gas. The flame accelerates to velocities that cause compression waves to form in front of the deflagration combustion wave. Shock waves may form, and the combustion process may transition to a detonation wave.

A mixing pump was installed in Tank 101-SY. The mixing action of the pump allows a gradual release of the hydrogen gas that is generated by the tank contents, preventing the buildup of this potentially explosive and flammable gas.

Forty-five SSTs have been identified as watch-list tanks. Conditions in these tanks could lead to on-site or off-site radiation exposure through an uncontrolled release of fission products. There are 4 categories of safety issues:

1) Tanks containing >1,000 g-mole of ferrocyanide (20 SSTs).
2) Tanks with potential for hydrogen or flammable gas accumulations above the flammability limit (17 SSTs).
3) Tanks containing concentrations of organic salts > 3 wt% total organic carbon (9 SSTs).
4) Tanks with high heat loads (> 40,000 Btu/hr) (10 SSTs).

Tank 241-101, located at the Department of Energy Hanford Site, has periodically released up to 283 m (sup 3) of flammable gas. This release has been one of the highest priority DOE operational safety problems because of potential consequences if the gas were ignited during one of these releases. The gases include hydrogen and ammonia (fuels) and nitrous oxide (oxidizer). There have been many opinions regarding the controlling mechanisms for these releases, but demonstrating an adequate understanding of the problem, selecting a mitigation methodology, and preparing the safety analysis have presented numerous new challenges.

The first priority is the flammable gas tanks. Second in safety priority are those that contain a mixture of fuels and oxidizers—compounds such as ferrocyanide, ethylenediaminetetraacetic acid, citric acid, or glycolic acid, mixed with oxidizers such as nitrates and nitrites. At elevated temperatures, such mixtures can react exothermically. The third tank safety priority relates to those releasing toxic organic vapors.

Objectives to be accomplished include (1) elimination of accumulation and periodic venting of flammable gases; (2) determination and validation of the risks posed by FeCN compounds and reduction of risks to acceptable levels through monitoring and/or corrective actions, as necessary; (3) evaluation and implementation of programs to resolve other safety issues; (4) characterization of wastes in high-priority tanks; (5) upgrade of tank instrumentation; (6) review and upgrade of conduct of tank operations, including procedures, staffing, and training; (7) development of an integrated data base network; (8) continued implementation of the workshop program to upgrade all aspects of high-level waste tank operations; and (9) identification and evaluation of codes, standards, other requirements, and guidance applicable to high-level waste tank activities.

Electrochemical oxidation was tested on laboratory scale to destroy organics that are thought to pose safety concerns, using a non-radioactive, simulated tank waste. Minimal development work has been applied to alkaline electrochemical organic destruction. Most electrochemical work has been directed towards acidic electrolysis, as in the metal purification industry, and silver catalyzed oxidation. Alkaline electrochemistry has traditionally been associated with the following: (1) inefficient power use, (2) electrode fouling, and (3) solids handling problems. Tests using a laboratory scale electrochemical cell oxidized surrogate organics by applying a DC electrical current to the simulated tank waste via anode and cathode electrodes. The analytical data suggest that alkaline electrolysis oxidizes the organics into inorganic carbonate and smaller carbon chain refractory organics. Electrolysis treats the waste without adding chemical reagents and at ambient conditions of temperature and pressure. Cell performance was not affected by varying operating conditions and supplemental electrolyte additions.

Low-temperature hydrothermal processing (HTP) is a thermal-chemical autogenous processing method that can be used to destroy organics and ferrocyanide in Hanford tank waste at temperatures from 250 C to 400 C. With HTP, organics react with oxidants, such as nitrite and nitrate, already present in the waste. Ferrocyanides and free cyanide will hydrolyze at similar temperatures and may also react with nitrates or other oxidants in the waste. No air or oxygen or additional chemicals need to be added to the autogenous HTP system. However, enhanced kinetics maybe realized by air addition, and, if desired, chemical reductants can be added to the system to facilitate complete nitrate/nitrate destruction. Tank waste can be processed in a plug-flow, tubular reactor, or a continuous-stirred tank reactor system designed to accommodate the temperature, pressure, gas generation, and heat release associated with decomposition of the reactive species.

13.5 RETRIEVAL

13.5.1 Robotics

Emptying the USTs is a technically challenging task made difficult because of the hazardous nature of the tank contents. This waste material is chemically complex and includes physical forms ranging from thick, sticky sludge to a crystalline saltcake. The sludge has a consistency of soft mud and the saltcake approximates low-grade concrete. Most of the tanks also contain small amounts of liquid.

PNL, Oak Ridge National Laboratory, and Sandia National Laboratory are working together with OTD to develop an advanced robotics retrieval system that will use robots, that is, remote manipulators, to get into the tanks to break up and remove the sludge and solidified waste. Since the project is technically complex, and because hazardous materials are involved, the development team is creating a full-scale, realistic mockup of a tank structure and retrieval system.

The facility, located at Hanford, will be used to test and fine tune all major subsystems of tank retrieval robotics using harmless simulated waste forms. The facility will use a 75 by 100 foot wide self-supporting platform that sits over the ground surrounding the underground tank. The platform supports the "long reach manipulator," a robotics arm that positions and operates waste-dislodging tools within the waste storage tank. The test bed will be fully operational in 1996.

The project has several important objectives: (1) to explore the capabilities of retrieval manipulator systems and acquire data necessary to develop specific remediation equipment and techniques; (2) to provide performance guidelines for large manipulator-based retrieval systems; (3) to improve the productivity and safety of such systems by first using them in a non-hazardous environment; (4) to reduce costs for long-term national remediation requirements; and (5) to apply lessons learned from this testing to applications at other DOE nuclear waste sites.

Multiple robotics technology needs have been defined in collaboration with tank waste remediators. Each need forms the basis for a development task.

Existing openings in underground waste storage tanks are limited in size. Manipulator payload and reach requirements dictate significant mechanical flexibility, both static and dynamic, that must be mitigated. Accomplishments to date include development of active and passive methods to damp vibrations in long reach manipulators. These approaches have been tested successfully on simplistic manipulator geometries. Many potential kinematic arrangements for waste retrieval manipulators have been investigated to ascertain optimal configuration for waste retrieval and maximum stiffness.

Improved capabilities in teleoperation (man-in-the-loop) and robotic (autonomous) operation are needed. During waste retrieval operations, it is required to remotely supervise and direct large manipulators. Highly productive man-in-the-loop operations must be available that include an accurate and complete perception of the work space for the human operator. Autonomous (robotic) control modes must also be selectable for performance of routine, repetitive actions. Autonomous functions also are needed to oversee actions of the operator, limiting these actions to avoid unsafe conditions, such as collisions.

Significant improvements in man-machine-interface have been achieved through development and demonstration of a graphic-based supervisory control system. Mapping sensors have been demonstrated that provide a three-dimensional digital image of the tank work space. The graphical control system can access the image and the autonomous manipulator controller to plan tool paths and avoid collisions in the tank.

Needs also have been identified to improve reliability and productivity of complex mechanical systems operating in the hostile in-tank environment. For example, the environment includes high radiation, hazardous and corrosive chemicals, high moisture, high temperature and airborne dirt. Improvements are being sought in equipment decontamination and maintenance, reliable cable and hose management, and cost effective power and data transmission capabilities.

Also needed are improved waste retrieval methods that are compatible with tank access constraints, various waste morphologies and safety requirements. Waste dislodging devices are required to be configured as end-of-arm tooling (end effectors) for the retrieval manipulator. Since more than one waste form is present in any given tank, an ability to remotely change end effectors during waste retrieval operation is needed.

Remotely controlled waste retrieval devices are being developed jointly with the Robotics Program of the OTD, which are capable of dislodging the sludge and saltcake fractions of the wastes and conveying them outside of the tanks. These devices are designed to fit through 12 inch diameter access ports, and work in highly radioactive environments. Waste separation techniques will separate tank wastes into low-level, TRU and high-level fractions, thereby significantly reducing the volumes of high-level wastes requiring costly disposal. Site closure technologies, a variety of containment barriers designed to capture effluents escaping from the tanks due to leakage, are currently being developed by related programs; they will be adopted in the near future by the UST-ID for full-scale testing, once their feasibility has been satisfactorily demonstrated.

The robotics test bed comprises commercially available, standard robotic systems. These devices are useful for:

1) First assessments of functionality of generic, very large robotic systems:
2) Testing and demonstration of new concepts for control of large, physically flexible manipulator links;
3) Testing and demonstrating new concepts for graphical supervisory control; and
4) Initial testing and demonstration of waste retrieval end effectors.

Each of the 3 contributing laboratories has invested in stand-alone simulation and modeling capability, which include kinematic and dynamic modeling of proven merit in evaluating potential configurations for the robotic test bed. Other analyses have been performed as a basis for the reference design of a waste retrieval system to be used at Hanford, Washington.

13.5.2 Characterization

Before tank wastes can be removed, their physical and chemical properties must be assessed. Placing sophisticated sensors inside the large tanks requires a manipulator with long reach (9 feet or more), a capacity to lift at least 50 pounds, and the ability to fit through the small 12 inch diameter access port.

Existing manipulators do not fully meet these requirements, nor do they conform to the stringent safety requirements at each site. The new light-duty utility arm will place instruments that allow evaluation of waste properties without removing the waste from the tanks. This provides data much more rapidly, over a wide area of the tank, without exposing workers to a radioactive sample.

The arm will deliver characterization tools such as optical sensors and physical measurement devices into the tanks. The arm also will place hardware inside the tanks as well as test-retrieval techniques.

The main purpose of the light-duty utility arm (LDUA) is to provide safe, controlled access to underground storage tanks containing highly radioactive waste.

The LDUA is a remotely-operated, mobile system to deploy end effectors for waste characterization and tank inspection. This device brings together technologies developed within multiple DOE laboratories and industry into an integrated system for providing a spectrum of storage tank characterization capabilities. The technology will enhance existing capabilities that are limited to single axis deployment of instruments into tanks. The arm will provide 7 degrees of freedom with a 4.5 m (13.5 foot) reach for positioning end effectors in multiple tank locations.

The present process of tank waste characterization requires core samples to be removed from the tank, processed through a hot cell and then undergo extensive analysis. Because each core sample can take up to 6 months to process, a large backlog exists for characterizing the 332 underground storage tanks across the DOE Complex. An easily deployable, in situ method of analyzing safety-related waste characteristics will expedite these characterization activities.

Core samples do not provide information on the integrity of the tanks themselves. Current capabilities for performing tank inspection have the same limitations as characterization techniques. Cameras and sensors are inserted into the tanks through risers on fixed supports. These systems are limited to vertical deployment of sensing devices and can only operate directly below tank penetration. Tank wall integrity and dispersion of material laterally cannot be properly evaluated. Remote in situ characterization helps to ensure a minimum of risk to personnel performing the characterization operation.

Characterization of tank wastes has historically been very expensive, had has failed to obtain representative data for many tanks. Under the UST-ID Program, sensors are being developed which will decrease laboratory analytical sample handling, alleviating safety concerns associated with handling. For example, Fiber Optic Laser Raman spectroscopy would be applied towards the in situ identification of radioactive tank wastes. Low-level waste (LLW) treatment will produce waste forms which are chemically and physically durable, and will reduce waste volume.

In the waste characterization development program, the Raman spectrometer and infrared spectroscopy system have been demonstrated on real waste in a hot cell, setting the stage for the development of a multi-sensor system for placement in tanks. Retrieval tool development includes the completion of testing of the Soft Waste Dislodging Tool and air conveyance system on a waste simulant, demonstrated control of intelligent end effectors, and extensive tests of waste dislodging and conveyance end effectors for the first generation Long Reach Arm.

Prior to deployment of in situ sensors, and during the retrieval operations, it will be necessary to know the topography of the waste in tanks. Two methods are being developed to address that need. The laser range finder (LRF) will provide quick, but coarse mapping of the tank interior. A structured light technique will be used for detailed mapping of the tank interior.

The characterization and surveillance instruments will be deployed and positioned within the tanks using a remotely operated, robotic arm. This Light Duty Utility Arm (LDUA) will operate as a versatile platform in a HLW environment, and therefore need to be hardened against the hostile (radiation, caustics, water vapor, etc.) elements present in the tank surroundings.

13.5.3 Techniques

At least 3 options may be capable of retrieving SST saltcake and sludge: (1) mechanical retrieval, (2) pneumatic retrieval, and (3) hydraulic retrieval. Methods for mitigating leaks from tanks may be required with hydraulic retrieval where the use of water is necessary to support retrieval operations.

Mechanical Retrieval: Mechanical handling properties of the SST waste range from those of a dry crystalline material that is nearly as hard as concrete, to those of mushy wet solids that have no structural integrity. The waste characteristics that must be known to support design of waste retrieval equipment are particle size distribution, bulk density, radiation levels, penetration resistance, shear strength/shear rate, shear and compressive strength, and abrasiveness. Currently, the waste retrieval equipment is being used on waste stimulants because, to date, there are very little data on the physical characteristics of the waste.

The central problem in removing the saltcake, sludge, and slurry in the tanks is the issue of generating secondary waste. Although the easiest way would be to pump in water to help loosen and fluidize the material, that approach would create more waste, and a greater potential for leaks during retrieval. So instead, methods must be found to mechanically remove the material.

Pneumatic Retrieval: The Mucksucker works like a giant vacuum cleaner to pull the claylike simulated sludge from the tank. The waste is first cut into fist-sized chunks by rotating blades. It is then sucked through a 6 inch diameter hose and collected in a retrieval bin at the rate of 20 to 30 gallons per minute.

Hydraulic Retrieval: A hydraulic retrieval system would use slurry transfer (pumping) to move the tank waste out of the tank. The equipment would include high-pressure, high-volume water jets, associated pumping and supply systems, slurry accumulation tanks, and sluicing water recirculation systems. The sluicing action of the water jets would dislodge and mobilize the waste, dissolve or disperse it in a slurry, and wash the waste to a slurry pump where it would be pumped to the surface and to the accumulation tanks. Here the material would be staged for recirculation of the decanted aqueous phase and storage or treatment of waste components.

Two concepts involving hydraulic retrieval are under consideration. They differ only in how the water jet would be maneuvered within the tank. Limited sluicing relies on an arm-based system to achieve precise maneuvering of a jet/or nozzle, while the other concept, called large-volume sluicing or traditional sluicing uses individual, riser-mounted devices with more limited maneuverability. Two types of sluicers are suggested. The first is a traditional sluicer with only vertical and horizontal nozzle rotation. The second type is an enhanced sluicer, which offers both rotation, translation, and other movements.

The riser-mounted system was the method successfully used in past retrieval campaigns. As with pneumatic retrieval, hydraulic retrieval cannot remove large debris items.

Early campaigns used sluicing and slurry pumping for tank waste retrieval. The equipment and technologies used were based on mining industry practices and adapted for use in radioactive service. Equipment failures occurred and process limitations were experienced, but overall, the campaigns generally were successful and achieved a high overall removal efficiency. In most tanks, sluicing was terminated when it was no longer cost-effective to continue operations to gain a few additional inches of storage space. Freeing up tank space for storage of newly created waste was an important goal of these historic campaigns. Leaks that occurred during sluicing in 2 underground storage tanks led to the termination of waste retrieval activities in those tanks.

The materials retrieved during sluicing were a variety of sludges. Saltcake, which constitutes more than two-thirds of the current SST inventory, was not a waste form involved in these early retrieval efforts.

It has been found at PNL that waterjets at a pressure of 10,000 psi can effectively cut the material which has been chosen to simulate the hardened saltcake within the storage tanks. Based on a parameterization test it has thus been calculated that an inlet flow volume of approximately 30 gallons per minute will be sufficient to excavate 30 gallons per minute of waste from a tank. In order to transport the resulting slurry from the tank, a modified jet pump has been developed and has demonstrated it s capability of conveying fluid and waste particles, up to 1 inch in diameter, to a height of more than 60 feet. Experiments were conducted to examine different configurations to achieve the production levels required for waste removal and to clean the walls of residual material. It was found more effective to clean the walls using an inclined angle of impact rather than a perpendicular angle of impact in order to provide a safeguard against driving the water through any cracks in the containment. It was demonstrated that excavation can take place with almost total immediate extraction of the water and debris from the cutting process. The results have qualitatively shown the potential of a medium pressure waterjet system for achieving the required results for underground storage tank waste retrieval.

There are several basic research needs, including understanding the hydrodynamics and surface properties of radioactive particulates, that are associated with hydraulic sluicing of insoluble radioactive wastes from underground tanks. Such sluicing operations are performed to suspend the solid wastes, some of which is very finely divided, so that it can be transported into appropriate facilities for separation from the liquid phase and subsequent processing for final disposal. Sluicing operations are complicated by the need to use remotely operated equipment in large-diameter tanks with many internal structures. Sluicing, suspension, and transport of finely divided radioactive solid wastes can involve complicated chemical and physical phenomena that have not yet been studied adequately.

Separation of finely divided solids from carrier liquids is a very challenging problem. Very high decontamination factors are typically required so that the clarified liquid will not need to be treated, for example, as a TRU waste. In some cases, remaining traces of radioactive solids will behave differently from the larger bulk concentration. In other cases, the last traces of a contaminant may be adsorbed on colloidal particles with high surface areas. An understanding of the hydrodynamics and surface properties of radioactive particulates needs to be provided.

The underground storage tanks at Hanford contain 3 basic material types, both individually and in combination: liquid supernatant, sludge, and hard saltcake. Removal of the sludge and saltcake has presented a technological challenge. A high pressure waterjet can be used to cut up and dislodge the tenacious sludge and saltcake. Combined with a conveyance system operating simultaneously, this confined sluicing can be used to effectively remove and convey waste from the tanks.

The University of Missouri-Rolla, in conjunction with Sandia National Laboratory, has been developing a confined sluicing technique to dislodge and convey difficult wastes from the underground storage tanks. Confined sluicing uses high pressure (70 Mpa or 10,000 psi) to cut the material in the tank into small pieces and then sucks the material out using a high pressure (50 Mpa or 7,000 psi) jet pump. All the water and debris is removed without significant water loss to the tank. The device is attached as an end effector to an articulated arm that enters the tank through an existing riser. The result of the process is a steady flow (at around 1.9 liters/second (30 gpm)) of extracted material from the tank as an aqueous slurry. This minimizes handling problems and converts the tank wastes to a form that can be more easily treated.

Confined sluicing reduces the water needed to clean the tanks, and therefore reduces the quantity of waste that must be processed.

Recycling the water results in waste minimization because the volume of waste water generated by the sluicing process is minimized. At the end of the entire tank farm remediation program, the waste water would be treated and reclaimed or suitably disposed of.

It is projected that CSEF (Confined Sluicing End Effector) will be able to excavate and remove all types of tank waste, including hard cake, sludge, and bulk supernate. The projected removal rate is 30 gpm with a water-to-solid ratio of 2:1 to 4:1. It is projected that the water expelled by the waterjets and retrieved by the pneumatic conveyor will be "close-looped" and recirculated throughout the entire tank farm reclamation process.

Hydraulic Impact End Effector: Lawrence Livermore National Laboratory (LLNL) has teamed with Quest Integrated, Inc. to develop an efficient method of breaking up large blocks of hard saltcake that have developed in underground storage tanks within the DOE complex. These remaining wastes often surround tank risers and equipment, making their removal doubly difficult. LLNL and Quest are developing a water cannon rubblizer as a hydraulic tool capable of fracturing the hard saltcake.

The system used ultra-high pressure (276 MPa, 40,000 psi) to generate a powerful hydraulic shock to fragment the monoliths. The resulting fragmentation is comparable to that achieved by explosive charges without the hazard of "fly" rock or toxic fumes and without the precise positioning required for waterjet cutters. The resulting fragment size varies with material. The current tool uses water as the working fluid, with only about 200 ml (one-half pint) per blast. The control console monitors the pressurization of the tool and controls the discharge of the tool through the control valve assembly. The end effector can be fired repeatedly with 5-10 seconds between blasts. The end effector is remotely operated, and the design incorporates several features to provide "fail safe" operation.

Development tasks are varied. The design can be refined to reduce the amount of additional processing required for the fragments. Alternative fluids are being evaluated which either vaporize or gel in the tank after discharge to limit the addition of water to the tank. Reduction of the poppet valve opening time will increase the shock energy rate. Finally, the end effector is being radiation-hardened and will be capable of accommodating remote decontamination.

The HIEF discharges 200 ml volume of water compressed to 40 kpsi before discharging. The end effector can be fired repeatedly with 5 s between blasts. The UHP power unit is locate 100 feet from the end effector and requires 480-VAC electrical power, 7-gpm cooling water, and 90-psi compressed air. The UHP has been designed to have a minimum fatigue life of 30,000 cycles. The flexible high-pressure hose is surrounded by a safety shield and has a typical burst pressure of 105 kpsi and a minimum blast pressure of 95 kpsi.

HIEF is applicable to monoliths and large boulders of the sodium nitrate and the sodium nitrite hard cake form of plutonium processing waste by-product. The system is designed for applicability to waste stored in underground storage tanks (0.5 to 1 M gallons in size) that may have limited access ports. Hydraulic impact methods have been used to break up natural geologic boulders and rocks in mining applications.

Soft Waste End Effector Air Conveyance System: The heavy sludge presents many difficult problems from the standpoint of its varying consistency. A system is needed that can adapt to the changing sludge and still effectively and efficiently remove and convey the material from the tanks. Currently, there is no baseline technology to deal specifically with this waste type.

The Sludge Dislodging End Effector Assembly uses a developmental tool which can be tested in a mechanical agitator configuration using blades and nozzles for water or air injection, or a scarifier configuration using only injection nozzles. The blades and injected air or water cut through the waste. The spray helps move the wastes to the air conveyance system inlet. The air conveyance system includes a water injection spray ring inside the conveyance hose, to facilitate waste transfer. The water that is injected by the conveyance system is immediately removed from the tank along with the waste, thereby causing no increase in risk of liquid leaking from the tank.

The system can be attached as an end effector to a long reach, remotely controlled manipulator arm. The Long Reach Arm will move the scarifier through the sludge until it is all successfully removed from the tank.

It is assumed that approximately 75% of all SSTs will be retrieved by hydraulic sluicing and the remaining tanks by arm–based methods.

13.5.4 Subsurface Barriers

The retrieval, demolition, removal, and transport of contaminated SST wastes and other materials creates the potential for highly dangerous levels of occupational and environmental exposures if the contaminated materials are not adequately contained. These activities may be conducted within surface confinement structures to minimize impact to the environment and personnel and to mitigate the undesirable effects of inclement weather. Poor weather conditions often causes delays in conducting outside operational and maintenance activities at the Hanford Site.

Subsurface barriers may be used to contain liquids used in or released from SSTs during tank waste retrieval options. They may also be used to complement soil flushing, a form of in-place treatment that involves flushing the soil with water to remove mobile contaminants. After all waste retrieval/treatment activities are completed, the remediated sites will be capped with a surface barrier. Capping will be necessary when residual contamination remains.

The potential for leaking contaminated liquid to the soil is a key issue with hydraulic retrieval or other retrieval methods that employ the use of water.

Subsurface barriers would be used in combination with retrieval methods that employ the use of water in the SSTs. The barriers would be placed throughout an entire tank farm in advance of sluicing activities. The barrier would stop or slow migration sufficiently to allow for remediation of the contaminated soil as part of closure. Soil flushing, contaminant immobilization, in situ vitrification, and soil removal are some of the soil remediation technologies under consideration to achieve closure.

Surface barriers would be used to prevent recharge of surface water and thus greatly slow the migration of contaminants. Surface barriers would be placed over an entire tank farm site as part of a final closure strategy.

Leakage from tanks may be minimized in most tanks by operating with a minimum free liquid depth (approximately 1 foot). Most SST leaks are believed to have occurred at elevations well above 1 foot where liquid/vapor interfaces existed in the past. The prevention or plugging of tank leaks is potentially the most desirable method of leak mitigation. For example, an injected barrier material placed next to the tank surfaces may effectively encapsulate a tank to prevent leaks.

Subsurface barriers must be cost-effective relative to other options, including mechanical retrieval techniques that do not employ water to aid in retrieving waste from the tanks and, hence, do not cause leakage to the soil.

The successful demonstration of subsurface barriers will enable the use of traditional, well-proven sluicing techniques for retrieval of wastes from the tanks. The combined subsurface barrier and traditional sluicing option may result in lowest overall risk to workers, the public, and the environment, and/or lowest cost when compared to other options.

The design of the barrier system must not preclude the ability to remediate excessive contamination that has leaked from the tanks.

The primary purpose of a subsurface barrier is to support the use of sluicing as the reference tank waste retrieval technology and thereby help to maximize the level of waste retrieval and enhance the probability of meeting a target limit of at least 99 percent removal. The barrier is intended to stop the migration of new and existing leaks and to facilitate clean up, if necessary, using in situ methods, such as soil flushing, where possible. The amount of residual contamination allowable in the soil following clean up has not been determined, but is roughly estimated to be equivalent to about 3,800 L (1,000 gallons) of saturated salt solution per tank.

Three subsurface barrier concepts have emerged for consideration: (1) injected or infused material barriers, (2) cryogenic barriers, and (3) desiccant barriers. These barrier types may be installed in 2 configurations: close-coupled (against the tank structure) and stand-off (with a soil layer between the tank and barrier).

13.6 SEPARATIONS

The Efficient Separations and Processing Integrated Program (ESPIP) was created in 1991 to identify, develop, and perfect separations technologies and processes to treat wastes and address environmental problems throughout the DOE system.

Billions of dollars could be saved if new separations technologies and processes could produce even marginal improvements. Treating essentially all

DOE defense wastes requires separation methods that concentrate the contaminants and/or purify waste streams for release to the environment or for downgrading to a waste form less difficult and expensive to dispose of.

In retrieving, processing and disposing of these wastes, separation technologies are the primary means to concentrate and chemically partition the wastes such that the volume of waste to be vitrified can be minimized. It is estimated that the volume of waste to be vitrified can be reduced by a factor of 10 to 200 using separation technologies.

Accordingly, the mission of ESPIP is to provide separations technologies to process, concentrate, and immobilize a wide spectrum of radioactive and hazardous defense wastes at DOE sites; coordinate separations technologies R&D within DOE's Office of Environmental Management; foster future expertise in separations technologies by encouraging university participation; and transfer separations technologies developed by DOE to the U.S. industrial sector to facilitate competitiveness of U.S. technology and industry in the world market. ESPIP provides the following categories of separations technologies:

1) Radionuclide removal, including Pu, Am, Sr, Cs, Tc, I;
2) Toxic materials removal
3) Conditioning and chemical treatment of wastes to enhance separations, including removal of Al, Cr, and P;
4) Solid/liquid separation; and
5) Organic destruction.

DOE is evaluating several high-level waste extraction technologies aimed at reducing the volume of high-level waste to be vitrified. These technologies include dissolution and solvent extraction, use of solid sorbents or precipitation, selected leaching, and calcination and dissolution.

Waste separations development areas include the following:

1) Removing radioactivity from the supernate;
2) Destruction of organics to prevent the buildup of hydrogen;
3) Ferrocyanide removal/destruction; and
4) Strontium removal from the sludge.

The major chemical constituents in the tanks are as follows: (1) nitrate and nitrite salts, (2) hydrated metal oxides, (3) phosphate precipates, (4) transuranics and (5) isotopes of strontium, iodine and cesium. The initial focus is on those technologies that can provide near-term benefits toward remediation of USTs.

Optimum treatment of high-level wastes in storage tanks at DOE facilities will require separation of several fission products. Cesium, strontium, and relatively short-life fission products are responsible for the largest fraction of the heat generated in HLW storage tanks. These components are also responsible for gamma radiation that creates the need to use extensive shielding and remote operations when handling the wastes. Removal of these elements would decrease the difficulties and hazard of handling or treating these materials. Removal of these materials from the liquid supernate in many tanks would allow remaining materials to be treated as low-level waste, so disposal would be both less costly and less hazardous. Small quantities of these elements may occur in wash or leach liquors from washing or leaching of the solid precipitates (sludges) in the tanks. These are also likely to be alkaline solutions with high sodium concentrations. If the tank solids (sludges) are dissolved for more extensive separation and concentration of the various sludge components, traces of cesium, and, especially, strontium may need to be separated from the other dissolved components of the sludge. Such separations are likely to be from acid solutions.

Selected long-lived components must be separated waste components that need to be sent to LLW. Technetium is the most notable example. This element can exist in different valence forms, is often distributed to different streams during waste processing, and has a high mobility from many waste forms.

R&D on fission product separations must address conditions that exist in DOE tank wastes or could result during treatment of those wastes. This may involve highly alkaline and high sodium content supernates or wash/leach liquors, as well as acid solutions that would result from possible dissolution of the solid sludges. In evaluating any selective separation agent, it is important to test the methods in the presence of all potential competing components. This is especially important when the target components, such as fission products, exist at only trace concentration levels, but other elements that can behave similarly are present at much higher concentrations. Removal methods need to be highly specific for target components and must remove them in a concentrated form.

Chemical separation processes involve contacting a waste stream with a separating agent (e.g., sorbent, extractant) in some physical support (e.g., packed bed, immiscible phase) and then physically separating the waste stream from the support, typically by flow. The separated species are left behind in the support and are typically removed so they can be routed to a waste form. However, in at least 2 cases, it is undesirable to remove the species from the physical support. In the first case, waste can be minimized by converting the physical support (e.g., a packed bed of the separating agent) directly into a waste form that incorporates the separated species. In the second case, the physical support is left in place as part of in situ treatment of groundwater. Sequestering agents developed under ESPIP will be utilized by the ISRIP for the development of in situ technologies to remove and immobilize contaminants from groundwater in situ.

Separation processes are needed that exchange, sorb, or sequester fission products (Cs, Sr, Tc, I, Ni, Sm, etc.) from a variety of waste streams such that the physical support, separating agent, and fission products can be formed directly into a waste form suitable for long-term storage or disposal.

The resulting waste form must endure irradiation and chemical changes associated with decay of radionuclides to decay products. Candidate waste streams include:

1) Acid or alkaline high ionic solutions removed from HLW tanks or resulting from processing of sludge removed from such tanks.
2) Complexed Sr in high pH supernatant in tank wastes.
3) Decontamination waste streams.
4) Miscellaneous aqueous waste or process streams.

Retrieval and pretreatment operations will require a substantial amount of double-shell tank space that currently does not exist. DOE's plans are based on the assumption that up to 18 double-shell tanks will be available during retrieval and pretreatment operations for storage and in-tank processing. According to GAO, DOE has no assurance, however, that any of these tanks will be available. The tanks currently contain waste that DOE plans to treat as low-level waste. Objections from Washington State about DOE's plans for treating this waste have already delayed DOE's schedule by 3 years and could prevent tanks from being available when they are needed.

DOE has initiated a project to construct 4 new double-shell tanks primarily for resolving tank safety issues that will also be used for retrieval operations. These tanks, scheduled to be completed in 1999, will not provide sufficient space for planned retrieval operations. DOE told GAO that the number of new tanks needed depends on the methods DOE will use to retrieve and pretreat the waste. DOE stated that as many as 70 new double-shell tanks may be needed, but it believes there is sufficient time to construct new tanks to support the treatment of Hanford wastes.

The highest priority isotope separation need is for cesium isotopes. Fission-product cesium, following one or more decades of decay, contains stable Cs-133, long-lived Cs-135, and short-lived Cs-137. Transmutation of Cs-135 may be desirable from a disposition perspective, but isotope separation is a prerequisite. Development of cesium isotope separation concepts is an important long-term basic research need.

If long-lived fission products other than Tc-99, I-129, and potential, Cs-135 require transmutation, some the them may also benefit from isotope separation operations prior to irradiation. In addition, isotope separation may be a prerequisite to beneficial use of some partitioned elements. For example, partitioning of fission-product palladium may be cost-effective if isotope separation can adequately remove long-lived Pd-107. Isotope separation may also be important in eliminating material damage due to the formation of activation products. For example, steel isotopically enriched in Ni-60 would eliminate the (n, alpha) reaction of Ni-59.

Within the mission framework, the main focus of ESPIP at present is to minimize the quantity of HLW and transuranic waste (TRU) destined for geologic storage and to minimize the quantity and environmental impact of low-level waste (LLW) generated when HLW and TRU are processed.

Priority is being given to separation technologies and processes (STPs) for the transuranic elements, such as neptunium, plutonium, americium and curium; highly radioactive elements, such as strontium-90 and cesium-137; long-lived soluble fission products, such as technetium-99 and iodine-129; and soluble activation products tritium and carbon-14. Since current plans are to incorporate the most serious radioactive components into a glass waste form, materials deleterious to the glass stability must be removed; thus STPs for aluminum, chromium and phosphorus are another priority for ESPIP.

13.6.1 Ion Exchange

Ion exchange is one of the proposed technologies to remove cesium and strontium from high-level waste found in underground storage tanks. Due to the somewhat unique chemical characteristics of the tank wastes, most common ion exchangers are not suitable for their processing.

Ion exchange is the second step in pretreating neutralized current acid waste (NCAW). The supernate, or liquid waste that remains near the top of the tank during sludge washing, will contain cesium 137, a highly radioactive material that must be separated through further pretreatment. To separate the cesium 137, DOE plans to use an ion exchange process in which the supernate liquid containing cesium passes through tubes containing ion exchange resin. As currently planned, the separated cesium will be sent to the vitrification plant for treatment as high-level waste. DOE has used this process before at Hanford to extract cesium from radioactive waste.

Zeolites have been utilized for selective removal of isotopes, however new materials are being explored such as silicotitanates, resorcinol-formaldehyde resins, potassium cobalt hexacyanoferrate, etc. It is interesting to note that a new process developed by PNL consists of a titanium-coated zeolite for removing plutonium ions from highly alkaline liquid wastes, and has been successfully demonstrated at West Valley, NY.

Technetium, Iodine,and Nickel: Technetium is a major fission product resulting from nuclear reactors and by nuclear fission of plutonium. At the present rate of production, Tc-99 will reach 170,000 Kg by the year 2000. Tc, as TcO_4, is a very mobile species in the environment. This characteristic, along with its long half-life (213,000 years) causes technetium to be a major contributor to the long-term risk. Additionally, incorporation of technetium into currently planned waste forms may pose unusual chemical and engineering problems during vitrification. For example, the Hanford glass frit mix contains formic acid, which may reduce technetium to TcO_2. At processing temperatures, this can disproportionate into the metal (which runs the risk of shorting out the vitrifier's electrical heater) and volatile Tc_2O_7.

The amount of nickel in the Hanford single-shell tanks exceeds the goal set by the "Clean Option" strategy making it necessary to evaluate technologies for nickel removal from HLW streams.

Several iodine transmutation design alternatives have been proposed to replace the original design that used an iodine-filled rod system. Calculations and engineering considerations for both a static-solid and dynamic gaseous system are being developed. Optimization considerations for thickness, heat transfer, etc., to develop figure of merits for transmutation conditions are being made. The literature survey for nickel removal technologies is nearly complete. First effort selection of technologies for nickel removal has been made. Some experimental work has been initiated where nickel removal from solutions is 99%.

One project at Los Alamos is divided into 3 separate initiatives addressing the separation needs for Tc-99, iodine-129 and nickel (both stable and nickel-63). For technetium separation, the approach being followed is a re-examination of the 2 baseline technologies, ion exchange and solvent extraction, in the light of recent work. In the former case, a new anion-exchange resin containing the pyridinium-functionality, Reillex -HPQ, which has shown superior stability towards radiation and nitric acid and is being tested as a replacement for the baseline Dowex resin.

For the solvent extraction, pyridinium-type extractants, or a liquid ion-exchange extractant, such as Aliquat 336, are under investigation. The use of water soluble chelating polymers containing quaternary amine functionalities specific to technetium will also be examined.

For iodine, the main effort will be to investigate the systems that would be needed if transmutation of waste were to become a disposal option.

For nickel, the 3 major objectives are to investigate the disposition of nickel in existing flowsheets, evaluate the technologies applicable to nickel removal from radioactive waste streams, and evaluate the behavior of nickel in the posited hydrothermal destruction of organic compounds (since much of the nickel present in Hanford wastes may be present as a nickel cesium ferrocyanide complex).

Titanate Ion Exchangers: A project at Sandia is focused on the development and demonstration of advanced, efficient separation technologies to selectively remove Cs-137, strontium, and other radionuclides from a wide spectrum of radioactive defense wastes. Crystalline silicotitanates (CST) and amorphous hydrous titanium oxide (HTO) ion-exchange technology will be developed and demonstrated for the removal of the radioactive materials from Hanford-type supernatant solutions and saltcake. The project objectives will be accomplished within 4 tasks:

1) The evaluation of the ion exchange properties, especially the effects of pH and sodium content on the ability of crystalline silicotitanate to remove Cs-137;

2) The assessment of the feasibility of regenerating the crystalline silicotitanate ion exchangers;
3) The measurement of the radiation stability of the crystalline silicotitanate materials; and
4) The delivery of the materials to Pacific Northwest Laboratory (PNL) for testing on actual Hanford wastes.

A variety of CSTs have been synthesized to evaluate the effect of structure and composition on ion exchange properties. Transmission electron microscopy micrographs show cuboidal crystals ranging from 20 to 50 nm and preliminary X-ray structural studies indicate a tetragonal structure. Cs adsorption was measured at various pHs and the distribution coefficient (K_d) was calculated. From pH 2 to 10, the K_d exceeds 10,000 ml/g. However, at a pH of approximately 10 the K_d decreases markedly and drops to 100 mL/g at pH 14.

A second generation of CST ion exchangers was prepared and is undergoing testing and evaluation. The main effect of the structural and compositional modification has been to increase the K_d to 1,000 ml/g in 2.5 NaOH to pH 11. Stability tests have been conducted on CST in high pH solutions for 100 days at 40°C and ambient temperature by measuring the Cs concentration. No change in Cs concentration and the K_d for Cs adsorption was measured.

Resorcinol-Formaldehyde Resin: The Savannah River Site (SRS) has developed a resorcinol-formaldehyde ion exchange resin for cesium removal. The resin has been demonstrated to give excellent performance in testing. Other cesium removing materials either have much lower capacity or are incompatible with the high pH and aluminum concentration that is found in the Hanford and SRS tank wastes. In particular, the resorcinol-formaldehyde resin is found to have 10 times the capacity of Duolite CS-100, a phenol-formaldehyde resin.

The resorcinol-formaldehyde resin has been demonstrated to give excellent performance with tank wastes; its high capacity will also allow a notable reduction in the size of process equipment (CPU). High-level supernatant from a waste tank will be processed in a CPU. Cesium will be sorbed onto the resorcinol-formaldehyde resin and then eluted with dilute acid, resulting in a concentrated Cs stream that can then be sent for vitrification. The treated stream will be processed further (e.g., removal of strontium in a CPU) before it is disposed of as low-level waste. This will produce a significant reduction in the volume of waste to be vitrified.

The Resorcinol-Formaldehyde Ion Exchange (ReFIX) resin is applicable to high-level waste streams containing cesium-supernate salt solutions. Radioactive cesium is a fission product found in waste produced by reprocessing fuels from nuclear power reactors. The highest concentrations of this isotope are found in alkaline high-activity wastes, a mixture primarily of sodium nitrate and sodium hydroxide that is called the supernate. This technology is a selective ion exchange resin (specific sorption of cesium ions) that has 10 times the capacity of the baseline Duolite CS-100 phenol-formaldehyde resin. Columns of ReFIX resins will be packaged in a standardized module to fit the Compact Processing Unit (CPU) waste processing module specifications. One specific benefit of ReFIX resin is that it is essentially unaffected by changes in temperature. However, high concentrations of competing sodium and potassium ions reduce the cesium sorption capacity and diffusion efficiency of the ReFIX resin.

High-level supernate from a Hanford waste tank will be processed through and appropriate number of ion exchange columns in a CPU processing module. Cesium will be removed by sorption onto the ReFIX resin in the processing columns. When a column becomes saturated, it will be temporarily removed from service so that the cesium can be eluted from the resin with acid (most likely nitric acid) in a concentrated stream that can be sent for vitrification. Once eluted, the newly generated column will be placed in service when another column is removed for elution. The treated streams from these columns may have to be processed with another series of columns containing resin specific for strontium removal before the stream can be incorporated in cement for final storage. Spent resin can be subjected to rigorous elution before disposal to lower its radionuclide content. Carefully eluted resin can then be stored or disposed of by incineration or chemical destruction before incorporation into cement.

Sodium-4-Mica: This material developed by Professor Komarneni at Pennsylvania State University is highly selective for strontium, and holds on to it much more effectively than zeolites. It forms a new crystalline compound in which the mica "collapses" around the hazardous pale-yellow metallic element, immobilizing it.

Compact Processing Units for Cesium Removal: Compact Processing Units (CPUs) or "Modular Waste Treatment Units," are relatively small mobile equipment modules that would be located near the waste storage tanks or in a conveniently located diversion box in the Hanford waste transfer system. They perform unit chemical process operations. The CPUs allow rapid deployment of technologies for the treatment of radioactive wastes in underground storage tanks. The modules would be manufactured off-site by commercial vendors and moved into place using trucks or special transports.

The CPU is designed to permit relocation using a construction crane and a transport trailer. The concept of having standardized modules is based on the notion that various radioactive waste treatment subsystems could be standardized to match the CPU hardware package, leading to more rapid, cost-effective deployment. The cost benefits are realized even when multiple units are deployed to achieve greater processing rates. The modular design concept will also allow for reuse of CPU components for different unit processes or process deployments. The CPU consists of 4 major subsystems: the containment system (safety), the process system (e.g., ion exchangers), the control system, and the process interface subsystem (which includes solid/liquid separations and waste stream routing).

The ion exchange CPU will pump undiluted liquid tank waste from an underground storage tank or receive liquid waste from a waste retrieval system for treatment. The CPU will filter this waste to remove solids. The solids removed will be transferred to a holding tank for further analysis and processing. The filtered tank waste will be adjusted to optimize the waste composition and temperature for maximum efficiency of the ion exchange process. The waste will then be pumped through 3 ion exchange columns in series to remove the cesium from the waste. The waste will be returned to the tank farms after the cesium is removed.

The ion exchange columns will use a new formaldehyde resorcinol ion exchange resin formulation developed at Savannah River Laboratories. The loaded ion exchange resin will be regenerated, using nitric acid to remove the cesium. This high-concentration cesium waste will be neutralized and transferred to the tank farms as a waste feed stream for the vitrification process. The waste stream with the cesium removed will be suitable for disposal as low-level waste.

This waste treatment technology is targeted for radioactive process liquids, sludges, and slurries. The CPUs are designed to incorporate waste treatment modules that could potentially have application to all Department of Energy (DOE) radioactive liquid tank wastes. The CPU waste treatment hardware system is applicable to high-level, low-level, and transuranic chemical separations technologies. The prototype CPU includes a process module of cesium-specific ion exchange resin columns that are selective to cesium ions.

The baseline technologies are large, centralized facilities for supernate waste treatment. The advantages of CPUs over the baseline technology are in 4 areas: cost reduction, schedule improvement, reduction of technical uncertainty, and reduction of process deployment uncertainty.

Potential commercial applications include waste treatment, separation, and volume reduction operations for reclamation of radioactive waste liquids, sludges, and slurries stored in underground storage tanks, and process effluent pretreatment before appropriate disposal.

Hanford has identified a project at the Cesium Demonstration Unit (CDU) which will have the following capabilities:

1) Demonstrate a deployable facility concept.
2) Separate solids from the feed stream.
3) Remove cesium to a high DF (DF = 1,000 to 10,000).
4) Collect and transfer HLW and LLW streams.
5) Test various ion exchange resins for removal of cesium (elutable and non-elutable resins).

The CDU will demonstrate the CPU facility concept and its viability for application to full-scale LLW pretreatment. The CDU will provide an early demonstration of pretreatment of supernate stored in the Hanford Site double-shell tanks for future immobilization.

The CDU will be capable of exploring key pretreatment issues such as solid/liquid separation, organic fouling of ion exchange resin, ion exchange column configuration, ion exchange material selection, etc. Early examination of these issues will reduce overall technical and cost risks in designing and constructing the LLW pretreatment facility.

13.6.2 Extraction

TRUEX Process: The TRUEX process is a solvent extraction procedure that can very efficiently separate transuranic (TRU) elements (e.g., Np, Pu, Am, and Cm) from aqueous nitrate- or chloride-containing wastes. These wastes are typically generated in reprocessing plant operations or in plutonium production and purification operations. The resulting solutions after extraction may be sufficiently free of TRU elements to warrant their disposal as non-TRU, low-level wastes. Furthermore, plutonium can be recovered and purified by this process. Treatment of stored wastes by the TRUEX process will lower the costs of final disposal significantly; treatment of waste streams as they are generated will allow recycle of streams and avoidance of future waste treatment and disposal costs.

The key extractant in the TRUEX process is octyl-(phenyl)-N, N-diisobutylcarbamoyl-methylphosphine oxide (CMPO). It is combined with tributyl phosphate (TBP) and a diluent to formulate the TRUEX solvent. The diluent is typically a normal paraffinic hydrocarbon (either a C_{12}-C_{14} mixture or n-dodecane) or a non-flammable chlorocarbon such as tetrachloroethylene.

The Center for TRUEX Technology Development at Argonne National Laboratory (ANL) continually performs research and development (R&D) to broaden the applicability of the TRUEX process in the treatment of high-level waste and TRU-containing waste streams.

If the TRU content of TRU waste streams can be lowered to below 100 nCi/g of solid, the waste can be classified as non-TRU. Additionally, if the radioactivity of other isotopes, such as Cs-137 and Sr-90, is reduced to an acceptable level, the wastes will be eligible for near-surface disposal. Use of the TRUEX process to treat TRU waste will greatly reduce the volume of high-level waste, resulting in high cost savings during disposal.

The TRUEX process being developed to pretreat about 5.5 million gallons, or about 75 percent, of the high-level waste at Hanford could permanently damage existing B Plant waste pipes. A report issued by the DOE Hanford Waste Pretreatment Technology Review Panel in December 1990 stated that the chemicals used in the TRUEX process would cause extensive corrosion to B Plant's embedded pipes. The report characterized the pipes as non-replaceable because they are embedded in the building's concrete structure. In addition to TRUEX chemicals, fluorides contained in double-shell tank wastes could also corrode B Plant's pipes.

According to Westinghouse engineers, technology to reduce the corrosiveness of TRUEX chemicals and fluorides in the high-level radioactive waste is still being developed. Westinghouse engineers told GAO that recent small-scale tests, which used an aluminum solution to alter these substances, did not adequately decrease their corrosiveness. Moreover, the aluminum solution inhibited the TRUEX waste separation process. Although the Westinghouse engineers emphasized that the problem is solvable, they could not estimate when the technology could be fully developed. They said that even after the technology is developed and demonstrated on a small scale, nothing can ensure that it will be successful on a large scale.

SREX Process: This solvent extraction process can be used to separate Sr-90 from acid-dissolved sludge wastes.

TOREX Process: The purpose of this work at Argonne is to develop new advanced solvent extraction and recovery processes in support of the "Clean Option" strategy that can reduce the complexity and cost of the chemical pretreatment of dissolved sludge to produce raffinates and effluent streams that will meet specifications of Class A low-level waste. The 3 objectives are to minimize the number of processes needed to achieve "Clean" status, to minimize the number of times that the initial volume of dissolved sludge must be handled, and to concentrate the product streams so as to reduce the scale of the operation to the smallest possible level. The requirements for an advanced chemical separations system that must meet this goal are that it must readily achieve the required decontamination factors, that it must have sufficient chemical and radiolytic stability, that it should not use highly hazardous substances, that it should not significantly increase waste volume, and that an engineering scale-up of the process must be feasible. The new separation scheme comprises a series of novel processes designed to extract and recover thorium, uranium, TRUs (neptunium, plutonium, americium, curium), Sr-90, and Tc-99 from dissolved sludge waste in the Hanford storage tanks.

For example, a combined Strontium Extraction Process/Transuranic Element Extraction Process (SREX-TRUEX) is being examined to extract strontium, technetium, uranium, and TRUs and to partition uranium and technetium from strontium and TRUs; a combined SREX-PUREX is being examined to separate strontium, neptunium and plutonium from americium, lanthanides and barium; a new technetium extraction process is being examined to separate technetium from uranium; an actinide/lanthanide resin is being examined to separate americium from lanthanide; and Diphonix ion exchange resin is being examined to separate barium and lanthanides from thiocyanate solution.

The advanced chemical separation processes (TOREX flowsheet) will be applicable in the chemical pretreatment of waste retrieved from storage tanks at DOE defense establishments (e.g., Hanford, Savannah River, etc.). The objective of these processes is to minimize the amount of waste that must be vitrified by reducing the level of alpha activity and reducing the concentrations of Sr-90, Cs-137, and Tc-99 in the dissolved sludge waste.

A process solvent has been successfully developed that is capable of removing in a single process thorium, uranium, neptunium, plutonium, americium, the lanthanides, strontium and technetium from synthetic dissolved sludge waste from the Hanford single- and double-shelled storage tanks. The TRU elements (neptunium, plutonium, and americium) and strontium are then selectively stripped from the uranium and technetium. The latter point is of interest since one does not then need to vitrify uranium. A technetium specific resin has been developed to remove technetium from uranium, thus the uranium fraction is suitable for recycle.

The concentrated TRU-strontium product stream is then treated by the SREX-NEPEX (strontium extraction-neptunium/plutonium extraction) process. A process solvent, similar to the front end process solvent that separates strontium, neptunium and plutonium from barium, the lanthanides and americium, has been developed. A major feature of this process is that barium can now be separated from strontium; this is necessary to minimize the number of glass canisters needed to hold the waste in the geologic repository. The raffinate (barium, the lanthanides and americium fraction) is now a suitable feed for the americium/lanthanide separation system.

TALSPEAK Process: The cost of disposing of massive volumes of both sludge and supernate in the HLW tanks will be prohibitive unless there are treatments to remove some of the radioactive and inert components. It is advantageous to reduce both the complexity and cost of the chemical pretreatment of the HLW to achieve sufficient decontamination that the product streams will meet Class A LLW specifications.

The objectives of this work by Martin Marietta are to develop treatment capabilities for radioactive supernatants and sludges that are compatible with advanced processing flowsheets such as the "Clean Option." The work is divided into 5 subtasks. The first investigates the extremely difficult separation of actinides from lanthanides by the TALSPEAK process, an advanced, low-waste generation chemical process which co-extracts the 2 groups of elements into di (2-ethyl-hexylphosphoric) acid (HDEHP) from a carboxylic acid solution and then partitions the actinides into an aqueous phase by stripping HDEHP into a carboxylic solution containing diethylenetriamine pentaacetic acid (DTPA).

The second task will explore the use of continuous ion exchange to remove fission products. This contrasts with the array of conventional fixed-bed sorption processes which, though possessing many desirable features, are inherently batch operations and suffer drawbacks that compromise efficiency, reliability and safety. The third task will investigate treatments for the off-gas streams likely to result from advanced processing options. The final 2 tasks consider the Accelerator Transmutation of Waste (ATW) option. The first of these will perform an indepth review of the proposed blanket concepts and configurations for the ATW to identify important problems and potential resolutions. The second will determine the characteristics of the aqueous slurry that would be used in the ATW blanket.

The influence of zirconium and molybdenum on the TALSPEAK process was determined and maximum concentrations of the impurities were established. Possible process improvements to the TALSPEAK process, such as alternative acid systems and operation at higher temperatures have been pointed out.

Bench-scale work using rare earth elements as a stand-in for waste constituents has demonstrated the feasibility of continuous ion exchange system. Measurement of the critical parameters were made to permit scale-up of the process and provide data for an economic evaluation.

The investigation of off-gas treatment schemes for advanced processing scenarios has identified several sources of emissions in processing the Melton Valley Storage Tank Waste (MVST).

The TALSPEAK process (Trivalent Actinide Lanthanide Separations by Phosphorous-reagent Extraction from Aqueous Complexes) is one of the few means available to separate the trivalent actinides from the lanthanides. The method is based on the preferential complexation of the trivalent actinides by an aminopolyacetic acid. Cold experiments showed that by using citric acid the deleterious effects produced by impurities such as zirconium are greatly reduced.

Aqueous Biphasic Separation: This project at Northern Illinois University seeks to develop aqueous biphasic separation (ABS) systems for the pretreatment of supernatant solutions from underground storage tanks. The ABS permit the extraction and recovery of dissolved inorganic ions from high ionic strength, acid or alkaline electrolyte solutions. The ABS systems can be generated by the addition of polyethylene glycols to high ionic strength electrolyte solutions. The primary objective is to remove long-lived radionuclides such as Tc-99, K-129, and Se-79 from alkaline solutions. The selective removal of these species by aqueous biphasic separation could be coupled with ion exchange processes to remove cationic species, such as Cs-137 and TRUs, from alkaline solution. This would permit near-surface disposal of the bulk of the alkaline supernatants and dissolved saltcake from underground storage tanks. The extractions could be carried out without pH adjustment and the only reagent added to the waste stream would be a minimum amount of water to ensure dissolution of the saltcake.

Near-surface disposal of single-shell tank supernatants, after conversion to grout, would require that long-live isotopes such as I-129, Tc-99, and Se-79 be removed. Ion exchange technology for removal of Cs from alkaline solutions is well developed, while little or no technology is available for recovery of iodine, technetium, or selenium from high-ionic strength alkaline solutions.

Acidifying the supernatants would be required for recovery of Tc and Cs by TRUEX and SREX processes and would thereby result in significant increases in waste volume. The process effluents would then have to be made alkaline again before conversion to grout. In addition, Tc, I, and Se are not well extracted by TRUEX, and recovery of Cs requires the use of a crown ether extractant in the SREX solvent.

Several tank waste supernatants have been prepared. Aqueous biphasic formation has been demonstrated with SY-101, NCAW, and SST wastes at 25° and 50°C. Aqueous biphase formation has been obtained with polyethylene glycols (PEGs) having average weights of 1,500, 2,000, and 3,400. Partition co-efficients for TcO_4 have been measured giving values that range from 12 to 50 depending on the choice of waste simulant. Irradiation of the PEG phase up to 20 Mrad had no detectable effect on partition co-efficients. This dose approximates the yearly dose that would be received by the solvent in processing tank waste from SY-101. They have also irradiated PEG-1500 and PEG 3400 to 75 Mrad. At this higher dose, PEG-3400 began to show evidence of gel formation due to polymer crosslinking, but the lower molecular weight PEG showed no noticeable change in solution viscosity. The lower molecular weight PEGs thus appear to be highly resistant to damage by radiolysis.

Other: The removal of strontium from HLW can be accomplished by crown-ether extractants. The process is simple and highly selective.

At the Los Alamos National Laboratory Plutonium Facility, an ion exchange is used to recover plutonium from nitric acid solutions. Although this approach recovers >99%, trace amounts of plutonium and other actinides remain the effluent and require additional processing. Currently, a ferric hydroxide carrier precipitation is used to remove the trace actinides and the resulting sludge is cemented. Because it costs approximately $10,000 per drum for disposal, they are developing an additional polishing step so that the effluent actinide levels are reduced to below 100 nCi/g. This would allow the resulting waste sludge to be disposed as low-level waste. They are investigating various solvent extraction techniques for removing actinides. The most promising are chelating resins and membrane-based liquid-liquid solvent extraction.

Magnetic separation techniques are being developed at Los Alamos, Argonne, and Oregon State University.

Finally, a technical interchange with the Commissariat à l' Energie Atomique (CEA) in France will examine the technical problems associated with underground storage tanks. This exchange will initially focus on the French ACTINEX process which uses diamides as extractants for the actinides; this process will be evaluated against comparable US extraction technologies, such as the TRUEX process.

13.6.3 Sequestrants/Complexing Agents/Matrices

Fixed Extractants: The purpose of work at PNL is to develop high-capacity, selective fixed extractants for the removal of cesium, strontium, chromium, silver and noble metals from nuclear wastes.

In this study, the primary objective will be the identification of solid-based ion exchange materials capable of removing Cs-137 and Sr-90 nuclear wastes. Materials with the ability to selectively absorb Cs-137 and Sr-90 will be developed and incorporated into matrices designed to produce flow rates much greater than those attainable with the bare materials.

Using simulated wastes, the composites will be tested to demonstrate that the ion exchange properties remain in tact after fabrication. At that point, the synthesis of the materials will be scaled up and samples will be provided to facilitate testing of the material on actual DOE wastes.

The task will consist of these phases. Phase I will involve testing the capacity, selectivity, and stability of selected sequestering agents and substrates and radiation environments. Phase II will involve scaling up the synthesis of the materials and testing the industrial technology with radioactive nuclides, while Phase III will involve further up-scale and pilot scale testing.

Inorganic Minerals: In this first project, managed through PNL, the materials are specifically designed inorganic minerals developed at Texas A&M University and engineered into a useable format by the Allied-Signal Corporation. Initial studies showed the best candidate materials to be sodium titanates for the recovery of Sr, and zirconium arylphosphonate phosphates for the recovery of Cs. The concept is to use layered materials (in which the spacing between the layers can be specifically engineered), to separate the target contaminants from aqueous solutions of high ionic strength and at a wide variety of pHs.

The layer spacing can be tailored to the ionic hydration-sphere radius by using rigid spacer molecules bonded to, but separating the clay layers. If a background material has too large an ionic hydration-sphere radius it can not enter the mineral lattice, and if it is too small a radius then it will not be selectively retained. Screening and initial testing are to be performed by Texas A&M University, with scale up, pelletization, testing, engineering and design being performed by Allied-Signal; testing in radioactive environments will be performed by PNL.

Membrane-Like Materials: In a second project (which, like the first, is managed through PNL) a collaboration of the 3M Company and IBC Advanced Technologies is working to develop membrane-like materials that will selectively remove Cs and Sr from DOE wastes. The 3M Company's Empore" membrane technology can incorporate a wide range of particles into a sheetlike membrane. The particles are physically retained within the sheet by a mesh of small fibers or tendrils enveloping the particle while leaving the particles fully exposed and accessible to the liquid passing through the membrane. The resulting sheet is flexible, but the particles are closely packed (over 90% of the membrane is particle) so that high loading capacity is achievable. IBC Advanced Technologies has developed a series of macrocycle and cryptand based ligands that can efficiently extract contaminants from waste solutions.

The size, shape and chemical nature of these ligands can be designed to reject all but the targeted Cs and Sr materials. These ligands are then chemically anchored to particles which can be incorporated into the 3M Company's Empore" membrane. The resulting membranes are highly selective toward the 2 target materials, effectively removing them, but little else. Testing of the products in radioactive environments will be performed by PNL.

Cobalt Dicarbollide: Los Alamos and Savannah River Laboratory are developing a separations system for removing cesium and strontium from radioactive aqueous waste using cobalt dicarbollide (CoD_2). The cobalt dicarbollide molecule was first synthesized in the United States in 1965, but investigations of its application as an extractant in nuclear waste management was first undertaken in the 1970's in the former Czechoslovakia and the former Soviet Union. Cobalt dicarbollide has been described as a nearly ideal hydrophobic anion for the extraction of cationic species from aqueous solutions into an organic solution, and the literature indicates that under the appropriate conditions, cesium and strontium can be extracted with very high selectivity and yield.

In addition, cobalt dicarbollide is believed to possess superior radiation stability. Most of the earlier work has focused on the use of the molecule in solvent extraction systems. The thrust of this work is to explore the incorporation of cobalt dicarbollide (or its derivatives) into polymers to produce a material with the benefits of a solid-based sequestrant. This program builds on a considerable body of experimental and development work performed at the Nuclear Research Institute in Rez, the Czech Republic.

This technology addresses a specific problem within part of the DOE Complex. It will be examined as a substitute for the tetraphenyl borate precipitation used in the Defense Waste Processing Facility (DWPF) at the Savannah River Site (SRS). The tetraphenyl borate precipitation has encountered serious difficulties since it contains 4 benzene functionalities per molecule. These threaten to stall the operation of the facility. A replacement is needed and the dicarbollide concept offers an alternative that can be rapidly evaluated.

Natural Complexing Agents: Efficient chemical processes for the selective removal of actinide elements are needed for the treatment and minimization of wastes such as those found at the Hanford site. To accomplish this objective, new metal complexing agents capable of withstanding harsh chemical and radiation environments must first be developed and then modified for use in practical extraction systems.

The objective of a task at Lawrence Livermore is to develop a separation system using a highly selective complexing agent derived from a natural material to remove plutonium (and perhaps other TRUs) from the high ionic strength waste waters which vary in pH depending on the exact flowsheet. In nature, bacteria and other micro-organisms produce siderophores, low molecular weight multidentate iron chelators, to scavenge the ferric ion from their environments. One of the major siderophores is desferrioxamine B, and derivatives of this material show promise in fulfilling the requirements of a substance that binds plutonium in the presence of other metals and in solutions ranging from highly-acid to highly-basic.

13.6.4 Other Techniques

Magnetically Assisted Chemical Separations: Work at Argonne will explore a process combining the selective and efficient separation afforded by chemical sorption with the magnetic recovery of ferromagnetic beads. The objective is to develop a compact, economic, in-tank or near-tank process for the removal of contaminants such as Cs, Sr, and TRUs from aqueous wastes stored at DOE sites. The magnetically assisted chemical separation (MACS) uses magnetic beads coated with either a selective ion exchange material, an organic complexant containing a solvent for contaminant removal, or solvents for the selective separation of TRUs.

The beads are formed from a magnetic material coated with organic polymer or bonded ion exchange resin. Organic solvents can be adsorbed onto the polymeric surface by contacting the beads with a solution of the solvent in a volatile diluent. The coatings extract the contaminants and the beads are then magnetically removed. The beads can either be directly added to the vitrification slurry, or the contaminant can be removed by a stripping procedure to regenerate the extracting beads.

Magnetic particles coated with several compositions of octyl (phenyl)-N, N-diisobutylcarbamoylmethylphosphine oxide/tributyl phosphate (CMPO/TBP) mixtures have been prepared; optimization of the coating composition and coating process continues. In tests of the ability of these substrate-coated particles to remove americium from 2 molar nitric acid, a reduction in the americium concentration by a factor of 7 was achieved using 3-10 g of particles per liter of solution. This is equivalent to a K_d value in the 600-4,000 range. Similar tests with plutonium gave concentration reductions of a factor of 50-120 (K_d value in the 3,500-6,000 range). Stripping tests using alcohol and a variety of stripping agents show a reduction in K_d values to less than 10.

Magnetic adsorbents can be applied to the treatment of waste water in various physical forms. For example, barium ferrite has been used successfully as powder, granules or pellets. Iron ferrite, or magnetite, a naturally occurring ore, can also be used in much the same manner. However, natural magnetite needs activation to have the same capacity as freshly prepared ferrite. Furthermore, ferrites have been used solely in a batch mode because of their finely divided nature. Work at Rocky Flats has uncovered a synergistic effect in using the magnetic resin in a column mode in conjunction with an external magnetic field for concentration of plutonium and americium from wastewater, greatly surpassing a batch model.

Hanford is evaluating cesium removal from tank supernatant using magnetic particles (MAG*SEP) coated with various absorbents.

Other Actinide Separations: The goal of this project at Los Alamos is to develop actinide separations capabilities that could be used in an advanced processing flowsheet, such as the "Clean Option" for the Hanford Tank Waste Remediation System or some other alternative. Since there is a wide variety of waste streams needing treatment, (such as aqueous acid waste, aqueous basic waste, low-level neutral wastewaters, contaminated soils and handling materials), this project has several subtasks. These include (1) investigation of selective inorganic precipitation of metals (molybdenum, technetium, ruthenium, palladium, cadmium, americium, and curium) to replace processes requiring the use of organic solvents, (2) investigating the use of liquid ion exchangers (LIXs) as alternatives to Aliquot 336 for actinide processing, (3) investigating diamides as alternatives to CMPO for TRU removal, (4) developing soft ion donor extractants for the separation of trivalent actinides from lanthanides, (5) investigating the use of water-soluble chelating polymers, and (6) investigating microporous hollow-fiber membranes for dispersion free liquid/liquid extraction.

Precipitation and separation of the lanthanides and transition metals from the actinides through a process of complexing the actinides with carbonate and precipitating the other metals has been achieved. Interesting differences between tracer studies and higher loading studies were observed.

In the research into LIX alternatives to Aliquot 336, 3 new pyridinium compounds have been synthesized and preliminary thorium extraction studies are in progress.

Two new dialkyldithiophosphoric compounds have been synthesized and tested for trivalent actinide/lanthanide separations. Initial results indicate enhanced extraction of americium over europium.

A commercially available pyrazoylborate was tested and found to have potential for separation. The instability of the compound has let to synthetic modification studies to improve both organic solubility and stability.

Collation of data has begun on the investigation of water soluble chelating polymers for waste stream treatment.

Plutonium Uranium Extraction Process (PUREX) testing on available equipment for the dispersion-free liquid-liquid extraction modular unit has been completed and data is being evaluated for comparison with other engineered systems.

Calcination/Dissolution Pretreatment: This process at Oak Ridge combines calcination and dissolution in a new and innovative manner in order to thermally destroy organic and inorganic constituents. A large-scale plasma arc heats the tank waste in excess of 800°C to destroy organic compounds, ferrocyanide, nitrate, and nitrite and to separate the strontium and transuranic waste into a small volume via water dissolution.

Target process flow is about 20-80 gpm for a typical operating unit. Previous attempts at calcining high sodium waste material using conventional calcining concepts, such as rotary kiln, spray tower, and fluidized bed, have encountered problems due to plugging of the reactors by molten material.

A large-scale plasma arc demonstration at the Westinghouse Science and Technology Center successfully calcined 3,000 pounds of simulated Hanford tank waste continuously without plugging. This demonstrates that large-scale, high calcination of high sodium wastes is possible. Future plans include a second demonstration of plasma arc calcination of a 101-SY tank simulated waste. This longer test will couple calcination and dissolution operations with enhanced sample analysis.

Thermal treatment by calcination offers several benefits for the treatment of Hanford Site tank wastes, including the destruction of organics and ferrocyanides and an hydroxide fusion that permits the bulk of the mostly soluble non-radioactive constituents to be easily separated from the insoluble transuranic residue.

13.6.5 Transuranic Element Removal

There is a need to remove transuranic elements and other selected components from high-level wastes (HLW) in underground storage tanks at Hanford and other DOE sites. This would allow concentration of those components into a minimum volume of HLW for ultimate disposal in a geologic repository. Present plans include incorporation of the HLW into borosilicate glass that will be sent to the repository. This is expected to be the most costly waste form per unit volume for tank wastes. Concentration of materials that must go to the HLW repository could have important cost benefits. Sufficient separation of these components would allow the remaining materials to be incorporated in a low-level waste form, which would be less costly and conserve space in the repository. The TRUEX solvent extraction process was discussed earlier.

Transuranium elements may need to be removed from the alkaline supernates from some of the tanks to permit these materials to be incorporated in LLW. Sludges (solid precipitates) in HLW storage tanks are also expected to be washed or leached to remove selected materials, and the washing or leaching may dissolve sufficient transuranium materials that the liquors would be unacceptable for incorporation in a LLW. It would then be necessary to remove the transuranium components from the leach liquor so it can be returned to the HLW. The potential need for removal of transuranium materials from leach liquors may depend upon how "aggressive" the leach step will have to be to remove sufficient fractions of other materials that hinder formation of borosilicate glass.

Optional approaches to minimizing the HLW produced by alkaline sludge treatment include actual acid dissolution of the sludge, followed by a series of separation steps to separate and concentrate the components that need to be incorporated in the HLW from those that can go to the LLW forms. Since dissolution is likely to involve use of nitric acid, separation methods for removing transuranium elements from dissolution liquors would need to operate with acid solutions.

It is important that R&D on transuranium separations be related to a potential scenario or flowsheet that could be used with HLW in DOE storage tanks. This means that proposed work needs to include pH and compositions of solution that exist or could result from processing or material in DOE tanks.

Transuranic (TRU) elements exist in many DOE wastes. These include:

1) Sludges precipitated from alkaline tank wastes, either acid or alkaline solutions formed by leaching or dissolving tank sludges, and complexed TRU species in high pH supernatant in tank wastes.
2) Decontamination waste streams.
3) Solid wastes, including mixed waste and buried waste.
4) Miscellaneous aqueous waste or process streams.

Consequently, innovative methods are needed that remove:

1) TRU elements from acid streams (e.g., 1M HNO_3); new processes should surpass existing solvent extraction processes.
2) TRU elements from high ionic strength (up to 10M) alkaline (pH up to 14) streams, including complexed species.
3) Very small concentration of TRU elements from waste waters and process streams.
4) TRU elements from solid wastes.

In all cases, methods are needed that:

1) Are efficient (remove virtually all TRU elements from waste and release all TRU elements to effluent streams), selective (do not remove transition metals, ideally do not remove lanthanides, etc.), and rapid (can be implemented at an industrial scale).
2) Generate minimal secondary wastes and do not require challenging separations to further treat effluent streams.
3) Are economical (especially for treatment of waste water and miscellaneous waste streams) and are simple to maintain and operate.

Development of the following 7 technologies is being sponsored by DOE to address the problems posed by TRU elements and other radionuclide components of HLW:

1) Solid Sequestering Agents for the Removal of Transuranics from Radioactive Waste.
2) Actinide Separations for Advanced Processing of Nuclear Waste.
3) Advanced Chemical Separations for the Clean Option.
4) Advanced Separations, Systems Development and ATW Blanket Analysis.
5) Derivatives of Natural Complexing Agents for the Removal of Plutonium from Waste Waters.
6) Aqueous Biphasic Systems for Radioactive Waste Pretreatment.
7) Fission Product Chemistry.

Cesium and Strontium Removal Technologies: Over the past 5 decades, a very large amount of highly radioactive nuclear waste has been accumulated as a by-product of the nuclear processing activities of DOE and its predecessor organizations. For example, the DOE inventory of wastes in the Weapons Complex includes over thirteen and a half million cubic feet of HLW. The destiny of this waste is to be turned into a glass waste form and then stored underground in a deep geologic repository. Due to the large volumes that must be processed, this destiny threatens to be extraordinarily expensive. If the components of this waste can be separated to produce streams that are amenable to less expensive disposal, a huge burden will have been lifted from the taxpayer.

Actually, most of this volume is composed of non-radioactive material, such as water and comparatively harmless salts, such as nitrate and nitrite. The source of the radiation, the radionuclides, are typically only a few tenths of one percent of the volume. Cesium-137, and strontium-90, are just 2 of these radionuclides, and although they in turn occur in comparatively small amounts, their physical properties dictate that they are of great importance in waste management considerations. These 2 substances dominate the fission product radiation, contributing over 90% of the thermal energy and penetrating radiation during the first 30 years after irradiation. These properties lead to major problems in handling the waste in the intermediate processing stages between reclamation of the wastes from the tanks and production of the final waste form, and may have a severely deleterious effect on the behavior of the final waste form. If these 2 components could be selectively removed from the waste, there would be an easing of the handling requirements and an improvement in the quality of the glass waste form.

Development of the following 6 technologies is being sponsored by DOE to address the serious problem of cesium and strontium removal:

1) Separation of Cesium and Strontium from High-Level Radioactive Waste.
2) Waste Separation and Pretreatment Using Titanate Ion Exchangers.
3) Magnetically Assisted Chemical Separations.
4) High-Capacity, Highly Selective Solid Sequestrants for Innovative Chemical Separation: Pillared Clay Approach.

5) High-Capacity, Highly Selective Solid Sequestrants for Innovative Chemical Separation: Membrane Supported, Particle Bound Sequestrant Approach.
6) "Carbollide"/Cobalt Dicarborane Anion Process Development for Cesium-137 Decontamination.

13.7 NITRATE DESTRUCTION/REMOVAL

A wide variety of nitrate-containing aqueous mixed wastes are produced and/or stored at various DOE facilities. These wastes generally have very high concentrations of nitrates (either sodium nitrate or nitric acid), high levels of radionuclides and heavy metals and/or solvents. The nitrates in the wastes will generally increase the volume and/or reduce the integrity of all the waste forms that have been proposed for ultimate disposal, so nitrate destruction prior to solidification is expected to be beneficial.

Nitric acid-based aqueous solutions have always been employed in DOE fuel reprocessing plants. Neutralization of the acidic waste solutions from such reprocessing operations produced very large volumes of aqueous sodium nitrate wastes which are stored in underground tanks at various DOE sites, principally the Hanford, Savannah River, and Oak Ridge sites. Over time, radiolysis has converted a substantial amount of the nitrate ion to nitrite ion.

Nitrate ion is an EPA-listed toxic chemical. Near-surface disposal (in grout form) of aqueous nitrate waste solutions, after prior removal of selected radionuclides, is complicated by the need to restrict water leaching and transport of toxic nitrate ion to permissible levels.

Abundant and useful technology already exists for destroying nitrate and nitrite ions in DOE radioactive and mixed wastes.

Several nitrate-destruction technologies are being investigated by DOE, each having advantages and disadvantages. A study by Oak Ridge resulted in a recommendation that development work continue on the NAC, electrochemical destruction, and hydrothermal processes. More information is needed on the capabilities and potential problems of these processes before a rational decision can be made regarding the best process for treating a particular waste stream.

Nitrates and nitrites constitute the largest fraction of wastes at some DOE sites, and these components can increase the volume requirements for disposal and/or storage. Nitrates and organic compounds complicate waste processing and are not easily incorporated into some waste forms; they can even increase the leaching of other components from waste forms. Organic compounds can include solvents that act as ligands for some metals or radioactive components of the waste, and reduce the effectiveness of some methods for removing those metals from the waste.

New and improved methods are needed to destroy nitrates and/or selected organic compounds. The methods should have significant potential advantages over current technologies. For instance, improved methods would operate on solid as well as liquid wastes, would be suitable for liquids with high concentrations of solids, would more easily operate in highly radioactive environments, or could offer greater safety or robustness for a variety of wastes. Where nitrates can be used (such as sodium nitrate or nitric acid) or where organic compounds can be recycled, effective recovery methods will have advantages over destruction methods. Most current approaches involve destruction of these materials. It is important to direct the R&D focus toward a current, specific DOE problem.

13.7.1 Nitrate to Ammonia and Ceramic Process (NAC)

The NAC process addresses the primary liquid and sludge leachate component at the Hanford and the Melton Valley Storage Tanks in Oak Ridge. Underground storage tanks at both sites contain large amounts of highly alkaline, nitrate based mixed liquid waste. Current plans are to place the liquid phase in grout and the acid washed solids (TRUEX process) in glass. The NAC project will support these objectives by generating innocuous gaseous and liquid products and a low volume, chemically stable solid waste form.

The Nitrate To Ammonia and Ceramic (NAC) process involves chemical reduction of the nitrate contained in the mixed (hazardous and radioactive) tank waste to gaseous ammonia utilizing a new technology developed at Oak Ridge National Laboratory (ORNL). The NAC process can achieve concentrations of nitrate below drinking water standards.

It is expected that radioactive species such as plutonium and strontium will enter the solid ceramic phase during the reduction of the nitrate anion. The alumina-based ceramic host matrix will undergo calcination, pressing and sintering which will generate a solid waste form capable of binding nearly all elements. Sodium will be in a nepheline ceramic phase. In the process, radioactive-contaminated scrap aluminum from various DOE sites could be shredded and used as feed to the NAC reactor. Therefore, the need for decommissioning and decontamination of such metal will be eliminated.

Laboratory experiments have shown that a decontaminated solid nitrate waste stream can be solidified by using a catalytic material to decompose the sodium nitrate to innocuous gas and a liquid secondary waste stream containing only trace amounts of nitrate. Additional advantages of the process are solidification of the radioactive fraction of the waste at rather low temperatures (50-90°C; unlike vitrification) and a final waste volume reduction of 55 to 75 vol% (as compared to a 30% to 40% volume increase by grouting).

The overall efficiency of the use of aluminum metal reactant has been lately increased from 50% to 91% by operating the NAC reactor in a continuous mode. This operation mode will:

1) Reduce processing costs by decreasing the amount of aluminum, a costly reactant, down to approximately $2/kg of nitrate converted; and
2) Operate with waste solution feeds containing nitrate at near saturation.

Sintering temperatures of the final ceramic product have been reduced by 150°C to 1,350°C, while sintering time was cut in half.

A patent is pending and its ownership was awarded to Martin Marietta Energy Systems, Inc. In virtue of its adaptability to alkaline supernates, numerous DOE sites could make use of the NAC process technology, including Savannah River, Oak Ridge, and Richland. Whereas no industrial partners are currently involved in the NAC process technology development, Florida International University worked to aid and support the development effort in FY94.

13.7.2 Electrochemical Reduction

The aim of this project at Savannah River is to develop and evaluate the various electrochemical methods for 1) destroying nitrates, nitrites and organic compounds in HLW and LLW, and 2) removing radionuclides and hazardous metals from alkaline waste solutions. In an electrochemical cell, nitrate and nitrite can be reduced to nitrous oxide, nitrogen and ammonia; these gases have very low solubilities in alkaline solutions and are effectively separated from the waste. Metal ions can be reduced producing solid phases or deposits on the electrodes which can once again be removed. In addition to electrochemical reduction, the corresponding electrochemical oxidation reactions can be brought about to oxidize organics to carbon dioxide and water.

This work will include examination of divided parallel-plate, packed bed, and fluidized bed electrochemical cells, and identification of electrocatalysts for these destruction reactions. The program consists of 5 major tasks. The first is the determination of optimum reactor conditions for the destruction and removal of hazardous waste components. The second is the development of engineering process models. The third is laboratory scale tests with radioactive materials. The fourth is the pilot scale tests, and the fifth is analysis and evaluation of the testing data.

The advantage of this method is that in the case of nitrate and nitrite reduction, one equivalent of nitrate destroyed results in one equivalent of hydroxide being produced; thus, there is no addition of any material to the waste stream. Further, considerable reduction (possibly up to 75%) in the volume of waste requiring disposal could be realized by recovery and recycle of the sodium hydroxide produced. Similarly, for organics the destruction produces innocuous materials without the addition of any material to the system.

Electrochemical reduction destroys nitrates without any chemical addition and the process can be used on acidic or alkaline wastes. However, other ions may plate out on the electrodes, release of off-gases requires further evaluation, the process requires significant quantities of makeup water, and NaOH by-product must be disposed.

13.7.3 Microbial Processes

The Biological Destruction of Tank Waste (BDTW) Process uses bacteria to reduce the nitrates in waste to nitrogen gas and separate the metals by a combination of biosorption (adsorption onto bacteria) and precipitation. Some degradation of organics is also anticipated.

The process uses a mixed culture of natural bacteria isolated from the Great Salt Lake and the Death Valley area. These bacteria are able to grow and reduce nitrate in the very high salt concentrations found in the tank wastes. The bacteria are grown in a bioreactor and then recycled to a biosorption tank where they are mixed with the incoming waste. The high radioactivity and metals concentration in this tank may kill the bacteria, but dead bacteria biosorb metals equally well.

The bacteria and any chemical precipitates that may have formed are removed by filtration, generating a biomass sludge waste containing the metals and radioactivity. In most cases this sludge will be a low-level radioactive waste, and will be dried and sent for final disposal by grouting or vitrification. The liquid containing the nitrate, organics and very low levels of metals flow into the bioreactor, where it is mixed with acetic acid as a carbon source for bacterial growth.

In the bioreactor, the nitrate is reduced to innocuous nitrogen gas that is released to the atmosphere after being filtered. Any remaining metals adsorb onto the growing bacteria, but their concentration is now too low to inhibit bacterial metabolism. The effluent from the bioreactor, after filtration, is a concentrated solution of non-radioactive, non-hazardous salts in which nitrate has been replaced mainly by bicarbonate.

This technology is currently being developed by EG&G at Idaho National Engineering Lab. It has potential application to treat waste streams from metals reprocessing facilities in addition to those from nuclear fuels processing and reprocessing facilities.

This biological process is applicable to treat the highly saline underground storage tanks of the Hanford Site, which contain various radionuclides, transuranic and toxic metals, and organic materials. The organic materials are principally saltcake, consisting mainly of nitrate salts and lower levels of metals, and concentrated supernatant whose composition is in equilibrium with the waste sludge and saltcake. It would be applicable to treat similar waste of other tank farms. The process should work on most tank waste, but a bench-scale treatability study would be needed for each tank.

The Biocatalytic Destruction of Nitrate and Nitrite task at Argonne is to develop an enzyme-based reactor system for the reduction of nitrate and nitrite to $N_2$and H_2O. It will demonstrate the validity of using immobilized enzymes coupled with biphase partitioning to efficiently destroy nitrate and nitrite. The reducing equivalents are provided by a low-voltage electrical current, which transfers electrons from the cathode to the enzymes via an electron transfer dye. The biphase system is necessary to protect the enzymes from excessive concentrations of electrolytes, especially H+ and OH- which would result in enzyme inactivation, while simultaneously allowing the transfer to nitrate and nitrite from the waste stream to the catalytic chamber. The use of enzymes enables very large specific catalytic activity to be obtained without the need for additional chemical reagents or the production of secondary waste streams.

13.7.4 Other Processes

1) Lawrence Livermore is developing a process utilizing bipolar electrohydrolysis membranes for salt splitting, used in conjunction with anion and cation permeable membranes.
2) A French group is studying the use of supported liquid membranes. The supported liquid membrane (SLM) renders the use of expensive tailor-made extractant molecules like CMPO or crown ethers possible.
3) Hydrothermal processes can simultaneously treat nitrates and organics in both acidic and alkaline wastes producing nitrogen gas.
4) Alkaline nitrate solutions can be acidified and treated with sugar, formic acid, or paraformaldehyde to convert the nitrate (and nitrite) ion to NO_x.
5) Chemical reduction requires acidification of alkaline wastes, converts nitrates to nitrogen oxides requiring off-gas treatment, and the process is potentially unstable so safety controls will be required.
6) Calcination has been used at INEL for many years to solidify low sodium wastes; however, with alkaline wastes, which are primarily sodium nitrate, the sodium nitrate melts rather than decomposing at the operating temperature of 500°C. The off-gas must also be treated to remove NO_x. Demonstrated technology exists for catalytically converting NO_x compounds to N_2.
7) Freeze crystallization.
8) Reduction with ammonia, hydrazine, or organic compounds containing oxygen.
9) The Clean Salt Process (CSP) at Hanford. A selective crystallization process.

13.8 SLUDGE PROCESSING

Participants in Hanford's Tri-Party Agreement (TPA), wishing to minimize chemical risks and associated programmatic delays, recently decided to pursue alkaline washing of tank sludge. The suitability of this path will be confirmed by testing sludge washing processes using real waste. Full dissolution of sludge and treatment of the resulting acidic solution is not considered baseline technology of treatment of Hanford tank waste; however, since it remains an important option until alkaline wash is proven, activities on acid side and other treatments will be pursued.

The suitability of the new TPA strategy will depend largely on the volume of the high-level waste (HLW) that would result from vitrification of the sludge. Since the cost of HLW disposal is high, there is an incentive to minimize the fraction of the sludge going to HLW disposal. This leads to 2 closely related, but opposing goals: (1) to remove from the sludge sufficient amounts of constituents that can be sent to the low-level waste (LLW) form, thus significantly reducing the HLW form volume; or (2) to remove from the sludge sufficient amounts of those constituents that cannot be sent to the LLW form so that the remaining sludge can be disposed of as LLW. The first involves partial removal of abundant non-radioactive constituents, while the second requires dissolution followed by nearly complete removal of dilute radioactive constituents.

To reach either goal, innovative processes are needed that involve minimal technical risk and complexity and do not generate substantial volumes of secondary waste. The process should be efficient (involving minimum addition of process chemicals, particularly any that incinerate into forms other than gases), selective (removing only the needed constituents), and rapid (can be scaled up for practical treatment of millions of gallons of tank waste). For example, it would be desirable to remove bulk constituents, such as aluminum, iron, sodium, etc., from the sludge while requiring minimal separation of HLW constituents from the leachate, in order for it to be disposed of as LLW. That is, constituents such as transuranics, Sr-90, U, etc., would be left in the sludge. There is a particular incentive to remove phosphate from the sludge, because it is sparingly soluble in borosilicate glass and would require production of a large mass of glass if it is not leached from the sludge.

The success of any selective washing or leaching process will depend on the properties of the sludge. Therefore, eventual testing using real sludges will be required, and must be anticipated and defined in proposals for new technologies.

In-tank sludge washing consists of washing tank wastes with water using large pumps inserted into the top of the tanks. Sludge washing is not the same as sludge mixing, the retrieval method discussed earlier in this chapter. Unlike sludge mixing, it introduces large volumes of additional outside water into the tanks. This approach will allow solid, high-level waste to settle to the bottom of the tank, while lower-level waste will remain near the top where it will be pumped to holding tanks for further pretreatment. DOE will wash the waste in this manner at least twice to separate the high-level waste solids from lower-level waste.

In the baseline process for pretreating Hanford tank sludges, the sludge is leached under caustic conditions, then the solubilized components of the sludge are removed by water washing. The purpose of the caustic leaching is to remove certain sludge components (e.g., Al, Cr, and P) that impact the volume of high-level waste borosilicate glass produced in treating these wastes. Tests of this method have been performed with samples taken from 5 different tanks at Hanford. The tests show that leaching with several molar NaOH solutions at 100 (degrees) C results in marked improvement in the removal of Al, Cr, and P from the sludge when compared to washing with dilute NaOH solutions.

Several options are being considered at Hanford for the pretreatment process:

1) Sludge washing with water or dilute hydroxide: designed to remove most of the Na from the sludge, thus significantly reducing the volume of waste to be vitrified.
2) Sludge washing plus caustic leaching and/or metathesis (alkaline sludge leaching): designed to dissolve large quantities of certain non-radioactive elements, such as Al, Cr, and P, thus reducing the volume of waste even more.
3) Sludge washing, sludge dissolution, and separation of radionuclides from the dissolved sludge solutions (advanced processing): designed to remove all radionuclides for concentration into a minimum waste volume.

The actinides are constituents of major importance in Hanford tank sludges. Vitrification of the washed sludges with no pretreatment could be prohibitively expensive, so a method that can remove from the sludges the contaminants of greatest concern represents a significant economic benefit. There is at present a dearth of technologies dealing with both the treatment of sludges and solids, and of technologies that might be amenable to in tank application. Technologies that are at a level of development such that timely implementation is possible are also at a premium.

13.8.1 Selective Leaching by the ACT*DE*CON" PROCESS

The objective of this effort is to assess the ability of Bradtec's proprietary ACT*DE*CON" process to treat and recover various radionuclides from the waste sludge of the single-shell tanks at Hanford, thus leaving a material amenable to less expensive disposal. The ACT*DE*CON" process is unique in that it combines dissolution of the contaminants with non-hazardous and non-corrosive dilute selective solvents, recovery of contaminants and regeneration of solvents for a continuous recirculating treatment process. The ACT*DE*CON" solvent chemical utilizes well established carbonate recovery chemistry, a chelant (EDTA) and an oxidant (hydrogen peroxide).

Variations of the ACT*DE*CON" process have been used in full-scale fuel dissolution applications and a predecessor of the ACT*DE*CON" process is currently treating 200 kg per day of TRU contaminated sludge and debris at the Dungeness Nuclear Power Station in the United Kingdom. The ACT*DE*CON" process chemistry has been developed and enhanced by Bradtec to treat soils contaminated with actinides and certain radionuclides and heavy metals.

A 3 phased approach will be used in this project: lab scale testing with waste stimulants, lab scale testing with actual waste, and pilot scale testing with waste stimulants and actual waste. One approach to be tested will be the ability to dissolve sludge directly in a tank. The solvent, dissolved materials, and entrained solids will then be separated to recover non-dissolved sludge materials. Provided these materials meet low-level radioactive waste criteria they can be disposed of in grout. If the sludges contain levels of TRU waste greater than 10 nCi/gram, they can be treated in a contactor with the ACT*DE*CON" process and mechanical action to further break up the sludge and dissolve the TRU waste.

Anion exchange will recover the uranium, technetium, iodine, and TRU, and further processing will deal with these components. Cesium and strontium will be recovered by cation exchange.

13.8.2 Sludge Washing and Dissolution of Melton Valley Storage Tank Waste

The focus of this task at Oak Ridge is the performance of experimental and modeling research and development in support of the comprehensive sludge/supernatant processing flowsheet efforts being conducted for the Underground Storage Tank Integrated Demonstration. The primary emphasis is on Hanford tank waste, disposal of which will probably involve dissolution of the sludge prior to pretreatment. A knowledge of the compositions of the dissolving solutions will be required in order to plan further treatment strategies, such as the various extraction options, grouting and vitrification.

This work will use actual wastes from the Melton Valley Storage Tanks (MVST) at Oak Ridge National Laboratory for experiments on sludge washing and dissolution. The aim of the washing is to minimize the amount of TRU containing sludge through dissolution of the inert components; dissolution of TRUs at this stage is to be avoided. The sludge residue will then be subjected to a series of leachings in which it is possible that the TRUs can be removed leaving a non-TRU residue. To this end, a knowledge of the compositions and properties of both the solution and the sludge is necessary.

The DOE site tank waste disposal strategy should require dissolving the sludge from the tanks before further treatment. Knowledge of the compositions of the dissolving solutions is needed to plan subsequent treatment strategies such as TRUEX, grouting or any of the other high-level waste separation technologies.

The processes and technologies that will be tested and evaluated include:

1) Acid dissolution of sludge;
2) Removal of cesium and strontium from supernate using ferrocyanide precipitation, ion exchange, and solid extractant; and
3) TRUEX solvent extraction of transuranic elements.

13.8.3 Calcine Residue Leaching: High-Level Waste Sludges

This project at Hanford addresses the need to reduce the volume of HLW by use of calcination. Calcination destroys the organics and nitrates present in the HLW and subsequent leaching of the calcine residue will further reduce the volumes. Calcining HLW from the Hanford tanks is the only non-aqueous alternative proposed for the updated Tank Waste Remediation System strategy. New calcination methods being tested as part of the Underground Storage Tank Integrated Demonstration are expected to be able to handle the high sodium content of Hanford wastes. The thrust of this work is to develop methods for removal of non-radioactive constituents from the residues produced by calcination and aqueous leaching.

Calcine residue volume, which would directly control HLW, could be significantly reduced by dissolving out the major non-radioactive components. The chemical state of the calcine and leach residuals and responses to additional thermal and chemical treatments will be established. This knowledge will be used to identify methods to selectively leach the non-radioactive species, or to dissolve broad ranges of constituents for separation. Extraction processes applied to calcine residues are expected to be more efficient than similar processes on raw tank waste because of previous elimination of sodium, aluminum, and nitrates.

The emerging Tank Waste Remediation System (TWRS) will call for several strategies for pretreatment of Hanford tank wastes. The strategies will include baseline aqueous processing as well as competitive, less developed approaches that promise significant improvement over the baseline. Calcination followed by aqueous leaching has been tentatively identified as one of the advanced processing alternatives to be included in the TWRS.

Equilibrium thermodynamic analysis and more complete work scope definition is in progress. Thermodynamic analysis of calcination of 5 wastes types—bismuth phosphate process waste, redox process sludge, plutonium finishing plant waste, complexant concentrate (101-SY), and composite Hanford tank wastes—has been performed. University support for the thermodynamic analysis of dissolution and residue treatment is being pursued and resources for residue treatment process development are being developed.

13.9 TRANSMUTATION

A number of concepts have been proposed that, if found to be technically and economically feasible, might reduce the volume and radioactive life of wastes destined for burial in a deep geological repository. These concepts involve transmuting (changing) constituents of the waste into elements with shorter radioactive lives or to non-radioactive elements through nuclear action in a reactor or an accelerator.

Actinide burning, also called waste partitioning-transmutation (P-T), is an advanced method for radioactive waste management based on the idea of destroying the most toxic components in the waste. It consists of 2 steps: (1) selective removal of the most toxic radionuclides from high-level/spent fuel waste and (2) conversion of those radionuclides into less toxic radioactive materials and/or stable elements.

Spent fuel contains a relatively small number of long-lived radioactive elements that are responsible for the long period that this waste is required to be confined in a repository. If DOE could transmute these elements to stable ones or ones with shorter radioactive life spans, it might reduce the long-lived hazards of the waste and increase the capacity of the repository. DOE could use a reactor's or an accelerator's nuclear reactions to transmute these long-lived elements. However, DOE would have to first reprocess the spent fuel to separate the long-lived elements and then incorporate them into new fuel (or a target for an accelerator to bombard). The fuel would be burned, reprocessed, refabricated, and burned again in a continuous cycle. Although the transmutation process might eventually produce a waste that has a much shorter radioactive life, residual high-level wastes and radioactive elements that cannot be transmuted would still need to be buried in a repository.

Although transmutation could be considered for treating defense-related nuclear waste, current plans call for DOE to separate the waste into high- and low-level components and dispose of the high-level component in a deep geological repository.

DOE managers who are responsible for the disposal of radioactive defense waste and commercial spent fuel are not in favor of transmuting waste before it is buried in a repository. They believe it unnecessary and costly and note that a repository will still be needed, even if transmutation of some of this waste is successful. It is expected that costs will be high and that any practical application of transmutation is decades away.

One of the highest priority isotope separation needs is for cesium isotopes. Fission-product cesium, following one or more decades of decay, contains stable Cs-133, long-lived Cs-135, and short-lived Cs-137.

Transmutation of Cs-135 may be desirable from a disposition perspective, but isotope separation is a prerequisite. Development of cesium isotope separation concepts is an important long-term basic research need.

If long-lived fission products other than Tc-99, I-129, and, potentially, Cs-135 require transmutation, some of them may also benefit from isotope separation operations prior to irradiation. In addition, isotope separation may be a prerequisite to beneficial use of some partitioned elements. For example, partitioning of fission-product palladium may be cost-effective if isotope separation can adequately remove long-lived Pd-107. Isotope separation may also be important in eliminating material damage due to the formation of activation products. For example, steel isotopically enriched in Ni-60 would eliminate the (n, alpha) reaction of Ni-59.

Five current transmutation concepts are discussed below.

13.9.1 The Advanced Liquid-Metal/Integral Fast Reactor

The advanced liquid-metal reactor (ALMR) actinide recycle system concept is a metal-fueled nuclear reactor that utilizes "fast" neutrons to produce the nuclear fission reactions on which its operation depends. The metal-fuel-cycle Integral Fast Reactor (IFR) program is a development of Argonne National Laboratory (ANL) near Chicago and its western branch, ANL-West, in Idaho. The purpose of the IFR project is to develop and demonstrate the essential features of a metal-fueled fast reactor and a metal-fuel-cycle process. The ALMR is a design project of General Electric Company at San Jose, California, to develop a commercial design for a modular nuclear power plant using the metal-fueled fast reactor.

This fast reactor concept differs in several ways from the light-water reactors (LWR) used in the current generation of nuclear power plants. The LWRs use slow or "thermal" neutrons instead of fast neutrons to produce the fission reactions. The LWRs are cooled by water instead of the liquid metal sodium, which is the coolant in the fast reactor. The LWRs cannot breed fuel—that is, they cannot produce more fuel than they use—while the fast reactors can. Finally, the fast reactors can transmute—burn up—minor actinides, a group of transuranic by-products produced in nuclear reactor operation that are major contributors to the long-lived hazards of radioactive waste. The LWRs cannot effectively transmute the minor actinides, although they can use the major actinide plutonium as fuel, if the spent fuel from LWR operations is processed. At the present time, LWR spent fuel is not processed in the United States.

The processing of metal fuel using a special pyrochemical technique is being developed by Argonne National Laboratory. The pyrochemical processing includes electrorefining the fuel in a molten salt, a process similar to the one used in the production of the metal aluminum from its ore. In its breeder version, the ALMR operates as a fast breeder reactor and can have a processing plant built as an integral part of the facility so that the spent fuel elements are processed and the new fuel is manufactured on-site. The fuel may alternatively be processed at an off-site facility to improve economics; that is, the facility might serve several ALMRs.

The ALMR can be operated as an actinide "burner" (transmuter) instead of a breeder. Breeding is not desired when the objective is to eliminate the existing actinide fuel materials in LWR spent fuel rather than to produce more fuel. In the burner mode of operation, the ALMR uses as fuel plutonium and minor actinides that have been extracted from LWR spent fuel at a separate processing facility. Transportation of spent or processed fuel will be necessary in this case because the LWRs will not, in general, be located at the fast reactor site.

Argonne National Laboratory and General Electric claim several advantages for this fast reactor concept as a long-lived radioactive waste transmuter. It separates the long-lived plutonium and minor actinide fuels from the rest of the radioactive waste and utilizes them to produce electricity. It provides convenient chemistry to incorporate some other long-lived radioactive materials—mainly, iodine-129 and technetium-99, which are not transmuted—into chemical forms that will be immobile when placed in a geologic repository. It maintains the fissionable materials in a form that is very hazardous and thus inhibits possible diversion or theft for nuclear weapons use. Finally, this technology is well advanced compared to any of the other transmutation concepts.

13.9.2 The Los Alamos National Laboratory's Accelerator

The Los Alamos National Laboratory (LANL) Accelerator Transmutation of Waste (ATW) concept is a project to use neutrons produced by a high-energy, high-current proton linear accelerator to transmute transuranic actinides and long-lived fission products. In the process, a large current (up to 0.25 amperes) of protons is accelerated to high energy (as high as 1,600 million electron volts). LANL has an aqueous and a non-aqueous version of the ATW. It is suitable for spent fuel, plutonium, etc.

The protons strike a target material and produce a shower of neutrons that slow to "thermal" energy in a tank of heavy water that surrounds the target. Most of the thermal neutrons are absorbed in transuranic actinides or long-lived fission products that flow in solutions or slurries through pipes located in the heavy water tank. Absorption of neutrons stimulates nuclear fission in the nuclei of actinide nuclei or alternatively converts long-lived iodine-129 and technetium-99 into short-lived or stable products. The nuclear fission releases heat, which is used to produce electricity by means of power-generating equipment that is coupled to cooling loops through heat exchangers. Part of the electricity produced in this way is used to supply energy to run the linear accelerator. The rest is available for sale to an electric utility. The cooling loops also include clean up elements that remove the short-lived and stable materials that are produced during the processing.

LANL claims several advantages for this transmuter concept. It is a subcritical system, which offers additional protection against criticality accidents. Also, the high concentrations of thermal neutrons produced from the linear accelerator target in the heavy water tank make possible rapid transmutation of actinides with much smaller actinide inventories in the transmuter than are required for the ALMR. This intense thermal neutron flux is a basis for the unique features of ATW. Furthermore, the thermal neutrons can transmute iodine-129 and technetium-99, which the ALMR cannot do efficiently. These radionuclides are major contributors to long-term risks associated with repository storage because they are more likely to be leached out of a repository than actinides. Finally, this process uses continual material feed and waste removal rather than batch refueling like the ALMR and thus can allow a smaller-capacity processing system.

None of the reactor-based plutonium burning systems demonstrated over the past 50 years of reactor development consume this material completely. Ultimately considerable unburned plutonium must be stored "forever" from those systems. Plutonium is considered to be dangerous both as a weapons material and as a health hazard. While properly stored plutonium might never make its way back by natural phenomena into the environment as a health hazard, stored plutonium is always accessible to recovery for malevolent purposes. It must be guarded wherever in the world it is stored for as long as it continues to exist. Complete destruction of the plutonium eliminates this material as a concern of future generations. Los Alamos National Laboratory accelerator-driven technology promises to allow safe and complete destruction of this material. Furthermore, it appears that in the process of destruction, the neutron rich features of the weapons, plutonium provides benefits to society that place a value on weapons plutonium exceeding that of highly enriched uranium.

13.9.3 The Brookhaven National Laboratory's Phoenix Accelerator

The Phoenix transmutation concept of Brookhaven National Laboratory comprises a linear accelerator (LINAC) with a subcritical target assembly as a transmuter for minor transuranic actinide constituents (neptunium, americium, curium) of LWR spent fuel waste. The LINAC is similar to the design proposed for the ATW, but uses less than half of the proton current required by the ATW. In Brookhaven's Phoenix design, the proton beam impinges on a subcritical sodium-cooled lattice of fuel rods containing oxides of minor actinides previously separated from spent LWR fuel. The protons interact with the heavy actinide nuclei to produce showers of neutrons that, in turn, cause additional nuclear fissions in other actinide nuclei. Each proton ultimately will lead to the fission of 170 to 350 actinide nuclei. The Phoenix target assembly also will include separate water-cooled targets containing iodine-129, which will be transmuted to stable forms of the element xenon by the neutrons.

Phoenix relies heavily on chemical separation processes to partition the LWR waste and to separate the constituents after transmutation. The reference Phoenix design uses the aqueous plutonium-uranium extraction (PUREX) and transuranic extraction (TRUEX) processes to prepare the accelerator target material from spent LWR fuel and to reprocess the targets after they have been irradiated. Phoenix does not transmute plutonium, uranium, or technetium. These materials (after separation from spent LWR fuel) are stored for eventual incorporation into current or future nuclear reactors. Phoenix also includes a waste stream of fission products destined for a geologic repository. Strontium-90 and cesium-137 are included in this waste stream, after an interim storage period to permit them to partially decay.

Brookhaven National Laboratory claims that 1 Phoenix proton-accelerator-subcritical lattice can transmute the minor actinides from 75 LWRs. The transmuter is proposed as part of a more general radioactive waste treatment system based on partitioning LWR fuel into a number of key components. If the spent fuel partitioning and transmutation are fully implemented, Brookhaven claims that the time required to reduce toxicity of the radioactive waste stream below that of uranium ore will be reduced from more than 10,000 years to approximately 30 years. In addition, Phoenix will be able to generate 850 megawatts-electric for sale.

Brookhaven has also developed the ADAPT concept for the rapid and complete burning of plutonium, employing the LINAC.

13.9.4 The Brookhaven National Laboratory's Particle-Bed Reactor

The Particle-Bed Reactor (PBR) nuclear waste burner is a nuclear waste transmuter concept proposed by Brookhaven National Laboratory. The fuel for the nuclear reactor consists of plutonium and minor actinides that have been extracted from LWR spent fuel. The fuel is contained in small graphite-coated particles that constitute the "particle bed" referred to in the reactor title. The structural components of the core of the PBR are made of materials like graphite that can withstand high temperatures. The core is cooled by helium, an inert gas. The fuel particles are not embedded in a matrix material, but are present in loose form so that they constitute a particle fluid that can flow into and out of cavities in the fuel elements. The PBR can achieve very high thermal neutron concentrations (high thermal neutron fluxes) and can therefore be an effective actinide and fission product transmuter.

Brookhaven proposes that an R&D program be undertaken to develop this concept as an actinide and fission product transmuter. If the R&D program is successful, Brookhaven proposes that PBR waste burners be built in modules producing 1,080 megawatts of thermal power. They may be used to produce electrical power, but the operating cycle is so short—about 20 days—that effectiveness in this application is somewhat questionable. Reprocessing the PBR fuel will require special techniques to separate the actinide and fission product constituents from the carbon particles. Brookhaven has not decided whether to undertake aqueous or non-aqueous processing of the fuel to achieve these separations.

Brookhaven claims that the PBR will destroy actinides and long-lived fission products from LWRs and defense wastes. The concept is attractive because it has low radioactive inventories, destroys both actinides and long-lived fission products, should be low in cost, and has various safety features.

13.9.5 Westinghouse-Hanford's Clean Use of Reactor Energy

The Westinghouse-Hanford Clean Use of Reactor Energy (CURE) concept is an integrated system of chemical processes and transmutation technologies for processing LWR spent fuel. It is designed to eliminate most long-lived waste components by partitioning and transmutation and thus to produce waste streams of low long-term disposal risk. CURE examines a variety of chemical processes and transmutation methods. The reference system comprises aqueous processing of LWR spent fuel, combined with fissioning of transuranic elements in an oxide-fueled fast reactor, which CURE calls a Cleanup Fast Reactor (CLFR). The CLFR differs in fuel type and fuel-processing technology from the ALMR/IFR, which uses metal fuel and non-aqueous pyrochemical processing. The CLFR can transmute technetium-99 and iodine-129 in special metal hydride cells that slow neutrons into an energy range where they interact strongly with these 2 fission products. The transmutation of strontium-90 and cesium-137 in a CLFR is not believed to be feasible.

The proponents believe that the CURE concept combines the superior transmuting properties of a fast reactor for transuranics with the potential for fission product transmutation in specially modified cells in the reactor. It also relies on the proven performance of oxide fuel in a fast reactor and aqueous processing methods for partitioning radioactive waste components. It includes extensive proposals for partitioning and disposing of problem nuclides in the waste.

13.10 IN SITU VITRIFICATION

In situ vitrification is normally considered a process for immobilizing radionuclides in soil, however the process has been adapted to vitrifying an entire tank.

Pacific Northwest Laboratory (PNL) has developed a remedial action technology for underground storage tanks through the adaptation of the in situ vitrification (ISV) process. The ISV process is a thermal treatment process that was originally developed for the stabilization of contaminated soil contaminated with transuranic waste at the Hanford Site in southeastern Washington for the Department of Energy (DOE). The application of ISV to underground storage tanks represents an entirely new application of the ISV technology and is being performed in support of the DOE primarily for the Hanford site and the Oak Ridge National Laboratory (ORNL).

A field scale test was conducted in September 1990 at Hanford on a small cement and stainless steel tank (1-m diameter) that contained a simulated refractory sludge representing a worst-case sludge composition. The tank design and sludge composition was based on conditions present at the ORNL. The sludge contained high concentrations of heavy metals including lead, mercury, and cadmium, and also contained high levels of stable cesium and strontium to represent the predominant radionuclide species present in the tank wastes.

The test was highly successful in that the entire tank and surrounding soil was transformed into a highly leach resistant glass and crystalline block with a mass of approximately 30 tons. During the process, the metal shell of the tank forms a metal pool at the base of the molten soil. Upon cooling, the glass and metal phases were subjected to TCLP (toxic characteristic leach procedure) testing and passed the TCLP criteria.

14

Mixed Waste Treatment

DOE faces major technical challenges in the management of low-level radioactive mixed waste. Several conflicting regulations, together with a lack of definitive mixed waste treatment standards hamper mixed waste treatment and disposal. Disposal capacity for mixed waste is also expensive and severely limited. DOE now spends millions of dollars annually to store mixed waste because of the lack of accepted treatment technology and disposal capacity. In addition, currently available waste management practices require extensive, and hence costly waste characterization before disposal. Doe is pursuing technology that will lead to better and less expensive characterization, retrieval, handling, treatment, and disposal of mixed waste.

Over 130,000 cubic meters of MLLW are stored at 48 DOE sites in 22 states. Much of this waste is highly heterogeneous and over 40% has insufficient process knowledge to adequately judge the contents of individual waste containers.

The Mixed Waste Characterization, Treatment, and Disposal Focus Area (referred to as the Mixed Waste Focus Area or MWFA) is one of five focus areas and three cross-cut areas targeted for implementation of a new approach. The MWFA deals with the problem of eliminating mixed waste from current and future storage in the DOE complex. Mixed Waste is various forms of waste that are contaminated by both hazardous and radioactive constituents. It includes mixed low-level waste (MLLW) and mixed Transuranic (TRU) waste. DOE has been storing mixed waste, mainly in steel drums, for years in violation of the Resource Conservation and Recovery Act (RCRA), because treatment capacity in the DOE complex or the private sector has been inadequate or non-existent. The Federal Facility Compliance Act (FFCAct) waived sovereign immunity for the DOE and required DOE to develop plans and facilities for achieving RCRA compliance. In many cases this compliance will be achieved by eliminating the hazardous constituents in the waste and stabilizing the waste for final disposal.

The MWFA is charged with developing the needed treatment system technologies to meet FFCAct goals. These systems require characterization of the waste prior to treatment, treatments that can handle a wide range of waste streams, off–gas and secondary waste treatment, waste handling, and waste disposal technologies.

Incineration, as well as new concepts for treating mixed wastes are indicated below.

14.1 INCINERATION

Although the historical operating experience base is still quite limited for radioactive and mixed waste incineration, the shrinking availability of publicly acceptable means of waste disposal and subsequent need to minimize waste quantities are generating increased efforts to use incineration to reduce the volume and hazardous chemical content of waste material. This section discusses some of the radioactive/mixed waste incinerators now in use or under consideration.

Operable incinerators are located at four U.S. Department of Energy (DOE) facilities. These are the controlled air incinerator at the Los Alamos National Laboratory (LANL), the Oak Ridge Toxic Substances Control Act (TSCA) incinerator, the Rocky Flats Plant (RFP) fluidized bed incinerator, and the Idaho National Engineering Laboratory (INEL) Waste Experimental Reduction Facility (WERF). Additionally, a new controlled air incinerator is planned for LANL, and a rotary kiln incinerator (the Consolidated Incineration Facility, or CIF) is planned for the Savannah River Site. The Savannah River Site Beta–Gamma incinerator, was shut down several years ago for equipment modifications. When the CIF was approved, modification of the Beta–Gamma incinerator was canceled, and there are no plans to restart this unit. DOE has discontinued work at the INEL Process Experimental Pilot Plant (PREPP) incinerator, while evaluating its future role in the DOE Waste Management Program. There is also an Advanced Nuclear Fuels (ANF) facility at Hanford.

A low–level radioactive waste incinerator owned and operated by the Scientific Ecology Group, Inc. (SEG) in Oak Ridge, Tennessee, began commercial operation in 1989. The SEG incinerator is an automatically controlled partial pyrolysis unit based on the Swedish Studsvik incinerator, which has been in service since 1976.

Advanced Nuclear Fuels, in Richland, Washington, operates a dual–chamber controlled–air incinerator for processing solid and liquid wastes contaminated with uranium. The incinerator has operated since October 1988. The wastes incinerated originate during the manufacture and recovery of nuclear fuel materials.

A commercial unit for thermal destruction of mixed waste was permitted in 1990. This unit, owned and operated by Diversified Scientific Services, Inc. (DSSI), in Kingston, Tennessee, is designed to use mixed waste, in fluid form only, as beneficial fuels in a boiler system. DSSI notes that most of the waste will come from hospitals and universities where various short-lived radionuclides are used, and that most of the radioactivity will have decayed away before the waste is processed and received by DSSI. The boiler system is designed for complete thermal destruction of the fuels and recovery and reuse of the energy produced.

Two electric utility companies investigated incineration for volume reduction of waste produced at their nuclear generating stations. Duke Power Company installed a fluidized bed incinerator at its Oconee Nuclear Station in South Carolina. Changes in station operating procedures made subsequent to the installation of the incinerator changed the potential incinerator feed material from that originally contemplated. The consequent need for design modifications, and associated delays, resulted in a decision to defer final completion and operation of the incinerator for an undetermined period. Duke Power uses the SEG incinerator described above for its incineration needs. Commonwealth Edison Company installed fluidized bed incinerators similar to the Duke Power unit at its Byron and Braidwood nuclear stations.

A number of smaller scale incinerators are used by medical facilities and other institutions to process radioactive waste. There are also a number of incinerators in at least 10 other countries.

Operations and maintenance practices vary with the particular type of incinerator. Fluidized bed, rotary kiln, and controlled air, are the most common. Specific practices pertaining to the various components of incinerator systems are beyond the scope of this report. However, a few major overall considerations are described below.

A fundamental characteristic of incinerators is that they are designed to function best under strictly controlled, predictable, steady-state conditions. Uncontrolled variations in the quantity and physical/chemical characteristics of the waste feed material can have a significant negative effect both on incinerator performance in terms of the combustion process and on the potential for air emissions. It is difficult, if not impossible, for incinerators to be capable of responding rapidly to wide fluctuations in the nature of the feed in such parameters as btu content, ash quality and quantity, pH of the off-gases, etc. Designing the unit for worst-case conditions it may encounter for each parameter will not be a satisfactory solution because optimizing for one condition will likely adversely influence performance in another area. For example, maintaining the upper limit of temperature for one type of waste will lead to slagging with other types of waste. Thus, analysis and control of feed material is a crucial aspect of operations. For radioactive waste, this involves the monitoring of the physical-chemical nature of the feed (sorting of low-level waste according to combustibility, shredding of dry material, etc.), as well as its activity levels and radionuclide content.

Process monitoring and control procedures are used to ensure the proper functioning of the actual incineration process. The need for attention to following proper procedures in monitoring, treatment, and handling of off-gases and solid residues (ash) obviously is of particular importance for radioactive and mixed waste incineration.

14.2 PLASMA HEARTH PROCESS

The fixed hearth plasma arc thermal treatment unit uses a direct current (DC) arc plasma transferred torch generated in a gas flowing between two electrodes. The term "plasma" refers to a highly ionized electrically conductive gas. Plasmas can be generated by a variety of techniques, over a wide range of pressures and energy levels. The type of plasma produced in the Plasma Hearth Process (PHP) application is a dc arc-generated thermal plasma and is created by a device known as a "plasma torch." The plasma torch used in the PHP operates in the transferred arc mode.

The transferred arc torch uses a flow of gas to stabilize an electrical discharge (arc) between a high voltage electrode (inside the torch) and a molten pool of waste (maintained at ground potential). Because of the very high resistance to electrical current flow through a gas, electrical energy is converted to heat. Additionally, energy is converted as the electric current passes through the melt, creating a Joule-heating effect in the molten pool.

Processing begins as complete drums of waste are fed to the plasma chamber, where heat from the plasma torch initiates a variety of chemical and physical changes. Complex organic compounds break down into non-complex gases that are drawn from the chamber, while the remaining inorganic material melts and separates into two phases: slag and metal. Actinides and oxidized heavy metals migrate to the slag phase which, after being removed, cools and solidifies into a glass-like, or vitrified, material. This high-integrity final waste form, similar to that selected for high-level radioactive wastes, has repeatedly shown the ability to meet or exceed disposal requirements instituted by the RCRA.

PHP thermal treatment technology is characterized by high-efficiency destruction of organics, encapsulation of heavy metals and radionuclides in the vitrified final waste matrix, maximum reduction of waste volume, low off-gas rates, and the capability of processing many waste types in a single-step process.

Materials that are not converted to a chemically benign phase by the plasma process will be converted into either a molten vitreous slag or a molten metallic phase. When these are removed separately from the furnace, they solidify into a physically and chemically stable compact waste form.

This program is a collaboration between Lockheed Idaho Technologies Company (LITCO), Argonne National Laboratory-West (ANL-W), Science Applications International Corporation (SAIC), and Retech, Inc. Patent rights are being investigated and both SAIC and Retech are interested in commercialization of the technology as appropriate. Retech is supplying the plasma torch equipment and the melter chamber.

14.3 STEAM REFORMING

The objective of this project was to demonstrate the steam reforming system developed by Synthetica Technologies to destroy organic and inorganic salts that decompose thermally (e.g., nitrates, nitrites, and carbonates) which are present in the following:

(1) An adsorbed aqueous organic liquid waste simulant,
(2) A high organic content sludge simulant,
(3) A cemented sludge/ash/solids simulant,
(4) A heterogeneous debris simulant,
(5) A laboratory pack simulant, and
(6) Trimsol coated machining waste.

In addition, scavenging of mercury vapor by molten sulfur coated on ceramic spheres was studied to determine whether a Synthetica ceramic sphere, in a packed moving bed evaporator liquid feed system can be used to scavenge mercury vaporized by steam gasification of mixed wastes.

Wastes are destroyed by this system in two steps. First, the organic components of a waste are gasified in the appropriate feed system by exposing the waste to superheated steam. Then, the gasified organic materials are destroyed by passage through the resistively heated high-temperature reaction chamber of the detoxifier, in which the mixture of steam and gasified organic fragments is heated to more than 1200°C.

This technology has been proven commercially on various types of hazardous waste streams by Thermo Chem and Synthetica Technologies.

14.4 VITRIFY-TO-DISPOSE

The purpose of this project is to further develop the "Treat-to-Dispose" vitrification technology. Vitrification involves converting wastes that are primarily inorganic in nature into glass. This is accomplished by using the Reactive Additive Stabilization Process (RASP), in which carefully chosen additives react chemically with potential glass-formers within the waste. The result of the RASP approach to vitrification is higher waste loadings and increased final waste-form homogeneity, which leads to decreased leachability.

Four specific wastewater treatment sludges (Oak Ridge Y-12 WETF sludge, LANL sludge, and SRSM-area sludge) and several flyash streams (including Oak Ridge TSCA flyash and LANL CAI flyash) have been identified for treatment. The study is also investigating advanced glass-making processes utilizing the expertise and pilot-scale resources at Clemson University.

The culmination of this work is a field-scale radioactive demonstration of vitrification technology initiated in FY95. The demonstration will utilize a fully integrated system including material handling, glass melting, off-gas treatment, and process control subsystems, all of which will be transportable in nature. The demonstration using the Compact Vitrification System (CVS) has been scheduled to begin in late FY95 with Oak Ridge K-25 Plant being the host site. The actual mixed waste stream selected for treatment is the Oak Ridge Y-12 Plant WETF sludge.

This project is a collaboration among SRS, Clemson University (which operates the DOE/Industry Laboratory for Vitrification Research), ORNL, and private companies that are providing equipment and technical support. EnVitCo, Inc. has a high waste loading, transportable melter, and StirMelter, Inc. has a high-rate, low-cost melter. RUST Remedial Services is providing chemical analysis services, waste form characterization, and engineering support. SRS is contributing its expertise on how the glass should be made using the Reactive Additive Stabilization Process (RASP). Westinghouse personnel will contribute glass formulation, process modeling, melter operations expertise, and off-gas treatment. This project is being supported by EM-50, SRS's High-Level Waste Program, the Savannah River Economic Development Program, and the DOE Office of Waste Management.

14.5 STUDIES ON WASTE STREAMS CONTAINING DIFFICULT TO VITRIFY COMPONENTS

This task is intended to investigate several waste streams that contain constituents known to be problematic for glass making. The constituents in question are organics, mercury, reduced metals, chloride, nitrate, and sulfate salts. The difficulties arise by either the constituent being insoluble in the glass melt (such as organics and reduced heavy metals) resulting in unacceptably low waste loadings, or by causing glass corrosion that over time reduces the structural stability of the glass and allows leaching from the glass matrix.

The vitrification processing envelopes are being defined for three waste streams: Hanford 183H solar pond sludge, Rocky Flats Saltcrete residues, and Oak Ridge TSCA incinerator residues. In addition, the application of direct vitrification of combustible material (rather than incineration followed by vitrification) has been studied.

The stabilization of mixed waste by converting it into a glass final form (e.g., vitrification) is complicated by the existence of certain "bad actors" in a given waste stream. One of the major advantages of a glass final waste form as compared to a grouted or cemented waste form is the decreased leachability of the glass final form. Glasses are also expected to be more physically stable over much longer periods of time than grouted or cemented forms, probably due to the excellent corrosion resistance of good glasses.

"Bad actors" for glass production do exist in many DOE mixed waste streams. Examples are the 183-H Solar Evaporation Pond sludge from Hanford (containing high levels of copper, sodium, and sulfur), and the Rocky Flats

Saltcrete waste (this waste is an example of the failure of cementation), which contains nitrates, sulfates, and chloride salts of sodium and potassium. This task is concerned with dealing with those "bad actors" in a manner that does not require their removal by pretreatment. Rather, the process involves the investigation of alternative glass systems in which the "bad actors" are acceptably soluble. Two other approaches to stabilizing problematic mixed wastes are presented below in the sections on phosphate-bonded ceramics and polymer encapsulation.

Both silicate and phosphate glass formulations have been developed for each of the referenced waste streams. Phosphate glasses are preferred, however, for two main reasons. Increased solubility of chloride ion in the phosphate glass melt has been observed, and the phosphate system allows evolution of sulfur oxide, which can be handled in an off-gas treatment train (sulfate typically is found in a separate molten phase on the surface of silicate glass melts, and upon solidification becomes available for leaching).

The glasses produced by this study on the crucible scale have been shown to be producible in joule-heated melters as a result of measurements of melt viscosity and electrical conductivity.

Joint participants include the Pacific Northwest Laboratory (PNL), Clemson University, Pacific Nuclear, MIT, and Washington State University. This project complements the treat-to-dispose vitrification technology development effort at SRS. Battelle-PNL is working with Vectra Technologies, Inc. in marketing vitrification technology for treating low-level waste produced in the commercial sector.

14.6 WASTE STREAM PRETREATMENT FOR MERCURY REMOVAL

A number of sites across the DOE Complex have mixed waste streams containing mercury compounds in various waste matrices. For an example of solid matrices, the Oak Ridge Y-12 site has approximately 363,000 kg of storm sewer sediments that were found to be not only radioactive, but to contain approximately 19,000 ppm of elemental mercury as well. Crushed fluorescent tubes and lamps are found at all sites (Oak Ridge alone has over 25,000 kg in storage). Examples of mercury-bearing liquids are the ICPP sodium-bearing acid waste (with approximately 1.5 million gallons in storage at Oak Ridge) and various leach solutions containing mercury as Hg^{2+} derived during this study from the acid-leaching of solid mixed wastes.

There is a need to separate mercury from the remainder of the mixed waste matrix when the waste is going to be treated thermally. The volatility of mercury is such that pretreatment for removal is needed in order to prevent the potential escape of the volatilized mercury from the off-gas treatment system. An alternate approach is to devise a process for capturing any volatilized mercury in the off-gas treatment system.

The successful removal of mercury and/or speciated mercury from waste streams will allow their treatment in thermal processes. Operators of thermal treatment units are reluctant to treat mercury-bearing wastes due to the problems presented by having to capture volatilized mercury.

Two of the methods are appropriate for leaching mercury from solid waste matrices and involve acid leaching in one case, and the patented GE KI/I_2 leaching process in the other. For the removal of mercury from liquids, two processes have been studied. The liquid technologies are sulfur-impregnated activated carbon and a process involving ion exchange resins and membrane technology.

This project involves researchers at ORNL and has participation by Nucon International, 3M Corporation, and the General Electric Corporation. 3M Corporation is providing services to the MWFA through their existing contract with the Efficient Separations and Processing Crosscutting Program.

14.7 MICROWAVE SOLIDIFICATION

The generation of nearly 1,200 cubic feet of mixed waste per year, coupled with a backlog inventory of 18,000 cubic feet, makes it one of the largest waste streams at RFETS. Radioactive and heavy metal components contained in liquid effluent are removed during a hydroxide co-precipitation process, and an insoluble sludge coagulates and is collected.

The current treatment of this waste material is accomplished by cement stabilization. Microwave solidification was identified as a potential replacement for cement stabilization and was selected for utilization on a variety of inorganic wastes.

Microwave solidification is a mixed waste treatment process being developed at the Rocky Flats Environmental Technology Site (RFETS). The process is applicable to homogeneous, wet or dry, inorganic solids. The process dries the waste, mixes it with a slice source and matrix modifier, transfers it to a processing container, and subjects the mixture to microwave energy to melt the materials.

The processed waste form then cools and solidifies. The RFETS process begins by connecting a 30 gal drum to the microwave unit. Waste and glass frit are placed in the drum while a turn-table moves continuously to ensure even distribution of the contents. Microwave energy is transmitted to the drum to raise the internal temperature of the material to 1000°C. The resulting waste form is a vitreous material that contains no free liquids, has limited releasable particulates, and is highly leach-resistant due to its vitreous nature.

Bench- and pilot-scale tests have been performed on both actual and surrogate wastes. The basic design of the current microwave system has been demonstrated with bench-scale tests on actual waste materials. Test results indicate that volume reductions of up to 80% are achievable over some other solidification technologies.

14.8 LOW TEMPERATURE THERMAL DESORPTION

Thermal desorption is a cost–effective way to treat smaller volumes of organically contaminated soils, sludges, and other solid matrices, and is being demonstrated at the Rocky Flats Environmental Technology Site (RFETS). The objective of the Low Temperature Thermal Desorption (LTTD) effort is to desorb and separate the hazardous contaminants without combustion of the waste matrix. Hazardous contaminants are separated from the mixed waste by heating the materials to temperatures no greater than 120°C. The waste is prepared and sized in a chilled environment to control the volatilization of the organic contaminants and to more accurately determine the separation efficiency of the process.

The waste material is then loaded into an indirectly heated, vacuum dryer equipped with agitator vanes. A heated nitrogen carrier gas is injected into the dryer and blankets the waste as it is agitated and brought to operating temperatures. When the desired temperature is reached, the waste is subjected to a vacuum for a predetermined period of time (residence time). Organic contaminants are driven off as vapors, which are either condensed and collected as liquids or are destroyed by flowing the gas stream through a non–thermal plasma (NTP) gas treatment system. The NTP reaction cells use electrical micro–discharges to break up organic molecules. The resulting LTTD/NTP products include decontaminated solids, organic condensate, and nitrogen–rich vapor. The vapor stream is further cleansed using a HEPA filter and granular activated charcoal (GAC) adsorption system prior to venting.

Removal of hazardous solvents from transuranic mixed waste and MLLW streams simplifies disposal of treated waste forms, and the corresponding volume reduction achieved separating out the hazardous compounds results in lower overall disposal costs. As a pretreatment system, LTTD renders solid waste more amenable to a final stabilization process, such as vitrification or polymer encapsulation. LTTD operates at a low temperature, typically around 120°C. Since the nitrogen atmosphere is inert, no combustion of organic material takes place.

Commercial organizations involved in the development program include EG&G Rocky Flats, Rust Remedial Services/Clemson Technical Center, and LANL.

14.9 BIOCATALYTIC DESTRUCTION OF NITRATE AND NITRITE

A wide variety of high nitrate–concentration aqueous mixed wastes are stored at various DOE facilities. The presence of nitrates and nitrites in these wastes presents several problems for many pretreatment and immobilization processes currently being considered for treatment of these wastes, including vitrification. These problems include an increase in final waste form volume or reduction in waste form integrity; generation of off-gases, such as nitrous oxides or ammonia, which will require expensive scrubbing facilities; the potential for unstable exothermic reactions during processing in the presence of organics; and the need to meet EPA guidelines on nitrate levels in aqueous discharges. Thus, nitrate destruction before solidification of the waste will generally be beneficial from the standpoint of cost and waste form performance.

This project is developing an enzyme-based reactor system that uses naturally-occurring reductase enzymes, to reduce nitrate and nitrite present in various aqueous wastes to nitrogen and hydroxide ions. The process involves a three-step reduction process: (1) reduction of nitrate to nitrite, (2) nitrite to nitrous oxide, and (3) nitrous oxide to nitrogen. The overall process requires three separate reductase enzymes, one for each reduction step. The reductase enzymes are co-immobilized along with electron-transfer mediators, such as functionalized bipyridinium complexes, onto the surface of an electrode. The enzymatic reactions proceed rapidly under mild conditions near ambient temperature and at redox potentials less than that required to electrolyze water (i.e., -0.828 V). The reducing equivalents are provided by a low-voltage electrical current, which transfers electrons from the cathode to the enzymes via the electron-transfer mediator. The immobilized enzymes are confined to the polyethylene glycol-rich phase of an aqueous biphase system to protect them from denaturization and inactivation due to the high ionic strength and radionuclide content of the mixed waste streams.

This aqueous biphase system allows selective partitioning of nitrate and nitrite from the waste stream to the polyethylene glycol-rich phase of the reaction chamber. Most other anions, such as hydroxide, carbonates, and sulfates and most of the cations present in the mixed waste stream remain in the aqueous phase. The use of enzymes enables high specific catalytic activity to be obtained without the need for additional chemical reagents or the production of secondary waste streams.

The reactor tests using simulated feeds were carried out jointly between ANL and the University of Iowa. Work at ANL focused on development of a biphasic extraction system and process integration. Researchers at the University of Iowa provided support in developing enzyme immobilization techniques and assays of activities.

14.10 FREEZE CRYSTALLIZATION

Freeze crystallization processes are based on the difference in component concentrations between solid and liquid phases that are in equilibrium. As an aqueous solution is cooled, ice usually crystallizes as a pure material, and dissolved components in the aqueous waste stream are concentrated in the remaining brine thereby reducing the volume of waste.

Freeze crystallization technology is capable of separating organic and inorganic contaminants in an aqueous waste stream by removing the bulk of the water as ice, and concentrating the contaminants in the remaining brine. This is a flexible technology that can be designed and operated for specific site needs for removal of inorganics, organics, heavy metals, and radionuclides from aqueous wastes.

The primary benefit of freeze crystallization is that it can produce a higher purity water than other processes, such as evaporation and membrane technologies, from aqueous solutions containing high percentages of organics and inorganics. Thus, this technology may be beneficial in applications where water recovered from a waste containing RCRA waste and radionuclides must be discharged directly to the environment. Based on previous tests, decontamination factors are expected to be in the range of 100 to 1,000. In addition, off-gas issues will not be as significant for freeze crystallization as with evaporation, because the low operating temperatures will keep volatile organics from vaporizing.

The engineering evaluation was performed by J.L. Humphrey and Associates, Austin, TX, an independent consultant in the field of separation technologies. Laboratory-scale and bench-scale tests have been performed by Wheelabrator HPD, Inc. in Naperville, IL. The principal investigator was Westinghouse Hanford.

14.11 SUPERCRITICAL CARBON DIOXIDE EXTRACTION

Supercritical Carbon Dioxide Extraction (SCDE) is a process that employs a flowing, non-combustible, non-toxic, environmentally safe fluid as a solvent. This process takes advantage of the enhanced ability of carbon dioxide to dissolve organic contaminants once it has been heated and compressed above 90°F and 1,080 psig. In waste cleanup applications, SCDE is used to dissolve the hazardous components and extract them from the substrate material. By lowering the temperature and pressure at the expansion vessel, the contaminants can be precipitated out of solution to allow separation and recycling of the carbon dioxide. This process is capable of producing a dry residual waste form that can be treated as radioactive, rather than mixed waste.

Successful development and implementation of organic removal technologies could remove selected waste streams from LDR status. Removal of hazardous solvents from TRU mixed and LLM wastes would simplify disposal of treated waste forms and result in cost savings. SCDE employs a noncombustible, nontoxic, environmentally safe fluid as the solvent.

Studies of SCDE at Rocky Flats are being conducted in collaboration with the University of Colorado Cooperative Institute for Research in Environmental Sciences (CIRES). Substantial industrial participation in studies of selected volatilization technologies is anticipated.

14.12 POLYMER ENCAPSULATION

Polymer encapsulation of mixed wastes encloses waste products in thermoplastic or thermosetting materials using commercially-available processing technologies. Two primary polymer processes are being tested for DOE mixed wastes.

In one process, micro-encapsulation, thermoplastic polymers such as polyethylene (a commonly-used plastic that is resistant to chemicals and moisture), are combined with dried waste in a commercially-available extruder, which melts the polyethylene and mixes it with the waste. The waste encapsulated in polyethylene is extruded into a drum, where it solidifies upon cooling. The process operates at a low temperature, requires no off-gas treatment, and generates no secondary waste. Since high loadings of waste may be incorporated into the polymer, a substantial reduction in volume may be possible relative to cementation, which has been used to immobilize wastes in the past.

A second process, macro-encapsulation, in which bulk materials (i.e., lead and "debris") are suspended in a drum and encapsulated with molten or liquid plastic, is also being investigated. The solidified polymer surrounds the waste and immobilizes hazardous contaminants. The use of recycled polyethylene is being investigated for this application. Thermosetting plastics (resins combined with hardeners, similar to epoxy) have also been evaluated for encapsulating wastes.

Polyethylene and modified sulfur cement encapsulation are two thermoplastic encapsulation processes developed at Rocky Flats Environmental Technology Site (RFETS) and Brookhaven National Laboratory (BNL) with demonstrated applicability to a wide range of mixed waste types. Bench-scale R&D has been completed for both processes, including application of waste form test criteria recommended by the U.S. Nuclear Regulatory Commission (NRC) in support of 10 CFR 61.

These technologies will provide improved waste form performance and result in reduced risk to human health and the environment. Polyethylene encapsulation of nitrate salt waste compared favorably with Portland cement grout solidification—both technically and economically. For example, use of polyethylene at Rocky Flats was estimated to result in up to 70% fewer waste drums for storage, transport, and disposal, resulting in annual net costs savings between $1.5 and $2.7 million. The Tanks Focus Area has estimated cost savings of $200 million over the life of the single shell tanks remediation project at Hanford. Similarly, modified sulfur cement and polymer-impregnated concrete can accommodate high-waste loadings, and thus reduce overall costs.

Polymer solidification development at RFETS is being conducted with the collaboration of researchers at the Colorado School of Mines and WHC.

Drying and extrudability studies are also being performed by several equipment vendors. Investigators at BNL are also participating in the development of this technology in conjunction with Pacific Nuclear Services.

14.13 PHOSPHATE BONDED CERAMIC FINAL WASTE FORMS

The purpose of this project is to exploit the attractive features of chemically bonded phosphate ceramics (CBCs) and develop superior waste forms for MLLW streams that cannot be handled by other established methods. Guidelines and assessments will be set up based on the waste stream with the best treatability performance, and that stream will be scaled up for pilot study. These include waste streams containing liquid mercury, mercury–contaminated aqueous liquids, toxic and heavy metal containing materials, salt cakes and processing salts, beryllium wastes, and pyrophorics.

Calcined MgO can be reacted with dilute phosphoric acid or dibasic phosphates (ammonium or sodium dibasic phosphate respectively) to form a stable ceramic form. Similarly, zirconium phosphates can be formed by reaction of zirconium hydroxide with phosphoric acid solution. These reactions occur at room temperature to form a dense ceramic that sets into a hard product, sometimes in a few hours. Surrogate ash waste streams, salt compositions, and cemented sludge were incorporated in phosphate ceramics with loadings up to 50%. These waste streams were spiked with RCRA metal nitrates (Cd, Cr, Ni, Pb, and CsCl). TCLP leaching tests were used to evaluate the performance of the final waste form.

CBCs have the potential for stabilizing several problem mixed–waste streams that have been identified by DOE. They are attractive for applications such as solidification and stabilization of these waste streams because the final waste forms can be fabricated at room temperature. The phosphate CBCs are pore–free, insoluble in groundwater, and stable at elevated temperatures. They form solid solutions with actinides and rare earths. In general, metals and metal oxides react with phosphoric acids to form stable CBCs at low temperatures. This method of stabilization can be exploited to incorporate various components of the mixed waste into solid monolithic forms of CBCs. The presence of highly volatile contaminants and pyrophorics in a waste stream makes it very difficult to stabilize these wastes with currently available technologies. Chemically–bonded ceramics, on the other hand, can accommodate these waste streams.

This work was performed in collaboration with the Center for Advanced Cement–Based Materials at the University of Illinois, Urbana–Champaign and the University of Dayton Research Institute.

14.14 CATALYTIC CHEMICAL OXIDATION

The CCO system uses both an iron catalyst and co–catalysts to degrade the organics in a strong acid solution. The system operates at temperatures much below those used in incineration and uses moderate pressures (expected operating conditions are approximately 150°C and 70 psig). Both solid and liquid wastes can be treated, and most metals are dissolved and concentrated in the reaction solution.

Delphi Research, Inc. has developed and patented a CCO system, called DETOX, which destroys hazardous organics at practical rates. This DETOX technology has been demonstrated at the bench–scale, with destruction efficiencies of 99.999% achieved for liquid hydrocarbons (including some chlorinated organics). Due to the strongly acidic nature of the reaction mixture, engineering development is focused on materials of construction, along with scale–up issues. Treatment of the spent reaction solution and system integration are also being studied.

Catalytic Chemical Oxidation (CCO) offers an alternative to incineration for the treatment of combustible MLLW. DETOX can treat combustible waste at a rate comparable to incineration and reduce the bulk volume of waste without the temperatures and off–gas associated with incineration. This technology has been selected for demonstration–scale development to provide information necessary to design and fabricate a production system.

14.15 HYDROTHERMAL OXIDATION

Hydrothermal Oxidation (HTO) involves bringing together organic waste, water, and an oxidant (e.g., air, oxygen, etc.) to temperatures and pressures above the critical point of water (374°C, 22.1 MPa). Under these conditions, the waste is treated at high–organic–destruction efficiencies of over 99.99% and the resulting effluents, which consists primarily of water and carbon dioxide, is relatively benign. In addition, HTO has the potential of being a highly cost–effective treatment process when compared to conventional technologies such as incineration. To date, some of the candidate DOE mixed waste streams for HTO treatment include: Spent solvent, oils, and other organic or aqueous liquids, sewage and organic–laden sludges, spent carbon, solvent–contaminated rags, explosives, and energetics.

The scope of work covered under this task is for the design, fabrication, permitting, operation, testing, and evaluation of the Hazardous Waste Pilot Plant (HWPP) and Mixed Waste Pilot Plant (MWPP) for treating DOE hazardous and mixed wastes. The HWPP demonstration will focus on identifying Hydrothermal Oxidation (HTO) technology development needs; providing technology improvements required to demonstrate that HTO is a safe, cost–effective technology; and demonstrating currently–available HTO technology using hazardous and surrogate mixed wastes of interest to DOE. Data generated in the HWPP demonstration will provide the basis for the decision of whether to proceed with the design basis of MWPP and the demonstration of the MWPP.

HTO technology holds promise for treating approximately 15% of DOE's mixed waste inventory. While this technology has been successfully demonstrated at the bench- and pilot-scales for a more limited number of wastes, numerous questions and risks remain in applying the process to DOE mixed wastes.

There are two significant problems, corrosion and precipitation of solids, which may limit the applicability of HTO in treating a variety of DOE aqueous waste streams.

Solids precipitation can lead to problems either through deposits that remain behind in the HTO reactor, or through relatively large volumes which may become contaminated with radioactive elements present in the waste stream and thus require disposal as solid low-level wastes. Acids are inherently formed in the HTO process through oxidation of organic compounds containing heteroatoms (e.g., chlorine, phosphorous, sulfur) at high temperatures to produce the acids or oxyacid of these compounds. Metal ions present in the waste stream may also contribute to the overall acidity for formation of hydroxy species as a result of hydrolysis at high temperatures. The use of in situ neutralization has the potential to significantly reduce the corrosion resulting from acids formed in the HTO system.

Successful design of a multi-layered ceramic material system is dependent on knowledge of material properties for each layer over a wide temperature range. Layered ceramic coatings can be designed to take into account the thermal stresses of the substrate and the stresses from the temperature flux. While many ceramics have demonstrated good corrosion resistance at HTO conditions, they have been unable to withstand the thermal cycles required for an HTO reactor. By using several layers of ceramics and matching their thermal stresses, a reactor may be designed with a corrosion-resistant material.

The process is being developed by Martin Marietta Energy Systems at Oak Ridge, and Lockheed Idaho Technologies Company at Idaho Falls.

14.16 WET OXIDATION

Delphi Research, Inc., under a DOE contract is developing a catalytic wet oxidation process for the treatment of multi-component wastes, with the aim of providing a versatile, non-thermal method which will destroy hazardous organic compounds while simultaneously containing and concentrating toxic and radioactive metals for recovery or disposal in a readily stabilized matrix. The DETOX process uses a combination of metal catalysts to increase the rate of oxidation of organic materials. The metal catalysts are in the form of salts dissolved in a dilute acid solution. A typical catalyst composition is 60% ferric chloride, 3 to 4% hydrochloric acid, 0.13% platinum ions, and 0.13% ruthenium ions in a water solution. Wastes are introduced into contact with the solution, where their organic portion is oxidized to carbon dioxide and water. If the organic portion is chlorinated, hydrogen chloride will be produced as a product. The process is a viable alternative to incineration for the treatment of organic mixed wastes.

14.17 SUPERCRITICAL WATER OXIDATION

Lockheed Idaho Technologies Co. studied Supercritical Water Oxidation (SCWO) technology under a DOE contract. Supercritical water oxidation is an emerging technology for industrial waste treatment and is being developed for treatment of the U.S. Department of Energy (DOE) mixed hazardous and radioactive wastes. In the SCWO process, wastes containing organic material are oxidized in the presence of water at conditions of temperature and pressure above the critical point of water, 374°C and 22.1 MPa. The study is a review and evaluation of tubular reactor designs for supercritical water oxidation of U.S. Department of Energy mixed waste. Tubular reactors are evaluated against requirements for treatment of 'U.S. Department of Energy mixed waste. Requirements that play major roles in the evaluation include achieving acceptable corrosion, deposition, and heat removal rates. A general evaluation is made of tubular reactors and specific reactors are discussed.

14.18 MEDIATED ELECTROCHEMICAL OXIDATION

Mediated Electrochemical Oxidation (MEO) is an aqueous process which oxidizes organics electrochemically at low temperatures and ambient pressures. The process can be used to treat mixed wastes containing hazardous organics by destroying the organic components of the wastes. The radioactive components of the wastes are dissolved in the electrolyte where they can be recovered if desired, or immobilized for disposal. The process of destroying organics is accomplished via a mediator, which is in the form of metallic ions in solution. Lawrence Livermore National Laboratory (LLNL) has worked with several mediators, including silver, cobalt and cerium. They have tested mediators in nitric as well as sulfuric acids, and recently completed extensive experimental studies on cobalt–sulfuric acid and silver–nitric acid systems for destroying the major organic components of Rocky Flats Plant combustible mixed wastes. The process was capable of destroying almost all of the organics tested, attaining high destruction efficiencies at reasonable coulombic efficiencies. The only exception was polyvinyl chloride, which was destroyed very slowly resulting in poor coulombic efficiencies.

14.19 ELECTRON BEAM TECHNOLOGY

Recently, an e–beam pilot plant, capable of treating an aqueous hazardous waste stream at a flow rate of 120 gpm, has been developed at Florida International University (FIU). This plant uses a 1.5 Mev continuous duty profile accelerator to produce doses in water approaching 1 Mrad. Studies at that plant used influent streams of potable water, and raw and secondary wastewater. Removal efficiencies range from 85% to greater than 99% for most common solvents. It could be attractive for solutions of mixed waste.

Electron beam treatment is presently cost-competitive with established commercial technologies for removing organics in water solution; new accelerator technology promises increased performance and reduced cost.

14.20 CATALYTIC EXTRACTION PROCESS

Molten Metal Technology's (MMT's) Catalytic Extraction Process (CEP) is an innovative and proprietary technology that allows organic, organometallic and inorganic waste streams to be recycled. The feed destruction and recycling capabilities of CEP have been demonstrated on a wide variety of materials ranging from simple organic surrogates to hazardous waste streams.

Quantum-CEP™ is an adaptation of the CEP technology for radioactive and mixed waste streams. Quantum-CEP allows both destruction of hazardous components and controlled partitioning of radionuclides. This leads to decontamination and recycling of a large portion of the waste components, as well as volume reduction and concentration of radionuclides for final deposit.

At the core of both CEP and Quantum-CEP is a molten metal bath which acts as a catalyst and solvent in the dissociation of the feed, the synthesis of products and/or the concentration of radionuclides in the desired phase. Upon introduction to the bath, feeds dissociate into their constituent elements and go into metal solution. Once in this dissolved state, addition of co-reactants enables reformation and partitioning of desired products. The partitioning control afforded by co-reactant addition is a distinguishing feature of CEP and Quantum-CEP.

Another development program underway is a joint effort between MMT and the Scientific Ecology Group (SEG), a subsidiary of Westinghouse, to process contaminated ion exchange resins from nuclear power plants. MMT installed a bench scale Quantum-CEP unit in February 1994 at SEG's facility in Oak Ridge.

15

Low-Level Waste Treatment

This chapter discusses various treatment technologies for low-level waste (LLW). Disposal and storage of low-level wastes are discussed in Chapters 2 and 5.

Sources of LLW include the following:

(1) Commercial power plant programs
(2) Medical/institutional facilities
(3) Industrial/laboratory facilities
(4) Naval reactor program
(5) Nuclear weapons program
(6) Uranium enrichment program
(7) DOE research and development programs

The operational definition of low-level waste is radioactive waste that becomes nonhazardous within several hundred years. Low level implies a hazard for a limited time. Low-level wastes may have very high initial levels of radioactivity. For radionuclides with half-lives less than 5 years, there is no limit of their concentration in LLW. For radionuclides with longer half-lives, the maximum allowable concentrations in LLW are less as the half-life and/or biological effects increase. For commercial wastes, there are three major classes of LLW—A, B, and C, where class A has the lowest and class C has the highest radioactivity. The class determines specific disposal requirements.

The boundary between LLW and TRU wastes is well defined (100 nCi/g TRU) but the boundary between LLW and HLW is less precise. In the U.S. regulatory system, there is a category of wastes called "Greater Than Class C (GTCC) LLW." This category (never fully defined) may include certain highly irradiated reactor components and cesium/strontium capsules. The rationale for this category of waste is that such wastes require greater waste isolation than needed for LLW in near surface disposal facilities. Based on current waste management practice, the quantities of such wastes are very small. The U.S. Nuclear Regulatory Commission (NRC) has recommended that these wastes go to the HLW repository since it is clearly uneconomical to build a specialized facility for such small quantities of wastes. The practical consequence is that the upper Class C LLW limit defines LLW and determines which wastes will be stored in the repository.

15.1 VOLUME REDUCTION TECHNOLOGIES

15.1.1 Waste Minimization

Industrial generation sources, waste minimization can sometimes be achieved by substituting different feedstock material, changing process technologies, or streamlining production processes. Waste minimization at most generation sources, however, is a question of management practices. The volume of waste generated can be effectively reduced through the implementation of new procedures, quality assurance programs, and programs to increase the general awareness of workers with regard to minimizing contaminated trash and avoiding cross-contamination of materials.

15.1.2 Compaction and Supercompaction

The volume of dry solid waste can also be reduced through the use of compactors, either hydraulic or pneumatic. The volume reduction achieved depends on the void space in the waste, its bulk density, its spring-back characteristics, and the force applied during compaction. Ordinary low-pressure compactors, where the applied force may be only a few tons, can provide a volume reduction factor of about 2 to 5 for contaminated trash waste. High-pressure compaction, also known as supercompaction, uses compaction forces of the order of 1,000 tons or more and can provide a volume reduction factor of about 6 to 10. It is also used for compacting waste-filled drums, which are flattened like pancakes. Compaction is a widely used technology at nuclear power plants, and it is also offered as a service by some companies through commercial mobile units.

15.1.3 Baling

Baling of solid LLW is generally carried out after compaction. Various types and sizes of baler units are available, however, rectangular bales are currently the most widely used technique for containerizing waste for storage or disposal. Baling of waste can also be used as an interim storage convenience for the combustible waste that will eventually be incinerated (for example, during shutdown periods of the incinerator).

15.1.4 Shredding

Shredding of contaminated paper, cloth, and plastic waste can result in a volume reduction of about 3:1. Shredding can also be used as a pretreatment step before incineration of combustible waste. Proper waste segregation is important because metallic pieces in the general trash have been often known to damage or break the cutting blades of shredders.

15.1.5 Cutting

Contaminated plant hardware can be cut into pieces for better packaging and storing. Often the discarded contaminated piping at nuclear reactor stations is sectioned with cutting equipment to fit into (and to reduce the empty volume of) transportation casks or storage containers. For this type of hardware, there is usually no intention of decontaminating and reusing the material. In addition to traditional saw cutting, specialized cutting technologies—such as oxyacetylene cutting and plasma arc torch cutting—can be used, as appropriate.

15.1.6 Incineration

If waste can be properly segregated at the source, incineration is perhaps the best option for dealing with combustible waste. It is estimated that about 50% of the solid waste generated at nuclear power facilities, the bulk of the waste generated as contaminated trash during research and development activities, and most of the biological waste (e.g., animal carcasses and contaminated laboratory waste) are combustible. A volume reduction of 100:1 or higher is not uncommon. The incinerator at Chalk River Nuclear Laboratories (CRNL) has achieved a volume reduction of about 170:1 (on as-received volume basis) for miscellaneous combustible uncompacted trash generated at the laboratories. Incineration of baled waste has also been successfully implemented at the CRNL incinerator.

Incineration technologies are well advanced, and various designs (such as starved-air incinerator, excess-air incineration, and pyrolytic or thermal decomposition) are commercially available. The ash or residue left from the incineration operation can be drummed for transportation and storage or can be immobilized through incorporation into matrices such as concrete or bitumen for disposal purposes.

The incineration of LLW generates a hot off-gas stream, with entrained particulate matter and fly-ash, that is filtered and scrubbed before release by high efficiency, multiple-stage, wet and/or dry treatment systems. Incineration also results in the volume reduction of the LLW to an ash by-product. The disposal of this incinerator ash, which is normally mixed with the filtered particulate matter, may be accomplished by shallow-land burial of the ash packed in high-integrity containers or permanently immobilized in cement, concrete, polymer or bitumen. Although the requirement for ash disposal can be characterized as a drawback of incineration, in reality it is a much better form of LLW for burial than was the original form since it is biologically and structurally stable, and usually consists of insoluble compounds.

Due to the high volume reduction achieved during incineration, a reciprocal increase in the specific activity takes place in the ashes. Care has to be taken of this phenomenon, especially as far as the radiation exposure to operational personnel is concerned.

Several process steps are involved in the safe combustion of radioactive wastes:

(1) Waste should be sorted before incineration. Sorting assures more homogenous feed, resulting in smooth and complete incineration, control of combustion and reduced risk of accumulating unburnt material in the ashes and the off-gas system.

(2) The temperature of the furnace is maintained between 700° and 1100°C to ensure complete combustion of the waste material.

(3) The combustion gas must be cooled and cleaned to protect the environment from the release of noxious gases.

(4) Sufficient ash collection points must be provided to permit cleaning of the entire furnace system. The ash must be treated further to prepare it for storage and disposal.

Off-gases from incineration can be highly corrosive, requiring precautions and construction materials that can withstand corrosion. The combustion gases leaving the furnace generally are at temperatures near 1100°C. Cleaning of the off-gases normally starts with a cooling step, followed by a wet or dry or a combined cleaning procedure. The final purification is normally carried out by high-efficiency gas filters. The off-gas cleaning process is also used to separate entrained radioactive particulates from the off-gas.

Currently, incineration must be considered as one of the best available technologies for the treatment of specific GTCC LLW streams (organic liquids, absorbed liquids, alcohols, sludges, filters, rags, plastics, cloth, etc.).

15.2 CHEMICAL TREATMENT/CONDITIONING TECHNOLOGIES

15.2.1 Decontamination

Reusable equipment and hardware that is contaminated only on the surface, can be decontaminated, generally with various cleaning fluids. The big advantage of decontamination techniques is that, after decontamination, the equipment and hardware can be released for unrestricted use. This is even more important in the case of high-capital-cost equipment. The main disadvantage is the generation of liquid LLW that must then be appropriately managed. High-pressure water-jetting techniques for decontamination are relatively inexpensive. For specialized decontamination applications, proprietary technologies are available, such as CAN-DECON (CANDU-Decontamination), CORD (Chemical Oxidation Reduction Decontamination), and LOMI (Low Oxidation State Metal Ion Reagents). Dry-cleaning techniques can also be used, e.g., method using FREON and employing agitation by ultrasonic waves. Mechanical decontamination techniques include manual cleaning, vacuum cleaning, grinding, and machining.

15.2.2 Immobilization

Immobilization processes involve conversion of the waste (ash from incinerators, residues, liquid concentrates) into physically and chemically stable forms. Although volume reduction and pretreatment technologies—such as incineration of solid LLW, chemical precipitation from liquid waste, and evaporation to concentrate liquid waste—can be highly effective in achieving volume reduction, they also have the net effect of concentrating radionuclides in a small volume of the leftover waste. This concentrated radioactive waste material presents a higher potential for negative impacts on human health and the environment. This necessitates immobilization of these wastes into stabilized forms to reduce the potential for migration or dispersion of the radionuclides.

Generally, the waste is incorporated into a matrix from which the leaching of radionuclides can be expected to be negligible (under natural conditions) from either storage or disposal operations.

Cementation processes have been widely used in the U.S. and abroad. Portland cements are the most commonly used matrix, but the use of high alumina cements, as well as pozzolanic cements is also becoming more widespread. Cements can also be used for immobilizing sludge and miscellaneous solid waste, and for embedding spent ion-exchange resins, decladding hulls, and contaminated hardware.

Bitumen has been employed as an immobilization agent in Canada and Europe. The bitumenization process generally involves heating asphalt (bitumen) to over 150°C, mixing the waste in it, and allowing it to cool and solidify.

Incorporation of radioactive waste into **plastics** is a relatively newer technology. The main plastic materials used for this purpose are polyethylene, urea-formaldehyde, polyester, and polystyrene. For dry solid waste (generally the structural parts), a polymer-impregnated cement matrix has also been used as the embedding matrix.

Vitrification is the process of converting materials into a glass or glass-like substance through heat fusion. Vitrification is conceptually attractive because of the potential durability of the product and the flexibility of the process in treating a wide variety of waste streams and contaminants. These characteristics make vitrification the focal point for treating HLW, and a viable treatment for GTCC LLW.

Vitrification has three major advantages. The primary advantage is the durable waste glass that it produces. This waste glass performs exceptionally well in leach tests. The second major advantage is the flexibility of the waste glass in incorporating a wide variety of contaminants, and accompanying feed material in its structure without a significant decrease in quality. Lastly, vitrification can accommodate both organic and inorganic contaminants.

DOE is developing cost–effective vitrification methods for producing durable waste forms. However, vitrification processes for high–level wastes are not applicable to commercial low–level wastes containing large quantities of metals and small amounts of fluxes. New vitrified waste formulations are needed that are durable when buried in surface repositories.

The major limitation of vitrification is that it is energy intensive, and thus may be more expensive compared with other treatment technologies. Vitrification can accommodate organic contaminants. However, a second major limitation is the potential for these contaminants to volatilize. An off–gas cleaning system must be operable if organic contaminants are subjected to vitrification. Pre-treatment of waste materials containing organic contaminants is preferred. An example of pretreatment includes incineration. Incineration volatizes the organic constituents leaving an ash that is then vitrified for disposal.

Immobilization technologies, especially the cementation and bitumenization processes, are well developed. Extensive experience already exists for these processes with a variety of equipment, and the properties of the immobilized waste forms have been well studied in the United States, Canada, and Europe.

APPENDIX I

DEPARTMENT OF ENERGY LONG-TERM (DOE) EXPENDITURES

The Department of Energy (DOE) has published (March 1995) the first annual report on the activities and potential costs required to address the waste, contamination, and surplus nuclear facilities that are the responsibility of the Department of Energy's Environmental Management program. It covers a 75 year period from 1995 to 2070, and is entitled *Estimating the Cold War Mortgage, the 1995 Baseline Environmental Management Report (DOE/EM-0232).*

IT ASSUMES

1). Significant productivity increases
2). Meeting current compliance requirements
3). Use of existing technologies

IT EXCLUDES

1). Cleanup where no feasible cleanup technology exists (e.g., Nuclear explosion sites, most contaminated groundwater)
2). Cleanup of currently active facilities (e.g., Pantex, Labs)
3). Naval Nuclear Propulsion facilities cleanups handled by U.S. Navy
4). Activities during first 5 years of program ($23 billion)

ALTERNATIVE CASES (Evaluated the Effect of)

1). Land use: biggest potential cost impact

2). New Technologies
3). Waste Management Facilities configuration
4). Funding and Schedule
5). Residual Risk: inadequate data limited analysis

WHAT WAS LEARNED

1). Total projected environmental costs are comparable to total U.S. nuclear weapons production costs
2). Projected future land use will dramatically affect costs
3). Significant ($24 billion) projected costs to support ongoing programs could be substantially reduced through greater pollution prevention
4). Development of new technologies will reduce certain cleanup costs and make possible other cleanups that are currently infeasible
5). Minimum action to stabilize sites – $170 billion

ESTIMATES, NOT DECISIONS

1). The estimated costs do not reflect final Departmental decisions in many cases. The report is intended to provide a framework for constructive local and national debate about the future of the environmental management program.
2). Projected costs significantly exceed current budget targets. Bridging this gap will require renegotiating compliance agreements and some statutory changes, in addition to planned productivity improvements.

KEY QUESTIONS

The future course of the Environmental Management program will depend on a number of fundamental technical and policy choices, many of which have not yet been made. Ultimately, these decisions will be made on the basis of fulfilling congressional mandates, regulatory direction, and adequate stakeholder input. The cost and environmental implications of alternative choices can be profound. For example, many contaminated sites and facilities could be restored to a pristine condition, suitable for any desired use; they also could be restored to a point where they pose no near–term health risks to surrounding communities but are essentially surrounded by fences and left in place. Achieving pristine conditions would have a higher cost, but may or may not warrant the economic costs and potential ecosystem disruption or be legally required. Resolving such issues will depend on what the Nation wants to buy.

Other key questions that affect the cost of the program include the following:

1). What level of residual contamination should be allowed after cleanup?
2). Should projects to reduce maintenance costs (i.e., high storage costs pending ultimate disposition of materials) be given priority over certain low-risk cleanup activities? In other words, how should cost affect priorities?
3). Should cleanup and waste management proceed with existing technologies or is it prudent, in some cases, to wait for the development of improved technologies? What criteria should guide decisions on this issue?
4). Should waste treatment, storage, and disposal activities be carried out in decentralized, regional, or centralized facilities? How are issue of equity among states factored into configuration decisions?

ESTIMATING COSTS IN THE FACE OF LARGE UNCERTAINTIES

Estimating the cost of future activities requires making assumptions about what those activities will be and is inherently uncertain. The uncertainty stems from:

1). Lack of characterization of the problems. For example, of the 10,500 hazardous substance release sites addressed in this report, only one-fourth have been fully characterized.
2). Lack of knowledge about what remedies will be effective or considered acceptable to regulators and the public, or what level of human health and environmental protection is sought through these remedies.
3). Lack of fundamental economic, social, and defense related decisions that affect the future use of land and facilities. For example, policy decisions related to the role of sites for nuclear nonproliferation and defense readiness will define the future mission for the Department's nuclear weapons complex. These policy decisions will affect the continued operations of some installations, including future land-use options and the final disposition of nuclear materials.
4). Lack of technical remedies for the problems. The contamination of soils deep underground from nuclear tests in Nevada is one such case. The costs to remediate these types of sites were excluded from the cost estimate, not because of a departmental policy to ignore such problems, but because no effective remediation technology currently exists.
5). Lack of defined program duration. The length of the program—approximately 75 years—is sufficient to introduce a variety of uncertainties into any cost and schedule estimate.

LIFE-CYCLE COSTS

Congress requested an estimate of the total cost of the Environmental Management program, which is referred to throughout the Baseline Report as the life-cycle cost. Base Case life-cycle costs are incurred over approximately 75 years. This is because scheduling under the Base Case assumes most activities are completed by approximately 2070. The availability of more or less funding than assumed for this analysis would, however, affect the length of the program.

The Base Case cost estimate does not include costs expended before 1995 (approximately $23 billion since the Environmental Management program was established in October 1989). It also does not include costs projected beyond 2070 associated with monitoring and maintaining disposal sites and other restricted-access areas, estimated to be $50 to $75 million per year, and costs of managing wastes from ongoing activities (e.g., basic research and nuclear weapons maintenance), estimated at approximately $300 million annually.

KEY WASTE MANAGEMENT ASSUMPTIONS

1). **High-Level Waste**
   a). Continue storage in tanks at Hanford, Savannah River Site, West Valley Demonstration Project, and in calcine bins at Idaho National Engineering Laboratory
   b). Vitrify and dispose of all high level wastes in geologic repository (available beginning in 2015)
2). **Spent Nuclear Fuel**
   a). Continue storage at 10 sites with costs for new wet and dry storage facilities estimated
   b). No reprocessing
   c). Dispose in geologic repository
3). **Transuranic Waste**
   a). Continue storage at 10 sites
   b). Treat as necessary to meet disposal criteria at the Waste Isolation Pilot Plant (starting in 1998)
4). **Low-Level and Low-Level Mixed Waste**
   a). Storage until treatment at 34 sites to meet minimum disposal requirements
   b). Disposal at Hanford, Idaho National Engineering Laboratory, Los Alamos National Laboratory, Nevada Test Site, Oak Ridge Reservation, and Savannah River Site
   c). Western sites using shallow land disposal and eastern sites using engineered disposal techniques

RESULTS

The Base Case cost estimate begins in 1995 and ends in approximately 2070, when environmental management activities are projected to be substantially completed. The estimate does not include costs expended since the program's formal inception in October 1989—about $23 billion—or costs incurred before 1989. Nor does it include costs beyond 2070 for long-term surveillance and maintenance, which are estimated at about $50–75 million per year. These costs are assumed to continue indefinitely after a disposal site or restricted access area is closed.

Under the Base Case, the life-cycle cost estimate for the Department of Energy's Environmental Management program ranges from $200 to $350 billion in constant 1995 dollars, with a mid-range estimate of $230 billion. **Figure 1** graphically depicts the life-cycle cost profiles. This includes not only the $172 billion for dealing with the nuclear weapons complex legacy, but $24 billion for future wastes from nuclear weapons activities, and $34 billion for past and future wastes from other activities. The projected costs for treatment storage, and disposal of waste generated by ongoing defense and research activities is $19 billion. The significant projected cost for support for future ongoing programs indicates the value of vigorous pollution prevention efforts to reduce these costs and threats.

The range of the cost estimate varies depending on the assumed level of productivity over the life of the program as described below.

1). The mid-range total program estimate of $230 billion reflects a planned 20% increase in productivity and efficiency over the next 5 years, plus an annual 1% productivity improvement over the remaining life of the program.

2). The low-end estimate of $200 billion reflects a more aggressive efficiency and productivity improvement program—20% for the next 5 years as in the mid-range total estimate, and subsequent annual improvements of nearly 2% (a number commonly used by the private sector in today's business climate).

3). The high-end estimate of $350 billion reflects costs if current levels of inefficiency and productivity were sustained over the program's life.

These levels of efficiency improvement are not only needed and planned, they are attainable. The Environmental Management program already has achieved significant improvements in efficiency and productivity. From FY 1994 to FY 1996, the program will have saved more than $2.1 billion through greater productivity.

Although the total life-cycle estimate is derived from a 75-year program duration, more than 90% of the life-cycle cost estimate reflects activities projected to occur during the next 40 years. The remaining costs are primarily for the operation of large waste treatment facilities at a limited number of sites. In 2070, given the Base Case assumptions, access will be restricted at the large, isolated Department of Energy sites with existing burial grounds. These sites include certain sections of the Hanford Site, Idaho National Engineering Laboratory, Savannah River Site, Nevada Test Site, Oak Ridge, Los Alamos National Laboratory, and the Waste Isolation Pilot Plant. At smaller Department of Energy sites, such as the Mound Site in Ohio or the Pinellas Plant in Florida, where contamination has been contained in place, future use is expected to be limited to industrial purposes.

Small non-Department sites or sites near heavily populated areas or water sources are assumed to be released for residential or industrial use. Examples include the General Atomics Site at La Jolla, California, and Battelle Columbus Laboratories in Columbus, Ohio.

Figure 2 shows cost estimates for the Environmental Management program under the mid-range Base Case estimate. The cost estimate is divided among the five major elements of the program: waste management, environmental restoration, nuclear material and facility stabilization, program management, and technology development.

BASE CASE ESTIMATE BY STATE AND SITE

Further examination of projected costs by State and site shows where the mid-range Base Case would be incurred **(Table 1)**:

1). Washington, South Carolina, Tennessee, Colorado, and Idaho account for $170 billion over the life of the Environmental Management program (71 percent).

2). The most costly sites are the Hanford Site (Washington); the Savannah River Site (South Carolina); the Rocky Flats Environmental Technology Site (Colorado); the K-25 Site, the Y-12 Plant, and the Oak Ridge National Laboratory (Tennessee); and the Idaho National Engineering Laboratory.

ALTERNATIVE CASES

The alternative cases reflect ways the Base Case could change if certain policy decisions were made. The alternative cases analyzed four areas most likely to affect total cost, scope, and pace of the Environmental Management program:

1). **Land Use**—What are the ultimate uses for currently contaminated lands, waters, and structures at each installation?

2). **Program Funding and Schedule**—How might activities be prioritized, and how rapidly will this money be spent?

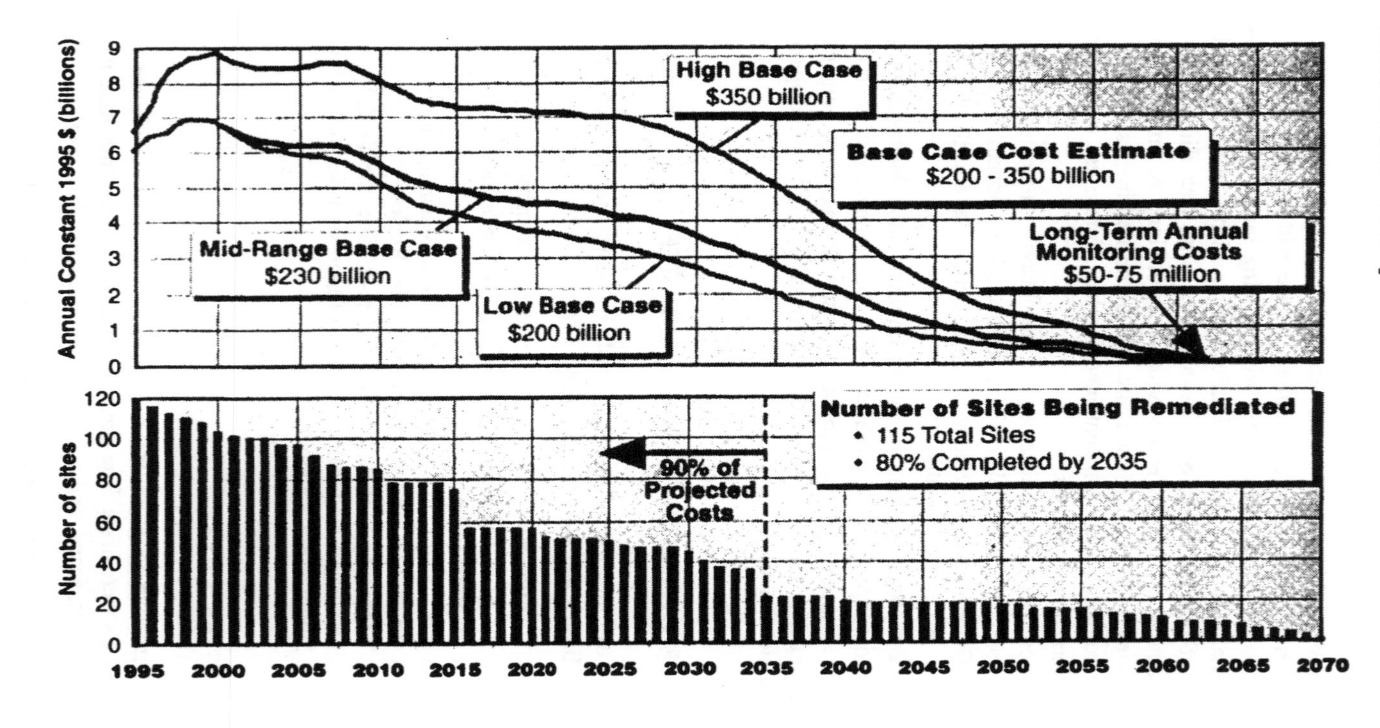

Figure 1: Base case cost and schedule estimate.

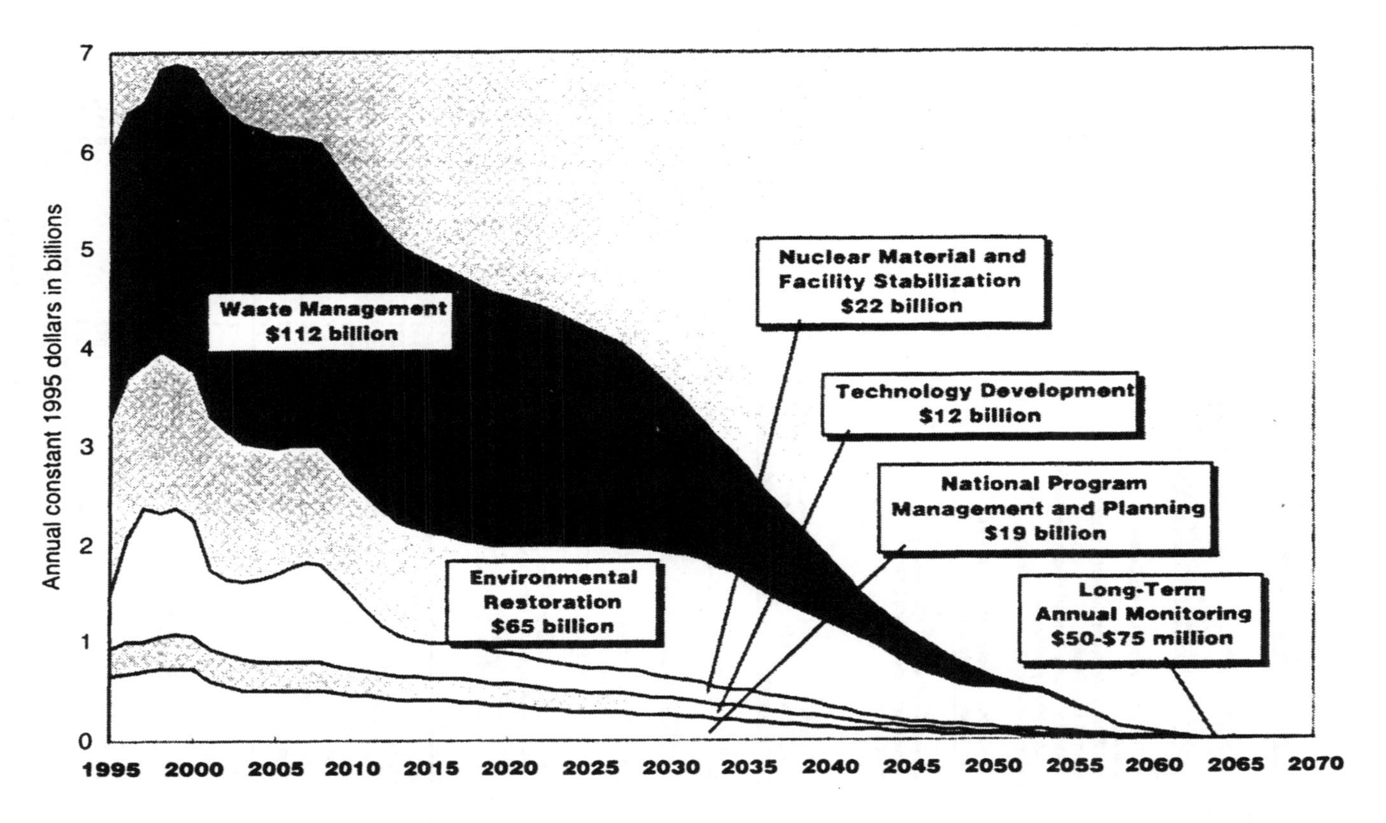

Figure 2: Mid–range base case cost profile for major elements of the environmental management program.

Table 1: Mid-Range Base Case Estimate by State and Site

Site	Mid-Range Base Case Cost (Constant 1995 $ in Millions)	Percentage of Total Mid-Range Base Case Cost
Alaska	*2*	*<1%*
Nevada Offsite* - Alaska	2	<.01%
Arizona	*139*	*<1%*
Completed UMTRA S&M** - Arizona	139	0.06%
California	*2,273*	*0.98%*
Energy Technology Engineering Center	249	0.11%
General Atomics	12	0.01%
General Electric Vallecitos Nuclear Center	18	0.01%
Geothermal Test Facility	6	<.01%
Laboratory for Energy Related Health Research	34	0.01%
Lawrence Berkeley Laboratory	208	0.09%
Lawrence Livermore National Laboratory	1,521	0.66%
Oxnard	13	0.01%
Sandia National Laboratories - Livermore	92	0.04%
Stanford Linear Accelerator Center	119	0.05%
Colorado	*23,294*	*10.10%*
Completed UMTRA S&M - Colorado	7	<.01%
Grand Junction Project Office Site	707	0.31%
Gunnison	14	0.01%
Maybell	23	0.01%
Naturita	26	0.01%
Rifle	34	0.01%
Rocky Flats Environmental Technology Site	22,455	9.74%
Nevada Offsite - Colorado	3	<.01%
Slick Rock	26	0.01%
Connecticut	*3*	*<.01%*
FUSRAP*** - Connecticut	3	<.01%
Florida	*189*	*<1%*
Pinellas Plant	189	0.08%
Idaho	*18,658*	*8.09%*
Argonne National Laboratory - West	229	0.10%
Completed UMTRA S&M - Idaho	<1	<.01%
Idaho National Engineering Laboratory	18,430	7.99%
Illinois	*612*	*<1%*
Argonne National Laboratory - East	527	0.23%
Fermi National Accelerator Laboratory	76	0.03%
FUSRAP - Illinois	1	<.01%
Site A/Plot M	8	<.01%
Iowa	*12*	*<1%*
Ames Laboratory	12	0.01%
Kentucky	*3,390*	*1.47%*
Maxey Flats	221	0.01%
Paducah Gaseous Diffusion Plant	3,368	1.46%
Maryland/District of Columbia	*30,143*	*13.07*
FUSRAP - Maryland	7	<.01%
Environmental Management Headquarters****	30,136	13.07%
Massachusetts	*14*	*<1%*
FUSRAP - Massachusetts	14	0.01%
Michigan	*1*	*<1%*
FUSRAP - Michigan	1	<.01%
Mississippi	*3*	*<1%*
Nevada Offsite - Mississippi	3	<.01%
Missouri	*1,074*	*0.47%*
FUSRAP - Missouri	388	0.17%
Kansas City Plant	312	0.14%
Weldon Spring Site Remedial Action Project	373	0.16%

*Nevada Offsite are locations where nuclear detonations occurred and environmental management activities are managed by the Nevada Operations Office.
** UMTRA S&M is the acronym for Uranium Mill Tailings Remedial Action projects with long-term Surveillance and Maintenance activities.
***FUSRAP is the acronym for the Formerly Utilized Sites Remedial Action Program.
****Approximately 71 percent of these costs are distributed across Environmental Management sites.

(continued)

Table 1: (continued)

Site	Mid-Range Base Case Cost (Constant 1995 $ in Millions)	Percentage of Total Mid-Range Base Case Cost
Nebraska	*<1*	*<1%*
Hallam Nuclear Power Plant	<1	<.01%
Nevada	*2,472*	*1.07%*
Nevada Test Site	2,443	1.06%
Nevada Offsite - Nevada	29	0.01%
New Jersey	*440*	*<1%*
FUSRAP - New Jersey	322	0.14%
Princeton Plasma Physics Laboratory	118	0.05%
New Mexico	*9,647*	*4.18%*
Albuquerque Operations Office	456	0.20%
Ambrosia Lake	<1	<.01%
Completed UMTRA S&M - New Mexico	3	<.01%
Inhalation Toxicology Research Institute	19	0.01%
Los Alamos National Laboratory	3,304	1.43%
Nevada Offsite - New Mexico	10	<.01%
Sandia National Laboratories - New Mexico	890	0.39%
South Valley Site	18	0.01%
Waste Isolation Pilot Plant	4,948	2.15*
New York	*4,003*	*1.74%*
Brookhaven National Laboratory	460	0.20%
FUSRAP - New York	273	0.12%
Separations Process Research Unit	112	0.05%
West Valley Demonstration Project	3,157	1.37%
North Dakota	*22*	*<1%*
Belfield/Bowman	22	0.01%
Ohio	*11,743*	*5.09%*
Battelle Columbus Laboratories	110	0.05%
Fernald Environmental Management Project	4,186	1.82%
FUSRAP - Ohio	197	0.09%
Mound Plant	1,539	0.67%
Piqua Nuclear Power Plant	<1	<.01%
Portsmouth Gaseous Diffusion Plant	5,575	2.42%
Reactive Metals, Inc.	135	0.06%
Oregon	*3*	*<1%*
Completed UMTRA S&M - Oregon	3	<.01%
Pennsylvania	*3*	*<1%*
Completed UMTRA S&M - Pennsylvania	3	<.01%
South Carolina	*48,174*	*20.90%*
Savannah River Site	48,174	20.90%
Tennessee	*24,812*	*10.76%*
Oak Ridge Y-12 Site	4,127	1.79%
Oak Ridge Reservation	277	0.12%
Oak Ridge K-25 Site	12,662	5.49%
Oak Ridge Associated Universities	18	0.01%
Oak Ridge National Laboratory	7,729	3.35%
Texas	*582*	*<1%*
Completed UMTRA S&M - Texas	21	0.01%
Pantex Plant	562	0.24%
Utah	*140*	*<1%*
Completed UMTRA S&M - Utah	8	<.01%
Monticello Millsite and Vicinity Properties	131	0.06%
Washington	*48,671*	*21.11%*
Hanford Site	48,671	21.11%
Wyoming	*25*	*<1%*
Completed UMTRA S&M - Wyoming	25	0.01%
[illegible]	[illegible] Billion	100%

3). **Technology Development**—How might future technologies influence the Environmental Management program?

4). **Waste Management Configurations**—Where and how will we treat, store, and dispose of wastes?

Land Use

How land will be used after environmental remediation dictates the type and extent of remedial approaches, and thus, total costs. The Base Case estimate in this report is based on a "bottom up" approach using large amounts of data and assumptions collected from field offices, rather than centralized estimating processes. This method resulted in more realistic land-use assumptions and, consequently, substantially lower costs than previous cost estimates. For comparison, total program costs were analyzed for a range of alternative future land uses, ranging from most to least restricted. **Figure 3** depicts a continuum of land use ranging from totally restricted to totally unrestricted use.

The most restricted case involves containing existing contamination in place and restricting public access thereafter. The least restricted land use requires removing or destroying contaminants in all parts of the environment, which would leave land clean enough for a wide variety of uses, potentially including farming and public recreation. Two other cases were also analyzed that were more reflective of the contractual and legal requirements accounted for in the Base Case analysis.

The life-cycle cost estimates for the range of land uses vary from approximately $175 billion to $500 billion depending on the level of cleanup assumed. This analysis indicates that future land-use determinations will have the single greatest impact on total program cost among the factors analyzed.

Each land-use case has its limitations. For example, containment rather than remediation is unrealistic across the Department of Energy complex because it would violate several existing cleanup compliance agreements. Also, in some cases, it is less costly to remediate contamination than to contain it. Establishing "green fields" at Department facilities nationwide is not realistic because it would preclude establishing any waste disposal areas, which must be located in restricted areas. Also, for certain contamination situations, technologies do not yet exist to remediate the environment to the level required for unrestricted use. For example, ground water beneath 150 square miles of the Hanford Site is contaminated with radioactive and chemical particles captured within a labyrinth of sediment and rock layers.

Residual Contamination Standards: Costs and schedules reported in the Base Case are based on each installation's best estimate of ultimate-cleanup levels. The site-specific land use assumptions in the Base Case result in significant restrictions on future land-use at many of the sites. Variations in residual contamination standards have little impact on costs because containment, rather than the removal of contamination, is assumed to be used. The Department

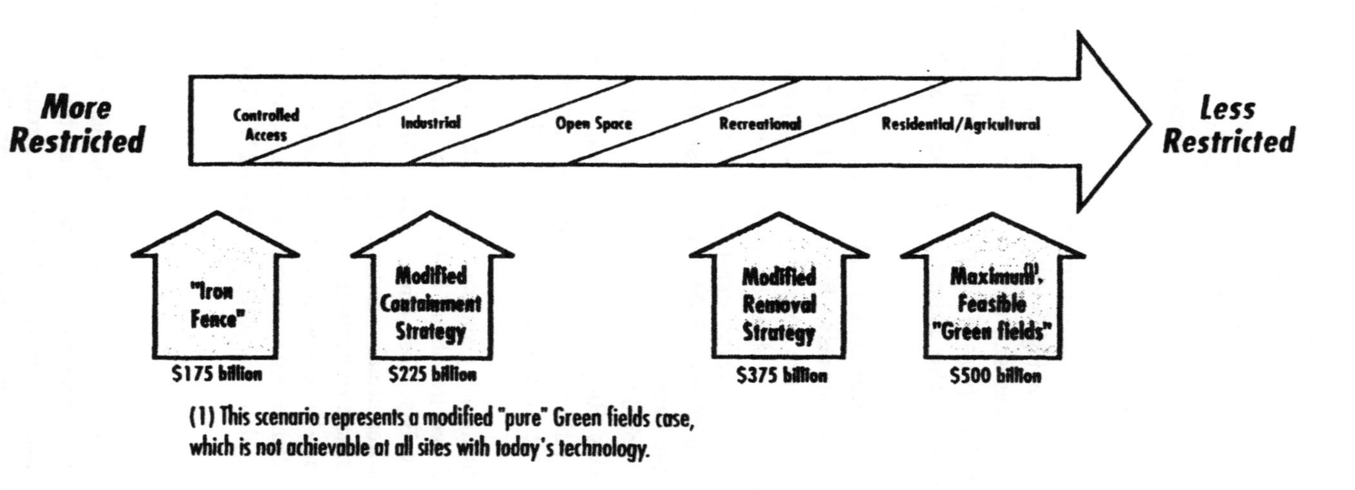

Figure 3: Conceptual illustration of land use continuum.

believes that more stringent cleanup standards will result in higher costs if more active remediation approaches are assumed. However, if less active remediation, such as containment is assumed, then little change in cost will occur from more stringent residual contamination standards. More information must be collected, and analyses need to be conducted before costs can be quantified nationwide.

Program Funding and Schedule

Another set of analyses addressed the impacts of more or less available funding for the program. Assuming additional funding, the impacts of accelerating stabilization activities and early closure of sites were analyzed. Assuming reduced funding, the impacts of reducing the scope of remediation and waste management activities are also addressed. Highlights of the scheduling analysis are shown below.

1). The life-cycle cost estimate for surveillance and maintenance could be reduced to approximately $500 million if pre-stabilization surveillance maintenance was reduced from 10 years (as in the Base Case) to 1 year. This is about 87 percent lower than the $4 billion in the Base Case. However, annual costs during the early years of the program would exceed the constant, or "flat," funding limit assumed for the Base Case.

2). Almost $5 billion would be saved if the Department closed the Rocky Flats Site, Oak Ridge's K-25 Plant, and the Fernald Plant substantially earlier (20–40 years) than currently scheduled. However, annual costs would exceed flat funding limits for several years.

3). If funding were significantly reduced beyond the year 2000, minimal action would require about $170 billion. This is about 27% lower than the Base Case through 2070. Minimal action would exclude environmental restoration, decontamination and dismantlement, and all treatment and disposal activities associated with future low-level, low-level mixed, and transuranic wastes. Annual surveillance and maintenance costs, however, would be as high as $500 million, compared with $50–$75 million projected in the Base Case.

Technology Development

Innovative technologies could make cleanup and other related activities more efficient and cost effective. More than 100 potential technology systems scheduled to be implemented by the year 2000 were screened based on the potential applicability to high-cost remediation projects. Of these, 15 were selected to evaluate potential cost savings.

Potential cost savings from implementing these new technologies range from

$9 to $80 billion, depending on future land use strategies, and assuming the technologies could be implemented by 2010.

Waste Management Configurations

The Department currently is examining alternative configurations (centralized, regionalized, and decentralized) for waste management facilities. This involves deciding where in the country wastes will be stored, treated, or disposed.

Alternative configurations, ranging from decentralized to centralized approaches, could increase costs by $9 billion or decrease them by $5 billion from the Base Case, because of the potential for economies of scale in building and operating fewer facilities. There is substantial uncertainty about the exact benefits of these economies. More analysis should be available for next year's version of the report.

APPENDIX II

DEPARTMENT OF ENERGY ADDRESSES AND TELEPHONE NUMBERS

U.S. Department of Energy, Headquarters (202) 586-5000
Forrestal Building
1000 Independence Avenue, S.W.
Washington, DC 20585

U.S. Department of Energy, Headquarters (202) 586-5000
Germantown
19901 Germantown Road
Germantown, MD 20585

- Assistant Secretary for Environmental Management (202) 586-7710
  - Principal Deputy Assistant Secretary (202) 586-7745
  - Executive Officer (202) 586-7709
  - Deputy Assistant Secretary for Compliance and Program Coordination (202) 586-8754
  - Deputy Assistant Secretary for Waste Management (202) 586-0370
    - Eastern Waste Management Operations (301) 903-7410
    - Program Integration (301) 903-7147
    - Management Systems Division (301) 903-7164
    - Waste Minimization Division (301) 903-1378
    - Waste Isolation Pilot Plant (301) 903-7201
    - Western Waste Management Operations (301) 903-7105
    - Hanford Waste Management Operations (301) 903-7170
    - Spent Fuel Management (301) 903-1450

Deputy Assistant Secretary for Environmental
Restoration (202) 586-6331
Eastern Area Programs (301) 427-1737
Program Integration (301) 427-1619
Northwestern Area Programs (301) 427-1757
Southwestern Area Programs (301) 427-1787

Deputy Assistant Secretary for Technology Development
(202) 586-6382
Technology Transfer and Program Integration (301) 903-7924
Research and Development (301) 903-7911
Demonstration, Testing and Evaluation (301) 903-8621

Albuquerque Operations Office, U.S. Department of Energy,
P.O. Box 5400, Albuquerque, NM 87185-5400 (505) 845-6049

Argonne Area Office, U.S. Department of Energy,
9800 South Cass Avenue, Argonne, IL 60439 (708) 252-2436

Argonne National Laboratory (East)
9700 South Cass Avenue, Argonne, IL 60439 (708) 252-3872

Argonne National Laboratory (West), Idaho Site,
P.O. Box 2528, Idaho Falls, ID 83403-2528 (208) 533-7000

Brookhaven National Laboratory,
P.O. Box 5000, Upton, NY 11973-5000 (516) 282-2772

Chicago Operations Office, U.S. Department of Energy,
9800 South Cass Avenue, Argonne, IL 60439 (708) 252-2110

Dayton Area Office, U.S. Department of Energy,
P.O. Box 66, Miamisburg, OH 45342-0066 (513) 865-3307

Fernald Area Office, U.S. Department of Energy,
7400 Willey Road, Cincinnati, OH 45030 (513) 648-3101

Golden Field Office, U.S. Department of Energy,
1617 Cole Boulevard, Golden, CO 80401 (303) 275-4778

Idaho Operations Office, U.S. Department of Energy,
785 DOE Place, Idaho Falls, ID 83401 (208) 526-5665

Idaho National Engineering Laboratory,
P.O. Box 1625, Idaho Falls, ID 83415 (208) 526-0111

Kirtland Area Office, U.S. Department of Energy,
P.O. Box 5400, Albuquerque, NM 87185-5400 (505) 845-4094

Lawrence Berkeley Laboratory,
1 Cyclotron Road, Berkeley, CA 94720 (510) 486-5111

Lawrence Livermore National Laboratory, University of California,
7000 East Avenue, P.O. Box 808, L–1, Livermore, CA 94550 (510) 422–1175

Lawrence Livermore National Laboratory, Nevada Test Site,
University of California, P.O. 45, Mercury, NV 89023 (702) 295–4080

Los Alamos Area Office, U.S. Department of Energy,
528 35th Street, Los Alamos, NM 87544 (505) 667–5105

Los Alamos National Laboratory, University of California,
P.O. Box 1663, Los Alamos, NM 87545 (505) 667–5101

Nevada Operations Office, U.S. Department of Energy,
P.O. Box 98518, Las Vegas, NV 89193–8518 (702) 295–3211

Nevada Test Site Mercury, Nevada, U.S. Department of Energy,
P.O. Box 435, Mercury, NV 89023 (702) 295–9060

Oak Ridge National Laboratory,
P.O. Box 2008, Oak Ridge, TN 37831 (615) 576–2900

Oak Ridge Operations Office, U.S. Department of Energy,
P.O. Box 2001, Oak Ridge, TN 37831 (615) 576–4444

Oakland Operations Office, U.S. Department of Energy,
1301 Clay Street, Oakland, CA 94612–5208 (510) 637–1810

Pacific Northwest Laboratory,
P.O. Box 999, Richland, WA 99352 (509) 375–2202

Pinellas Area Office, U.S. Department of Energy,
P.O. Box 2900, Largo, FL 34649 (813) 541–8088

Princeton Area Office, U.S. Department of Energy,
P.O. Box 102, Princeton, NJ 08542 (609) 243–3700

Richland Operations Office, U.S. Department of Energy,
825 Jadwin Avenue, P.O. Box 550, Richland, WA 99352 (509) 376–7395

Rocky Flats Field Office, U.S. Department of Energy,
P.O. Box 928, Golden, CO 80402–0928 (303) 966–2025

San Francisco Support Office, U.S. Department of Energy,
1301 Clay Street, Room 1060 North, Oakland, CA 94612–5219 (510) 637–1942

Sandia National Laboratory, U.S. Department of Energy,
P.O. Box 969, Livermore, CA 94551–0969 (510) 294–2211

Sandia National Laboratory, Albuquerque,
P.O. Box 5800, Albuquerque, NM 98185–5800 (508) 844–7261

Sandia National Laboratory, Nevada Test Site,
P.O. Box 238, Mercury, NV 89023 (702) 295–7477

Sandia National Laboratory, Tonopah Test Range,
P.O. Box 871, Tonopah, NV 89049 (702) 295–8313

Savannah River Operations Office, U.S. Department of Energy,
P.O. Box A, Aiken, SC 29801 (803) 725–2277

Waste Isolation Pilot Project Office, U.S. Department of Energy,
P.O. Box 3090, Carlsbad, NM 88221 (505) 887–8103

For information regarding Research Opportunity Announcements (ROAs), and Program R&D Announcements (PRDAs): contact U.S. Department of Energy, Morgantown Energy Technology Center, P.O. Box 880, MS 107, Morgantown, West Virginia 26507; (304) 285–4087.

For information regarding the DOE Small Business Innovation Research (SBIR) Program Office, and Small Business Technology Transfer Pilot Program (STTR): contact U.S. Department of Energy, Small Business Innovation Research Program Hotline, ER–16 GTN, Washington, DC 20585; (301) 903–5707.

The Offices of Research and Technology Applications (ORTA) serve as technology transfer agents for the federal laboratories. They coordinate technology transfer activities among laboratories, industry, and universities. ORTA offices license patents and foster communication between researchers and technology customers. ORTA Contacts:

Ames Laboratory	(515) 294–5640
Argonne National Lab	(708) 252–5361
Brookhaven National Lab	(516) 282–7338
Fermilab	(708) 840–2529
Idaho National Engineering Lab	(208) 526–1010
Lawrence Berkeley Lab	(510) 486–6467
Lawrence Livermore National Lab	(510) 422–7839
Los Alamos National Lab	(505) 665–9090
Morgantown Energy Technology Ctr	(304) 285–4709
National Renewable Energy Lab	(303) 275–3015
Oak Ridge Institute/Science & Ed	(615) 576–3756
Oak Ridge National Lab	(615) 576–8368
Pacific Northwest Lab	(509) 375–2789
Pittsburgh Energy Technology Center	(412) 892–6029
Princeton Plasma Physics Lab	(609) 243–3009
Sandia National Lab	(505) 271–7813
Savannah River Technology Center	(803) 725–1134
Stanford Linear Accelerator Center	(415) 926–2213
Westinghouse Hanford Company	(509) 376–5601

APPENDIX III

FOREIGN NUCLEAR WASTE MANAGEMENT ORGANIZATIONS AND ACTIVITIES

ARGENTINA

NUCLEAR POWER

Policy: Pressurized heavy water reactors (PHWR) with natural uranium and indigenous fuel cycle; currently government ownership and operation of all nuclear power plants – other options being evaluated; development of nuclear plants and services export capability.

INDUSTRIAL FUEL CYCLE

Policy: Develop all phases of the PHWR fuel cycle, gaseous diffusion capability for U enrichment (Pilcaniyeu), and D_2O production; may export Pu to nations with breeder reactors. Interim AR and AFR storage of spent fuel.

Waste Management Strategy: Options for reprocessing spent fuel analyzed, including vitrification of HLW and disposal of HLW glass canisters in granite host-rock repository, but no decision made.

Comision Nacional de Energia Atomica (CNEA)
Avenida del Libertador 8250
1429 Buenos Aires, Argentina

National Atomic Energy Commission—owns and operates all nuclear facilities.

AUSTRALIA

NUCLEAR POWER

Policy: No nuclear power installed; none planned. Large uranium reserves; uranium currently produced for export. Government sponsors nuclear waste management R&D.

Australian Nuclear Science and Technology Organisation
New Illawarra Road, Lucas Heights
Private Mail Bag 1
Menai NSW 2234, Australia

Fuel cycle R&D—HLW immobilization (SYNROC process development and waste form properties), mill tailings treatment, actinide transport, surface hydrology, and radionuclide release.

BELARUS

NUCLEAR POWER

Though Belarus currently produces no nuclear power, approximately 25% of its total electricity consumption is provided by nuclear power plants at Ignalina in Lithuania and Smolensk in Russia. Construction of a nuclear power plant at Minsk, with projected capacity of 2,000 MWe and planned additional capacity of 6,000 MWe, was halted in 1986 due to the events at Chernobyl. The current government has stated that nuclear power is a necessity in the future of Belarus.

INDUSTRIAL FUEL CYCLE

Policy: Because Belarus has no uranium natural resources, no uranium enrichment is foreseen, nor is fuel reprocessing.

Waste Management Strategy: A waste management concept and strategy for disposal of waste from the planned first Belarus NPP is now being developed, LLW generated during operation and from decommissioning of a research LWR (IPEP) was managed in accordance with known regulations in the former Soviet Union. This waste was stored in an engineered structure in an underground facility near Sosny that is also used for spent radioactive sources. Spent fuel from decommissioning the research LWR was sent to Russia for reprocessing.

Belarus Research and Design Institute of Power
Engineering Industry (BEL NIPI ENERGOPROM)
ul. Romanovskaja sloboda 5A
220048 Minsk, Belarus

Development of technical policy for electric power/energy resources and of electric power network installations.

State Committee on Supervision of Industrial/Nuclear Safety (GOSPROMATOMNADZOR):
ul. Chkalova 6
22039 Minsk, Belarus

Responsible for regulations, control, and licensing of nuclear installations and radiation-emitting facilities.

State Chernobyl Committee (GOSCOMCHERNOBYL):
ul. Lenin 14
220030 Minsk, Belarus

Regulate, control, and finance the National Chernobyl Program; licensing of decontamination/waste management activities for area affected by the Chernobyl fallout.

State Specialized Enterprise (GSP POLESJE):
ul. Karpovich 11
246017 Gomel, Belarus

Decontamination of affected zone in southern Belarus; treatment and conditioning of waste generated as a result of decontamination.

Institute of Power Engineering Problems (IPEP):
Belarus Academy of Sciences
Sosny
220109 Minsk, Belarus

Waste management R&D - LLW/ILW immobilization, liquid LLW treatment, thermal/chemical processing of radioactive wood waste.

Institute of Radio-Ecological Problems (IREP):
Belarus Academy of Sciences
Sosny
220109 Minsk, Belarus

R&D on radionuclide migration in biosphere, decontamination, conditioning of liquid LLW (generated after remediation of contaminated site), nuclear medicine, radiochemistry.

BELGIUM

NUCLEAR POWER

Policy: Produce base load electricity by nuclear and coal power plants. Decided against adding proposed eighth (1300 MWe) nuclear unit (at least during next few years).

INDUSTRIAL FUEL CYCLE

Policy: Well-rounded capability—uranium enrichment (share in Eurodif); MOX and UO_2 fuel fabrication; purchase of foreign reprocessing services; decision made to dismantle former Eurochemic reprocessing plant.

Waste Management Strategy (responsibility of ONDRAF): Vitrify HLW and store 50 years (investigation of HLW, ILW disposal in clay formations underway); treatment and immobilize other wastes; sea-dumping of LLW halted; shallow-ground disposal of LLW under investigation.

Belgonucleaire S.A.
Avenue Ariane 2-4
1200 Brussels, Belgium

Provide engineering services for nuclear power plants, nuclear fuel cycle facilities, and waste treatment plants; fabricate MOX fuels.

Belgoprocess
Gravenstraat 73
2480 Dessel, Belgium

Maintenance/dismantling of ex-Eurochemic reprocessing facilities and obsolete waste treatment facilities formerly belonging to CEN/SCK; treatment/conditioning of all categories of low-, medium-, and high-level waste; from 1986 to 1991 joint operation with WAK of Pamela vitrification plant, now being kept in standby for potential future vitrification of HLLW from WAK pilot reprocessing plant at Karlsruhe, Germany.

Organisme National des Dechets
Radioactifs et des Matieres
Fissiles (ONDRAF/NIRAS)
Place Madou 1, B.P. 24
1030 Brussels, Belgium

Define Belgian waste management policy and R&D requirements; responsible for transportation of radioactive materials, waste treatment, conditioning and interim storage, spent fuel AFR storage, waste disposal, fissile material storage.

Studiecentrum voor Kernenergie
Centre d'Etude de l'Energie Nucleaire Laboratories
Boeretang 200
2400 Mol, Belgium

Geologic waste isolation in clay formations, waste treatment (decontamination and recycling of boric acid, removal of plutonium from waste generated by fuel fabrication, etc), decommissioning (decontamination, dismantling, restoration) of nuclear facilities.

BRAZIL

NUCLEAR POWER

Policy: Complete nuclear industry with closed fuel cycle, based upon technology transfer from FRG and other countries.

INDUSTRIAL FUEL CYCLE

Policy: Development of full commercial capability for closed fuel cycle—U mining and milling; conversion of U_3O_8 to UF_6; enrichment; UO_2 fuel fabrication; fuel reprocessing.

Waste Management Strategy: Not yet defined for HLW; near-surface disposal for LLW, including the Cs-137 waste from the Goiania accident (1987).

Comissao Nacional de Energia Nuclear (CNEN)
Rua General Severiano 90
Botafogo ZC-82, CEP 22290
Rio de Janeiro, RJ, Brazil

Regulation, surveillance, and licensing of nuclear reactors, fuel cycle facilities and radiation-emitting installations; promotion of nuclear technology R&D and technology transfer to private industry; promotion and training of personnel. Controls four research institute: CDTN, IPEN, and IRD.

Instituto de Pesquisas Energeticas e Nucleares (IPEN)
Cidade Universitaria
Caixa Postal 11.049
Pinheiros, CEP 01000
Sao Paulo, Brazil

Nuclear physics, nuclear medicine, radiobiology, radiation health/safety, engineering/reactor technology/instrumentation, nuclear materials chemistry, isotope and radiation applications/production, nuclear waste disposal, nuclear metallurgy, radiochemistry.

CANADA

NUCLEAR POWER

Policy: Strong support for domestic use and export of the CANDU reactor system.

INDUSTRIAL FUEL CYCLE

Policy: Interim storage of used fuel for decades, pending an environmental assessment and review of a concept for the disposal of nuclear fuel waste (review process of concept has started).

Waste Management Strategy: Geologic disposal of nuclear fuel waste and spent CANDU fuel in a crystalline rock repository. Disposal of LLW in engineered, shallow-ground facility.

Atomic Energy Control Board (AECB)
P.O. Box 1046
270 Albert Street
Ottawa, ON, K1P 5S9, Canada
Federal nuclear control agency, answers to Parliament; responsible for health/safety regulation, compliance/licensing.

Atomic Energy of Canada Ltd. (AECL)
344 Slater Street
Ottawa, ON, K1A 0S4, Canada
Crown Corporation answers to Parliament via Ministry of Natural Resources. R&D; design, engineering and sale of CANDU and research reactors; proprietary rights on CANDU Nuclear Steam Supply Systems; waste management R&D at Whiteshell and Chalk River laboratories.

AECL Research
Chalk River Laboratories
Chalk River, ON, R0J 1J0, Canada
Development and operation of processes for the treatment of LLW and ILW using incineration, compaction, microfiltration/reverse osmosis evaporation, ion exchange, and solidification in bitumen. LLW/ILW repository consisting of prototype vault. Capacity of 2,000 m^3 radwaste in drums or bales, when full to

be covered with backfill, roofed with concrete, and mounded with earth. Waste can be retrieved from the IRUS vault until concrete cap is emplaced. Construction start, 1993.

Ontario Hydro
700 University Avenue
Toronto, ON, M5G 1X6, Canada

Provincial public utility. Owns/operates 20 CANDU nuclear power plants, 15,340 MWe total capacity; responsible for developing interim fuel storage/transport technologies. Process/store low- and intermediate-level radioactive waste from Ontario Hydro CANDU reactors and research/maintenance facilities.

CHINA

NUCLEAR POWER

Policy: Develop nuclear power as one of three major sources of energy to solve problems caused by uneven distribution of resources; be self-sufficient, but introduce foreign advanced technology.

INDUSTRIAL FUEL CYCLE

Policy: Activities include uranium mining, milling, and diffusion enrichment; isotope separation; fuel fabrication; future spent fuel reprocessing.

Waste Management Strategy: Interim storage of spent fuel in pools for 5–8 years if <1,000 tU, in transport/storage casks if >1,000 tU; interim storage, reprocessing, vitrification, and disposal all to be at one site, to be selected, located in northwest China or the Gobi Desert; final disposal in deep geologic formation; plan for a small pilot reprocessing plant, followed by a commercial-size facility, about 500 tU/a.

China National Nuclear Corporation (CNNC)
P.O. Box 2102
Beijing 100822, PRC

Conglomerate of over 200 enterprises and institutions. Plans to construct four regional final LLW/ILW disposal facilities in northwest (Gansu), east, south, and southwest China for waste from nuclear facilities, including Qinshan and Daya Bay nuclear power stations.

China Institute of Atomic Energy
P.O. Box 275
Beijing 102413, PRC

Large comprehensive nuclear R&D institute. FBR development. HLW vitrification; waste form characterization; pilot plants to be built.

FINLAND

NUCLEAR POWER

Operates 2 PWR and 2 BWR nuclear power reactors, by IVO and TVO.

INDUSTRIAL FUEL CYCLE

Policy: Purchase fuel and fuel-cycle services from other countries (spent fuel from Soviet-built reactors is returned to Russia).

Waste Management Strategy: Spent fuel from TVO's power plants will be stored for 40 years, then disposed of in crystalline bedrock; IVO returns spent fuel from its (Soviet-built) reactors to Russia. Operating wastes are conditioned, stored above ground, and disposed of in crystalline bedrock at the nuclear power station sites. Decommissioning wastes will be disposed of in extended operating waste repositories.

Nuclear Energy Commission (NEC)
Pohjoinen Makasiinikatu 6
FIN-00130 Helsinki, Finland

Advisory organization for general matters connected with nuclear energy; coordinated by the Ministry of Trade and Industry.

Finnish Centre for Radiation and Nuclear Safety (STUK)
P.O. Box 14
Laippatie 4
FIN-00881 Helsinki, Finland

Regulatory enforcement and inspection authority; research related to transport of radionuclides in biosphere.

Technical Research Centre of Finland (VTT)

VTT Energy
Nuclear Energy
Tekniikantie 4C, Espoo
FIN-2044 VTT, Finland

VTT Communities/Infrastructure
Rock/Environmental Engineering
Betonimiehenkuja 1, Espoo
P.O. Box 19041
FIN-02044 VTT, Finland

VTT Chemical Technology
Environmental Technology
Physics Building
Otakaari 3A, Espoo
FIN-02044 VTT, Finland

VTT Manufacturing Technology
Materials/Structural Integrity
Kemistintie 3, Espoo
P.O. Box 1704
FIN-02044 VTT, Finland

FRANCE

NUCLEAR POWER

Policy: Vigorous nuclear power program, scaled down recently to construction of less than one new reactor per year; commercialization of the breeder reactor; export of nuclear plants and services.

INDUSTRIAL FUEL CYCLE

Policy: Maintain full domestic fuel cycle capability and aggressive export of fuel cycle plants, equipment, and services (including uranium enrichment and spent fuel reprocessing).

Waste Management Strategy: HLW—vitrify and store in engineered storage facility for indefinite period, then emplace in geologic repository (granite or clay). LLW—immobilize in bitumen, concrete, or resin and dispose in engineered surface facility.

Agence Nationale pour la Gestion des Dechets Radioactifs (ANDRA)
Route Du Panorama Robert Schumann
B.P. 38
92266 Fontenay-aux-Roses Cedex, France

Design, site, construct, and manage long-term waste disposal facilities; establish radioactive waste packaging/disposal specifications and ensure compliance; contribute to R&D programs related to long-term waste disposal.

Commissariat a l'Energie Atomique (CEA)
Centre d'Etudes Nucleaires (CEN)
31-33, Rue de la Federation
75752 Paris Cedex 15, France
Responsible for R&D related to all areas of the nuclear fuel cycle through activities of several operational units (scientific directorates), research centers, and wholly/partially owned industrial concerns.

Centre d'Etudes Nucleaires de Cadarache (CEA/CEN-CA)
B.P. 1
13108 Saint Paul Lez Durance, France
Treatment of TRU wastes, LLW, and ILW; properties of non-HLW waste forms and waste isolation (radionuclide migration).

GERMANY

NUCLEAR POWER

Operates 14 PWR and 7 BWR nuclear power reactors.

INDUSTRIAL FUEL CYCLE

Policy: Full commercial capability—enrichment, fuel fabrication, plutonium recycle to LWRs; reprocessing is to be handled by foreign plants.

Waste Management Strategy: Vitrification of HLW (by foreign plants) and interim storage of HLW glass; disposal of reprocessing wastes in future salt-dome repository; interim storage of ILW/LLW wastes; future disposal of reactor and decommissioning wastes in abandoned iron mine or salt repository.

Bundesamt fur Strahlenschutz (BfS)
Postfach 10 01 49
38201 Saltzgitter, Germany
Execution of the federal responsibilities concerning testing/standards for radiation protection, nuclear safety, radioactive waste disposal, and transport/storage of radioactive materials; in particular, the responsibility for construction and operation of repositories.

Bundesanstalt fur Geowissenschaften und Rohstoffe (BGR)
Stilleweg 2
Postfach 510153
30655 Hannover, Germany

Responsible to BMWI for all geological/geotechnical aspects related to planning, construction/operation of a final repository for radioactive wastes; conduct special research for BMU.

Bundesministerium fur Forschung und Technologie (BMFT)
Heinemannstrasse 2
Postfach 200240
53175 Bonn, Germany

Responsible for R&D programs on fuel cycle and radioactive waste management.

Bundesministerium fur Umwelt, Naturschutz und Reaktorsicherheit (BMU)
Kennedyallee 5
53175 Bonn, Germany

Responsible for storage, transportation, and disposal of radioactive wastes; supervision of state licensing procedures; federal standards for nuclear safety and radiation protection.

Endlager fur Radioaktive Abfalle Morsleben (ERAM)
Am Schacht 105
39343 Morsleben, Germany

Final repository for LLW of the former East Germany, now operated by DBE under contract to BfS.

Gesellschaft fur Nuklear-Service mbH (GNS)
Zweigerstrasse 28
45130 Essen, Germany

Service to nuclear facilities, including waste treatment/conditioning, transportation of radioactive materials, shipping cask development, and facility dismantling.

GSF-Forschungszentrum fur Umwelt und Gesundheit, GmbH
Institut fur Tieflagerung
Theodor-Heuss-Strasse 4
P.O. Box 2163
38122 Braunschweig, Germany

Waste Management R&D: Development/testing of techniques for safe, final geologic disposal of radioactive and chemical toxic wastes; acquisition of data for planning, construction, and operation of underground repositories. Safety analyses of long-term performance for the post-operational phase of underground repositories.

INDIA

NUCLEAR POWER

Policy: Heavy dependence on nuclear power to augment the nation's electric power generating capacity. A three-phase program—first phase, reactors fueled with natural uranium; second phase, FBRs fueled with Pu produced by first-phase reactors; third phase, self-sustaining thorium-uranium-cycle reactors.

Due to resource and technical problems, it is doubtful that 1997 nuclear power forecasts (end of 8th five-year plan) can be met, including the commissioning of the 500 MWe FBR by the year 2000.

INDUSTRIAL FUEL CYCLE

Policy: Achieve self-sufficiency in CANDU-type and LWR fuel cycle—uranium mining and milling, conversion to UO_2, fuel fabrication, reprocessing (in small plants adjacent to power stations); if enriched UF_6 supply for India's BWRs is cut off; they may fuel with UO_2-PuO_2.

Waste Management Strategy: Vitrification of HLW, interim storage for at least 20 years and geologic disposal in a crystalline rock formation; disposal of LLW and short-lived ILW in near-surface engineered facilities; disposal of long-lived ILW will be in a deep geological repository.

Department of Atomic Energy
Chatrapati Shivaji Marharaj Marg
Bombay 400 039, India
Design, construction, and operation/maintenance of nuclear power stations; help realize nation's goal of having 10,000 MWe of nuclear power on-line by the year 2000.

Bhabha Atomic Research Centre, Trombay (BARC)
Bombay 400 085, India
BARC has five test reactors; radiochemistry and isotope laboratories; an isotope production and processing unit; pilot plants for production of heavy water, zirconium, and titanium; a thorium plant; a uranium metal plant; a pilot-scale fuel reprocessing plant; the Fuel Irradiation and Processing Laboratory; and

supporting facilities. Fuel cycle R&D includes fuel reprocessing, HLW solidification; treatment of alpha-emitting wastes (incineration, wet oxidation, decontamination, and immobilization of cladding hulls); D&D; and waste isolation in geologic formations.

ITALY

NUCLEAR POWER

Policy: The current national energy plan calls for abandonment of nuclear power and increased use of coal and natural gas for electricity generation; research into nuclear energy will continue but with a reduced R&D budget.

INDUSTRIAL FUEL CYCLE

Waste Management Strategy: Spent fuel from previous nuclear power plant operations is being reprocessed abroad; vitrified HLW will be returned, starting in 1995; canisters will be temporarily stored until a final repository is available (clay formations are being considered); dry storage on site is also presently considered. No site for disposal of LLW/ILW has been selected.

Ente per le Nuove Tecnologie,
l'Energia e l"Ambiente (ENEA)
Viale Regina Margherita 125
00198 Rome, Italy

Direct basic and applied research on energy and environment (mostly non-nuclear). Current nuclear-related work includes cooperation in international programs and is carried out in three departments: Fusion, Innovative Reactors, and Fuel-Cycle Plant Dismantling. Decommission facilities, including removal of stored nuclear material. Tasks: conditioning of liquid/solid radioactive wastes stored at the Eurex (Saluggia) and Itrec (Trisaia) plants and the Casaccia Center; removal (foreign reprocessing being considered) of spent fuel from reprocessing pilot plants; decontamination and dismantling of plants and laboratories, including plutonium oxide fuel fabrication laboratory.

Nucleco
Via Anguillarese 351
00060 Rome, Italy

Treat and dispose of LLW/ILW from hospitals, laboratories, industrial establishments, and nuclear plants; eventual plans include decommissioning work on nuclear installations.

JAPAN

NUCLEAR POWER

Policy: Strong nuclear power program to lessen dependence on foreign energy sources; install LWRs for near-term needs; develop advanced HWR (ATR); aim for commercial FBR operation ~2020–2030; supply domestic needs and build export business.

INDUSTRIAL FUEL CYCLE

Policy: Obtain ownership of foreign uranium resources; develop complete fuel cycle capability (enrichment, reprocessing, and waste treatment; buy foreign reprocessing services until domestic capacity is available); recycle Pu to FBRs, HWRs, and LWRs.

Waste Management Strategy: HLW – vitrify with borosilicate glass, store for 30–50 years, and dispose in geological formations; LLW – dispose in engineered structures in shallow-land facility and at sea, if politically feasible.

Atomic Energy Bureau (AEB)
Science and Technology Agency
2–1 Kasumigaseki 2–chome
Chiyoda–ku, Tokyo 100, Japan
Provide support to the Atomic Energy Commission.

Atomic Energy Commission (AEC)
2–1 Kasumigaseki 2–chome
Chiyoda–ku, Tokyo 100, Japan
Formulate national policy on nuclear energy R&D and utilization; advise Prime Minister.

Central Research Institute of Electric Power Industry (CRIEPI)
1–6–1, Ohtemachi
Chiyoda–ku, Tokyo 100, Japan
Transportation, storage, disposal of LLW; intermediate and long–term storage of spent fuel; long–term storage and disposal of HLW.

Government Industrial Research Institute, Osaka (GIRIO)
1–8–31 Midorigaoka, Ikeda–shi
Osaka 563, Japan
Alternatives for HLW solidification; waste form characterization.

Hitachi Engineering Co., Ltd.
2-1 Saiwai-cho, 3-chome
Hitachi-shi, Ibaraki-ken 317, Japan

Develop technology to reprocess spent LWR fuel; fixation, storage, and disposal of HLW; spent fuel storage; Pu fuel production; decommissioning.

Japan Atomic Energy Research Institute
2-2, Uchisaiwai-cho, 2-chome
Chiyoda-ku, Tokyo 100, Japan

Semi-government research organization implementing national long-term programs in nuclear energy, including joint projects and international cooperation. Waste management research at Tokai.

JGC Corporation
Nuclear and Advanced Technology
New Ohtemachi Building
2-1 Ohtemachi 2-chome
Chiyoda-ku, Tokyo 100, Japan

Design and construction of fuel reprocessing and radwaste treatment facilities.

JGC Nuclear Research Center
2205 Narita-cho, Oarai-machi
Higashi-Ibaraki-gun
Ibaraki Pref. 311-13, Japan

Wet oxidation (organic materials, e.g., spent ion exchanger resin) incinerator; waste solidification processes (cementing, bituminization, plastic solidification); regeneration waste recycle process; selective nuclide removal process; ash melting process.

Power Reactor and Nuclear Fuel Development Corporation (PNC)
Sankaido Building
1-9-13 Akasaka
Minato-ku, Tokyo 107, Japan

Develops D&D technology at the Oari Engineering Center; also incineration and vitrification technology at Tokai.

Radioactive Waste Management Center
Mori Building #15
8-10, Toranomon 2-chome
Minato-ku, Tokyo 105, Japan

R&D on safe and effective treatment and disposal techniques for radioactive wastes.

KOREA (SOUTH)

NUCLEAR POWER

Policy: Continue expansion of electric power capacity; reduce dependence on foreign oil by strong nuclear program with indigenous manufacturing capability; long-term goal—develop FBR capability.

INDUSTRIAL FUEL CYCLE

Policy: Develop long-term contracts for fuel supplies, holdings of foreign uranium resources; fabricate fuel for PWR and HWR (CANDU); "wait and see" on reprocessing and recycle of Pu for FBR, CANDU, and LWR.

Waste Management Strategy: LLW/ILW repository to be constructed by mid-1990s with emphasis on engineered barriers; candidate sites have been identified, but final decision is pending; utility surcharge of 2 mil/kWh to fund waste management; extended storage (~60 years) of SF planned, in AR and AFR facilities; no decision has been made on reprocessing or disposal of SF/HLW.

Atomic Energy Office (AEO)
Ministry of Science and Tech.
1 Chungang-dong
Kyonggi-do, Kwacheon 171-11, Republic of Korea

License nuclear power plants and fuel cycle facilities; manage nuclear waste fund; sponsor nuclear R&D.

Atomic Energy Commission (AEC)
1 Chungang-dong
Kwacheon 171-11, Republic of Korea

Decision-making body for policies regarding nuclear energy; R&D plan for nuclear energy applications; always chaired by Deputy Prime Minister; ministers of MOST and MTIE and president of KEPCO are required members.

Korea Atomic Energy Research Institute (KAERI)
150 Tukjin-dong
Daeduk-gu, Taejon, Republic of Korea

Develop reactor engineering and nuclear fuel cycle technology; assist government (MOST) with regulatory/licensing issues and in establishing national nuclear policy. Fuel fabrication, uranium ore processing and conversion, radioactive waste management, and post-irradiation examination.

Ministry of Science and Technology (MOST)
1 Chungang-dong
Kwacheon, Kyonggi-do, Republic of Korea
Authority over virtually all scientific and technological efforts in Korea.

NETHERLANDS

NUCLEAR POWER

Policy: Expansion of nuclear capacity is on indefinite hold as a consequence of events at Chernobyl.

INDUSTRIAL FUEL CYCLE

Policy: Use foreign services (fuel fabrication, reprocessing); participate with FRG and U.K. in URENCO (uranium enrichment consortium).

Waste Management Strategy: Utilize single centralized waste collection service; extend interim storage of all wastes (50–100 years) until decisions are made regarding disposal; studies on final disposal of all radioactive wastes in geological formations are executed in the framework of the national research program (OPLA); ocean dumping of LLW and ILW has been terminated; the Netherlands contributed to NEA feasibility study regarding subseabed disposal; feasibility of disposal within international or bilateral framework is also being explored.

Centrale Organisatie Voor Radioactief Afval (COVRA)
Spanjeweg 1
P.O. Box 34
4453 ZGs'-Heerenbroek, Netherlands
Responsible for collection, treatment, and storage of all waste (multi-funded: utilities, government, ECN).

Stichting Energieonderzoek Centrum Nederland (ECN)
Westerduinweg 3
Postbus 1
1755 ZG Petten, Netherlands
Organize and sponsor energy research and development (partially government-funded). Geologic waste isolation in salt dome repositories (conceptual design, thermo-mechanical, safety, and radionuclide migration studies), seabed disposal, actinide burning, and decontamination study of large components.

N.V. Tot Keuring van Elektrotechnische Materialen Arnhem (KEMA)
Utrechtsweg 310
Postbus 9035
6800 ET Arnhem, Netherlands

Research and consulting development; services for utilities; waste management R&D; characterization, quality assurance, volume reduction, and storage of radioactive wastes.

PAKISTAN

NUCLEAR POWER

Policy: Provide up to 50% of electrical power supply with nuclear.

INDUSTRIAL FUEL CYCLE

Policy: Develop complete domestic fuel cycle - uranium mining, milling, conversion, and enrichment; fuel fabrication; reprocessing.

Pakistan Atomic Energy Commission (PAEC)
P.O. Box 1114
Islamabad, Pakistan

Advocate increased nuclear energy generation to overcome serious energy shortages in a country substantially lacking in natural energy resources. In an effort to accelerate Pakistan's overall economic development, the Commission also promotes the use of nuclear technologies in other areas, such as enhancing agricultural production and for medical diagnosis/therapy. Has a centrifuge enrichment plant.

Pakistan Institute of Science & Technology (PINSTECH)
P.O. Nilore
Islamabad, Pakistan

Fuel cycle R&D activities, including analytical chemistry, nuclear materials, metallurgy, fuel development, digital electronics, control instrumentation, and computational physics; basic research facilities are open to scientists/engineers from universities as well as research organizations.

RUSSIA

NUCLEAR POWER

Policy: Major program to develop nuclear power to avoid transport of fossil fuels from east of the Ural Mountains to the more densely populated western areas.

INDUSTRIAL FUEL CYCLE

Policy: Complete domestic fuel cycle capability, including enrichment, fuel fabrication (UO_2 and MOX), and reprocessing; complete fuel cycle services, including SF storage and LLW/ILW disposal; shift to PWRs (since Chernobyl accident in 1986).

Waste Management Strategy: Spent nuclear fuels from PWRs are stored 3–10 years, followed by reprocessing to recycle fissile materials and separate a number of other specific radionuclides for beneficial uses and different disposition; HLW is vitrified for disposal in a future geologic repository; HLW partitioning processes are being developed to recover most long-lived radionuclides. SF from RBMK (Soviet acronym for light-water-cooled, graphite-moderated) reactors is stored, pending decision on ultimate disposition.

Liquid LLW from nuclear reactor operations is currently evaporated, incorporated into bitumen or cement, and stored and/or disposed of at disposal facilities at each reactor station. Thirty-six other regional facilities exist for medical, industrial, and radioactive waste disposal. Efforts are underway to decrease liquid LLW volumes and to recycle them in water and reactant circuits.

Solid LLW, compacted at each nuclear power station, is stored/disposed of at reactor sites; regional burial facilities are being considered to minimize transportation-related risks.

All-Russian Scientific Research Institute for Inorganic Materials
Rogov Str. 5a
123060 Moscow, Russia
R&D on SF reprocessing, radioactive waste processing/solidification (bitumenization/vitrification, etc.), off-gases.

All-Russian Scientific Research Institute for Nuclear Power Plants
Ferganskaya Str. 25
109507 Moscow, Russia
Processing and disposal of NPP radioactive wastes; decontamination of equipment/facilities; emergency situation studies.

GOSATOMNADZOR
Taganskaya Str. 34
109147 Moscow, Russia

Supervision of all safety aspects of Russian nuclear industry.

Research Production Association
V.G. Khlopin Radium Institute
2nd Murinski Ave. 28
194021 St. Petersburg, Russia

Development of SF treatment (reprocessing, thermal decladding, meltdown of hulls), improved HLW partitioning, waste immobilization, off–gas treatment, Kr–85 storage, waste disposal, geochemistry, studies on solidified waste properties; environmental remediation, protection and monitoring.

Russian Scientific Centre
Kurchatov Institute
Kurchatov Square 1
123182 Moscow, Russia

Nuclear power research; R&D on LLW/ILW.

Production Association 'MAYAK'
Lenin str. 31
454065 Chelyabinsk–65, Russian Federation

Nuclear complex with multitude of activities and facilities, including radiochemical processing, weapon materials production reactors, isotope production, special waste storage, and burial sites; produced first Soviet weapons–grade plutonium. Has an HLW vitrification plant.

Ministry for Atomic Energy of the Russian Federation (MINATOM)
Staromonetny per.26
109180 Moscow, Russia

Manages all aspects of nuclear power industry. Established 1/92 on an interim basis; successor to MAPI, the former USSR Ministry of Atomic Power and Industry.

Mining and Chemical Enterprise
Lenin Str. 53
660033 Krasnoyarsk–33, Russia

SF reprocessing, waste management, underground disposal.

Ministry for Ecology and Natural Resources
B. Gruzinskaya Str. 4/6
123812 GSP Moscow, Russia

Responsible for control and standardization of releases containing radionuclides.

Research Production Association RADON
7th Rostovski per. 2/14
119121 Moscow, Russia
Research-and-production association; disposal of institutional radioactive/hazardous waste, R&D on waste treatment/conditioning, engineering design/support services, environmental protection services, special accident-related emergency services/investigations. Has largest waste disposal facility at Sergiev Posad. Treatment/conditioning of wastes is by compaction, combustion, cementation, bitumenization, vitrification, and special immobilization in metal matrix; disposal is in engineered concrete in-ground structures.

Scientific Research Institute of Chemical Engineering
Griboyedov Str. 32
620010 Ekaterinburg, Russia
Develops waste vitrification technology.

All Russian Design/Research Association (VNIPIET)
Savushkin Str. 82
197228 St. Petersburg, Russia
Design plants/facilities for SF reprocessing, waste processing, storage/disposal; SF transport/storage; decontamination.

SOUTH AFRICA

NUCLEAR POWER

Policy: Expand electric power production capacity chiefly through coal-burning plants, but develop modest nuclear capability to complement coal, particularly post-2000.

INDUSTRIAL FUEL CYCLE

Waste Management Strategy: Interim storage of reactor LLW/ILW at the reactor, followed by disposal at two shallow-land disposal facilities; interim storage of spent fuel for ~40 years; plans for disposal not defined.

Atomic Energy Corporation of South Africa Ltd. (AEC)
P.O. Box 582
Pretoria 0001, South Africa

Overall responsibility for government nuclear activities including uranium conversion and enrichment, R&D, radioisotope production, radwaste disposal and repository; fuel fabrication. Operates land–disposal facilities at the Pelindaba site.

SPAIN

NUCLEAR POWER

Continue to operate existing nuclear power plants. Moratorium on new nuclear power plant construction has been in place for several years (confirmed 1991).

INDUSTRIAL FUEL CYCLE

Policy: Once though fuel cycle for LWRs; no domestic reprocessing and no further contracts for foreign reprocessing, except GCR fuel (Vandellos I).

Waste Management Strategy: Store spent fuels at the reactor sites for at least 10 years; reracking in some reactor pools and dry storage in dual–purpose casks planned to provide additional capacity until geologic repository is ready to receive HLW (SF); granite, salt, and clay are being considered as host rock for repository; shallow–land burial of LLW in fully engineered facility at El Cabril, province of Cordoba (in operation since 10/92).

Centro de Investigaciones Energeticas, Medio Ambientales
y Tecnologicas (CIEMAT)
Avenida Complutense 22
Ciudad Universitaria
28040 Madrid

Organized into four research institutes; nuclear technology (R&D on nuclear fuel cycle, decommissioning, material sciences, and safety analyses); fundamental research; radiological protection and environment; renewable energies.

Consejo de Seguridad Nuclear (CSN)
Justo Dorado, 11
E–28020 Madrid, Spain

Independent body, responsible to Parliament, with regulatory powers on nuclear safety and radiation protection matters.

Empresa Nacional de Residuos Radiactivos S.A. (ENRESA)
Emilio Vargas, 7
E–28043 Madrid, Spain

Provide waste management services and disposal facilities to all Spanish nuclear companies and radwaste producers; responsible to the Ministry of Industry and Energy; funded by CIEMAT (80%) and the National Institute of Industry (20%).

SWEDEN

NUCLEAR POWER

Policy: Phase out all nuclear plants by the year 2010 at the latest; changing this policy will require a new decision by Parliament.

INDUSTRIAL FUEL CYCLE

Policy: Direct disposal of spent fuel; no Pu recycle is planned; costs for waste management and future decommissioning of nuclear power plants are paid by fees collected from the nuclear utilities.

Waste Management Strategy: Store spent fuel for 30–40 years in an underground pool storage facility; encapsulate spent fuel in a highly corrosion-resistant canister; emplace in a deep geologic (crystalline rock) repository.

Svensk Karnbranslehantering AB (SKB)
Box 5864
102 48 Stockholm, Sweden

Coordinate and arrange for nuclear fuel supply and reprocessing services for all Swedish nuclear power reactors; manage and fund R&D for the back end of the fuel cycle; responsible for design, construction, and operation of all necessary storage and disposal facilities; demonstrate that SF and other long-lived wastes can be disposed of safely and permanently; provide transportation of SF outside reactor sites.

SWITZERLAND

NUCLEAR POWER

Policy: Federal government is in favor of nuclear power, but local opposition has delayed its expansion.

INDUSTRIAL FUEL CYCLE

Policy: Purchase most services from other countries, including reprocessing of spent fuels; recycle Pu to LWRs or FBRs.

Waste Management Strategy: Develop two waste repositories: a horizontally accessed rock cavern in a host rock with considerable overburden for LLW/ILW, and a deep repository in crystalline rock or sedimentary formations for HLW glass and unreprocessed SF elements and alpha wastes; interim storage of all waste at common center until repositories available; ocean-dumping of LLW discontinued in 1982.

NAGRA/CEDRA/CISRA (National Cooperative for the Disposal of Radioactive Waste)
Hardstrasse 73
5430 Wettingen, Switzerland
Provide for safe disposal of radioactive wastes produced by the Swiss nuclear industry; funded by utilities and government.

Paul Scherrer Institut (PSI)
Wurenlingen/Villigen
5232 Villigen, Switzerland
Federal(Department of Interior) institute for reactor and nuclear R&D. Incineration of TRU wastes; modeling of radionuclide migration through heterogeneous geologic media; chemical behavior of radionuclides during migration; transport of radionuclides through the biosphere; natural analogue studies; hydrological studies; sorption constants on different rocks; immobilization of LLW/ILW in cement; leaching rates on LLW/ILW forms; and long-term corrosion tests on waste package materials.

TAIWAN

NUCLEAR POWER

Policy: Plan for nuclear power to meet rapidly growing demand for electric energy; continue with nuclear power at about 1/3 of total electricity.

INDUSTRIAL FUEL CYCLE

Policy: Purchase fuel materials and enrichment; develop indigenous fuel production capability: UF_6 conversion; UO_2 pellet preparation; fuel hardware fabrication; fuel assembly.

Waste Management Strategy: Evaluating spent fuel/HLW interim storage options; may reprocess (in other countries); maximize existing SF pool storage capacity by reracking; build MRS facility at existing reactor site for interim storage until final disposal in geologic repository; LLW stored in National Waste Storage Facility on nearby Orchid Island; LLW/ILW will eventually be disposed of on the sea floor, if internationally acceptable, or in a shallow-land facility.

Atomic Energy Council (AEC)
67, Lane 144
Keelung Road, Section 4
Taipei 10772, Taiwan
Regulatory agency

Institute of Nuclear Energy Research (INER)
P.O. Box 3
Lung-Tan 32500, Taiwan
Solvent extraction technology; yellowcake conversion to UO_2; cement and thermoplastic waste forms for reactor wastes; HLW conditioning processes; burial of LLW.

UKRAINE

Ministry of Energetics & Electrification of Ukraine
Kotkiv 252 001
Kreshtatik 30, Kiev 931, Ukraine

State Committee of Ukraine for Nuclear Energy
Arsenal Ulitsa, 252 001
Kiev 931, Ukraine

NUCLEAR POLICY: Due to the disaster caused by the explosion of Reactor No. 4 at the Chernobyl site, no further RBMK nuclear plant construction has been contemplated. Unit No. 4 was encapsulated in close to 500,000 tons of concrete. Reportedly, two of the other three units at Chernobyl are currently operating, even though they are the same type as Unit No. 4 (RBMK models). Unit No. 2 was shutdown in 1991, after an extensive turbine fire. The government would like to shut the other two down, but needs the energy, particularly since the country has a heavy demand for electric heating purposes. These four plants at Chernobyl will be replaced by a gas-fired plant, under a 1995 agreement with a consortium led by Asea Brown Boveri (ABB). Ukraine also has PWRs.

UNITED KINGDOM

NUCLEAR POWER

Policy: Continue nuclear power as a significant element of total electricity production; substantially based, to date, on gas-cooled reactors, but now diversifying to PWRs; eventual active FBR pursuit expected.

INDUSTRIAL FUEL CYCLE

Policy: Reprocess and recycle U to AGR and LWR systems; develop and maintain complete fuel cycle capability (UF_6 conversion, enrichment, UO_2 and MOX fuel fabrication, spent fuel reprocessing); sell fuel cycle services abroad

Waste Management Strategy: Reprocess spent Magnox/AGR fuels as rapidly as plant capacity permits; reprocess other thermal reactor fuel after several years' cooling; vitrify HLW (French process); long-term interim storage of HLW glass for at least 50 years before disposal; shallow-land burial of LLW currently; future deep-land disposal of LLW and ILW.

AEA Technology
11 Charles II Street
London SW1Y 4QP, United Kingdom
Government-owned nuclear research and applications agency, since 1986 operating on a fully commercial basis; provides contract R&D, technical and engineering services to governments and companies in the U.K. and worldwide.

AEA Decommissioning and Radwaste
Winfrith Technology Center
Dorchester, Dorset DT2 8DH, United Kingdom
Decommissioning of all types of nuclear facilities; all aspects of radioactive waste storage, processing, transport, and disposal; decontamination technology and robotic handling

AEA Environment & Energy
Harwell Laboratory
Oxfordshire OX11 0RA, United Kingdom
R&D and consulting services to industry and regulatory bodies covering pollution control technology, waste management, and regional and global environmental impacts.

AEA Fuel Services
AEA Technology Dounreay
Caithness KW14 7TZ, United Kingdom

Fuel reprocessing, special fuel manufacturing and testing, laser enrichment, waste conditioning, R&D in radioactive equipment and safeguards. Conditions liquid wastes by cementation.

Atomic Weapons Establishment
Aldermaston, Reading RG7 4PR, United Kingdom
Waste management

British Nuclear Fuels plc
Sellafield, Seascale
Cumbria CA20 1PG, United Kingdom
Provide spent fuel management services, including storage, reprocessing and waste management; transport of SF/wastes and complete fuel cycle services. Utilizes a number of waste treatment, encapsulation, and disposal processes.

H.M. Inspectorate of Pollution (HMIP)
Department of the Environmental
Romney House, 43 Marsham Street
London SW1P 3PY, United Kingdom
Administer U.K. waste management programs; fund and coordinate waste treatment and waste isolation R&D at Harwell, BGS, and NRPB; regulate discharge of radioactive materials to the environment.

Ministry of Agriculture, Fisheries and Food (MAFF)
Ergon House, Room 231
c/o Nobel House
17 Smith Square
London SW1P 3JR, United Kingdom
Regulate, jointly with HMIP, management of waste prior to disposal.

U.K. Nirex Ltd.
Curie Avenue, Harwell
Didcot, Oxon OX11 0RH, United Kingdom
Commission/manage research and development to propose (to the government) a site suitable for a deep repository for LLW/ILW; construct and operate the repository; continue necessary R&D on long-term waste emplacement.

NOTE: The information contained in this Appendix (with the exception of Ukraine) was abstracted from the *International Nuclear Waste Management Fact Book* (1994), prepared by the International Program Support Office of the Pacific Northwest Laboratory, Richland, Washington. It contains substantial additional information that is of value to anyone interested in foreign nuclear activities. It can be ordered from NTIS (DE94018005/GAR).

Other countries with nuclear power reactors include:

Bulgaria
Cuba
Czech Republic
Hungary
Kazakhstan
Korea (North)
Lithuania
Mexico
Philippines
Rumania
Slovakia
Slovenia

NOTE: A "World List of Nuclear Power Plants, Operable, Under Construction, or On Order;" is contained in the March 1995 issue of *Nuclear News*.

APPENDIX IV

ACRONYMS

ABS	Aqueous Biphasic Separation
ACM	Asbestos Containing Material
AEC	Atomic Energy Commission
AECL	Atomic Energy of Canada Limited
AFR	Away-From-Reactor
AGR	Advanced Gas-Cooled Reactor
AL	DOE Albuquerque Operations Office
ALARA	As Low As Reasonable Achievable (or Allowable)
AMES	Ames Laboratory
ANDRA	National Radioactive Waste Management Agency (France)
ANL	Argonne National Laboratory
ARAR	Applicable or Relevant and Appropriate Requirement
ARPA	Advanced Research Projects Agency
ATR	Advanced Thermal Reactor
BDAT	Best Demonstrated Available Technologies
BNL	Brookhaven National Laboratory
BWID	Buried Waste Integrated Demonstration
BWR	Boiling Water Reactor
CAA	Clean Air Act
CANDU	Canadian Deuterium Uranium Reactor
CC&AT	Cross Cutting and Advanced Technology
CDU	Cesium Demonstration Unit
CEA	Atomic Energy Commission (France)
CEPOD	Catalytic Electrochemical Plutonium Oxide Dissolution
CERCLA	Comprehensive Environmental Response, Compensation, and Liability Act of 1980
CFR	Code of Federal Regulations
CH	DOE Chicago Operations Office
CMST	Characterization, Monitoring, and Sensor Technology
CMSTIP	Characterization, Monitoring, and Sensor Technology Integrated Program

CMU	Carnegie Mellon University
COCA	Consent Order and Compliance Agreement
CPU	Compact Processing Unit
CRADA	Cooperative Research and Development Agreement
CSTP	Conceptual Site Treatment Plan
CTMP	Comprehensive Treatment and Management Plan
CUA	Catholic University of America
CWA	Clean Water Act
CWCO	Catalytic Wet Chemical Oxidation
CWL	Chemical Waste Landfill
D&D	Decontamination and Decommissioning
DIAL	Diagnostic and Control Analytical Laboratory
DD&D	Deactivate, Decontaminate, and Decommission
DNAPL	Dense Non-Aqueous Phase Liquid
DOD	U.S. Department of Defense
DOE	U.S. Department of Energy
DOIT	Development of On-Site Innovative Technologies
DP	Office of Defense Programs
DPM	Defense Prioritization Model
DSTP	Draft Site Treatment Plan
DT&E	Demonstration, Testing, and Evaluation
DWPF	Defense Waste Processing Facility
EBWR	Experimental Boiling Water Reactor
EH	Environmental Safety and Health
EM	Office of Environmental Restoration and Waste Management
EM-30	Office of Waste Management
EM-40	Office of Environmental Restoration
EM-50	Office of Technology Development
EM-60	Office of Facility Transition and Management
EPA	U.S. Environmental Protection Agency
EPRI	Electric Power Research Institute
ER	Environmental Restoration
ES&H	Environmental Safety and Health
ESPIP	Efficient Separations and Processing Integrated Program
FBR	Fast Breeder Reactor
FEMP	Fernald Environmental Management Project
FMPC	Feed Materials Production Center
FN	DOE Fernald Operations Office
FR	Federal Register
FUSRAP	Formerly Utilized Sites Remedial Action Program
FY	Fiscal Year
GAC	Granulated Activated Carbon
GAO	General Accounting Office
GCR	Gas Cooled, Graphite Moderated Reactor

GDP	Gaseous Diffusion Plant
GJPO	Grand Junction Project Office
GOCO	Government-Owned, Contractor-Operated
GPM	Gallons Per Minute
GTCCLW	Greater Than Class C Low-Level Waste
HEPA	High-Efficiency Particulate Air
HLLW	High-Level Liquid Waste
HLW	High-Level Waste
HMCSP	Heavy Metals Contaminated Soils Project
HMTA	Hazardous Materials Transportation Act
HWPP	Hazardous Waste Pilot Plant
HWR	Heavy Water Reactor
HWVP	Hanford Waste Vitrification Plant
IAG	Interagency Agreement
ICPP	Idaho Chemical Processing Plant
ID	Integrated Demonstration
IIA	Innovative Investment Area
ILW	Intermediate Level Waste
INEL	Idaho National Engineering Laboratory
IOO	DOE Idaho Operations Office
IP	Integrated Program
ISV	In Situ Vitrification
IWIF	Idaho Waste Immobilization Facility
IWPF	Idaho Waste Processing Facility
KC	Kansas City (Plant)
LANL	Los Alamos National Laboratory
LBL	Lawrence Berkeley Laboratory
LDR	Land-Disposal Restrictions
LDUA	Light Duty Utility Arm
LLNL	Lawrence Livermore National Laboratory
LLW	Low-Level Waste
LWR	Light Water Reactor
MAWS	Minimum Additive Waste Stabilization
MD	Mound (Plant)
METC	Morgantown Energy Technology Center
MIT	Massachusetts Institute of Technology
MLLW	Mixed Low-Level Waste
M&O	Management and Operations
MOX	Mixed Oxide
MRS	Monitored, Retrievable Storage
MTRU	Mixed Transuranic Waste
MVST	Melton Valley Storage Tank(s)
MWIP	Mixed Waste Integrated Program
MWL	Mixed Waste Landfill
MWLID	Mixed Waste Landfill Integrated Demonstration

MWPP	Mixed Waste Pilot Project
MWTP	Mixed Waste Treatment Project
NAC	Nitrate to Ammonia and Ceramic
NAGRA	National Cooperative for Disposal of Radioactive Waste (Switzerland)
NAPLs	Non-Aqueous Phase Liquids
NARM	Natural and Accelerator Produced Radioactive Materials
NDA	Non-Destructive Assay
NEPA	National Environmental Policy Act
NMP	Nuclear Materials Production
NMP&M	Nuclear Materials Production and Manufacturing
NORM	Naturally Occurring Radioactive Materials
NPDES	National Pollutant Discharge Elimination System
NPL	National Priorities List
NRC	Nuclear Regulatory Commission
NREL	National Renewable Energy Laboratory
NRTS	National Reactor Testing Station
NSF	National Science Foundation
NTS	Nevada Test Site
NV	DOE Nevada Operations Office
NWC	Nuclear Weapons Complex
NWCF	New Waste Calcining Facility
NWPA	Nuclear Waste Policy Act of 1982
OCRWM	Office of Civilian Radioactive Waste Management
OER	Office of Energy Research
OMB	Office of Management and Budget
OR	DOE Oak Ridge Operations Office
ORNL	Oak Ridge National Laboratory
OTA	Office of Technology Assessment
OTD	Office of Technology Development
PCT	Product Consistency Test
PEG	Polyethylene Glycol
PEIS	Programmatic Environmental Impact Statement
PEL	Permissible Exposure Limit
PETC	Pittsburgh Energy Technology Center
PHWR	Pressurized Heavy Water Reactor
PI	Principal Investigator
PLWR	Pressurized Light Water Reactor
PNL	Pacific Northwest Laboratory
PPE	Personal Protection Equipment
PRDA	Program Research and Development Announcement
PSU	Pennsylvania State University
PUREX	Plutonium-Uranium Extraction
PWR	Pressurized Water Reactor

PX	Pantex (Plant)
RA	Remedial Action
R&D	Research and Development
RAD	Radiation Adsorbed Dose
RCRA	Resource Conversation and Recovery Act of 1976
RCS	Remote Characterization System
RDDT&E	Research, Development, Demonstration, Testing and Evaluation
RF	DOE Rocky Flats Operations Office
RFETS	Rocky Flats Environmental Technology Site
RFP	Request for Proposals
RFP	Rocky Flats Plant
RI	Remedial Investigation
RI/FS	Remedial Investigation/Feasibility Study
RL	Richland Operations Office
ROA	Research Opportunity Announcement
ROD	Record of Decision
RSL	Remote Sensing Laboratory
RSM	Radioactive Scrap Metal
RTDP	Robotics Technology Development Program
RW	Radioactive Waste Management
RWMC	Radioactive Waste Management Complex
S&M	Surveillance and Maintenance
SBR	Fast Breeder Reactor (Europe)
SBIR	Small Business Innovation Research
SB-TIP	Small Business Technology Integration Program
SCWO	Supercritical Water Oxidation
SF	Spent Fuel
SIP	Stabilization in Place
SITE	Superfund Innovative Technology Evaluation
SKB	Swedish Nuclear Fuel and Waste Management Company
SKI	Swedish Nuclear Power Inspectorate
SNL	Sandia National Laboratory—New Mexico
SNLA	Sandia National Laboratory—Albuquerque
SNLCA	Sandia National Laboratory—California
SNLL	Sandia National Laboratory—Livermore
SOS	Stabilization On Site
SR	DOE Savannah River Operations Office
SRP	Savannah River Plant
SRTC	Savannah River Technology Center
SRS	Savannah River Site
STL	Special Technologies Laboratory
SWMU	Solid Waste Management Unit
TBD	To Be Determined

TCLP	Toxicity Concentrate Leachate Procedure
TLV	Threshold Limit Value
TPA	Tri-Party Agreement
TPO	Technical Program Officer
TRU	Transuranic Waste
TRUEX	Transuranic Extraction
TSCA	Toxic Substances Control Act
TSG	Technical Support Groups
TTP	Technical Task Plan
TWRS	Tank Waste Remediation System
UMTRA	Uranium Mill Tailings Remedial Action Project
UMTRCA	Uranium Mill Tailings Radiation Control Act
UST	Underground Storage Tank
UST-ID	Underground Storage Tanks Integrated Demonstration
VOC	Volatile Organic Compound
WHC	Westinghouse Hanford Company
WIF	Waste Immobilization Facility
WIPP	Waste Isolation Pilot Plant
WM	Waste Management
WSRC	Westinghouse Savannah River Company
WVDP	West Valley Demonstration Project

BIBLIOGRAPHY

Major Sources of Information

Overall/General/History/Locations/Plans

1. Sites Contaminated and Potentially Contaminated with Radioactivity in the United States; S. Cohen & Associates, Inc. and Roy F. Weston; EPA68D90107; 2/91.
2. Environmental Management 1995, Progress and Plans of the Environmental Management Program; U.S. Department of Energy; DOE/EM-0228; 2/95.
3. Environmental Management 1994, Progress and Plans of the Environmental Management Program; U.S. Department of Energy; DOE/EM-0119; 2/94.
4. Complex Cleanup, The Environmental Legacy of Nuclear Weapons Production; Office of Technology Assessment; OTA-0-484; 2/91.
5. Environmental Restoration and Waste Management Five-Year Plan, Fiscal Years 1994-1998, Volume 1; U.S. Department of Energy; DOE/S-00097P Vol. 1; 1/93.
6. Environmental Restoration and Waste Management Five-Year Plan, Fiscal Years 1994-1998, Installation Summaries; U.S. Department of Energy; DOE/S-00097P Vol. 2; 1/93.
7. Environmental Restoration and Waste Management, Five-Year Plan, Fiscal Years 1993-1997; U.S. Department of Energy; DOE/S-0090P; 8/91.
8. Nuclear Weapons Complex Reconfiguration Study (no longer applicable); U.S. Department of Energy; DOE/DP-0083; 1/91.
9. Implementation Plan for the Nuclear Weapons Complex Reconfiguration Programmatic Environmental Impact Statement; U.S. Department of Energy; DOE/EIS-0161IP; 2/92.
10. Implementation Plan, Tritium Supply and Recycling Programmatic Environmental Impact Statement Revised; U.S. Department of Energy, Office of Reconfiguration; DOE/EIS-0161IPREV; 1/95.
11. Estimating the Cold War Mortgage, The 1995 Baseline Environmental Management Report; U.S. Department of Energy, Office of Environmental Management; DOE/EM-0232 Vol. I; 3/95.

12. Estimating the Cold War Mortgage, The 1995 Baseline Environmental Management Report; U.S. Department of Energy; DOE/EM-0232 Vol. II; Site Summaries; 3/95.
13. Radioactive Waste Disposal, An Environmental Perspective; U.S. EPA; EPA 402-K-94-001; 8/94.
14. Matching Leading DOE Waste Management and Environmental Restoration Needs with Foreign-Based Technologies, Sandia National Laboratories, Golder Associates, Inc. and Applied Sciences Laboratory, Inc.; TTP No. AL-2233-05; 10/92.
15. Environmental Consequences of, and Control Processes for, Energy Technologies; Argonne National Laboratory; Noyes Data; 1988.
16. Radioactive Waste Definitions, Standards, Criteria, and Approaches in the United States of America; Martin Marietta Energy Systems, Inc.; CONF-9006233-1.
17. Nuclear Terms Handbook 1993; U.S. Department of Energy; DOE-94016248; 1993.
18. Nuclear Science, Developing Technology to Reduce Radioactive Waste May Take Decades and be Costly; GAO; GAO/RCED-94-16; 12/93.
19. Nuclear Health and Safety, More Can be Done to Better Control Environmental Restoration Costs; GAO; GAO/RCED-92-71; 4/92.
20. Military Bases, Environmental Impact at Closing Installations; GAO; GAO/NSIAD-95-70; 2/95.
21. Nuclear Cleanup, Completion of Standards and Effectiveness of Land Use Planning are Uncertain; GAO; GAO/RCED-94-144; 8/94.
22. Nuclear Weapons Complex, Weaknesses in DOE's Nonnuclear Consolidation Plan; GAO; GAO/RCED-93-56; 11/92.
23. Nuclear Science, Developing Technology to Reduce Radioactive Waste May Take Decades and be Costly; GAO; GAO/RCED-94-16S; 12/93.
24. Nuclear Waste, Change in Test Strategy Sound, But DOE Overstated Savings; GAO; GAO/RCED-95-44; 12/94.
25. National Laboratories Need Clearer Missions and Better Management; GAO; GAO/RCED-95-10; 1/95.
26. Environmental Cleanup, Better Data Needed for Radioactively Contaminated Defense Sites; GAO; GAO/NSIAD-94-168; 8/94.
27. Nuclear Cleanup, Difficulties in Coordinating Activities Under Two Environmental Laws; GAO; GAO/RCED-95-66; 12/94.
28. Cleaning Up Inactive Facilities Will Be Difficult; GAO; GAO/RCED-93-149; 6/93.
29. Management Changes Needed to Expand Use of Innovative Cleanup Technologies; GAO; GAO/RCED-94-205; 8/94.
30. Uranium Enrichment, Observations on the Privatization of the United States Enrichment Corporation; GAO; GAO/T-RCED-95-116; 2/95.

31. Research and Agency Missions Need Reevaluating; GAO; GAO/T-RCED-95-105; 2/95.
32. DOE, Need to Reevaluate Its Role and Missions; GAO; GAO/T-RCED-95-85; 1/95.
33. Nuclear Weapons Complex, Establishing a National Risk-Based Strategy for Cleanup; GAO; GAO/T-RCED-95-120; 3/95.
34. International Nuclear Waste Management Fact Book; Department of Energy; PNL-9450-1; 5/94.

Specific Sites

35. Hanford Site Waste Management Plan; U.S. Department of Energy; DE90-008175; 12/89.
36. Hanford Site Waste Minimization and Pollution Prevention Awareness Program Plan, U.S. Department of Energy; DOE/RL—91—31; 5/91.
37. Nuclear Waste, Hanford Single-Shell Tank Leaks Greater Than Estimated; GAO; GAO/RCED-91-177; 8/91.
38. Nuclear Waste, Problems and Delays With Characterizing Hanford's Single-Shell Tank Waste; GAO; GAO/RCED-91-118; 4/91.
39. Nuclear Waste, Hanford's Well-Drilling Costs Can Be Reduced; GAO; GAO/RCED-93-71; 3/93.
40. Nuclear Waste, Hanford Tank Waste Program Needs Cost, Schedule, and Management Changes; GAO; GAO/RCED-93-99; 3/93.
41. Nuclear Waste, Pretreatment Modifications at DOE Hanford's B Plant Should Be Stopped; GAO; GAO/RCED-91-165; 6/91.
42. Richland Operations Office, Technology Summary; U.S. Department of Energy; DOE/EM-D171P; 5/94.
43. Researchers Take Up Environmental Challenge at Hanford; C&EN; 6/93.
44. Nuclear Health and Safety, Environmental Problems at DOE's Idaho National Engineering Laboratory; GAO; GAO/RCED-91-56; 2/91.
45. Evaluation of Technologies for Remediation of Disposed Radioactive and Hazardous Wastes in a Facility at the Idaho National Engineering Laboratory; U.S. Department of Energy; EGG-M—89371; 4/90.
46. Active Waste Disposal Monitoring at the Radioactive Waste Management Complex, Idaho National Engineering Laboratory; U.S. Department of Energy; EGG-WM-9276; 10/90.
47. Idaho National Engineering Laboratory Conceptual Site Treatment Plan; U.S. Department of Energy; DOE/ID-10453; 10/93.
48. Rocky Flats Transition Plan; U.S. Department of Energy; DOE/EM-0079; 7/92.
49. Nuclear Health and Safety, Increased Rating Results in Award Fee to Rocky Flats Contractor; GAO; GAO/RCED-92-162; 3/92.
50. Project Management at the Rocky Flats Plant Needs Improvement; GAO; GAO/RCED-93-32; 10/92.

51. Sandia Laboratories/Production Agency Weapon Waste Minimization Plan; Sandia Laboratories; SAND91-8225; 7/91.
52. Development and Pilot Demonstration Program of a Waste Minimization Plan at Argonne National Laboratory, Argonne National Laboratory; ANL/CP—73954; 9/91.
53. LLNL Waste Minimization Program Plan; Bechtel National, Inc.; UCRL—21215-Rev. 1; 9/89.
54. Oak Ridge National Laboratory Technology Logic Diagram, Volume 1, Technology Evaluation, Part B, Remediation Action; Martin Marietta Energy Systems, Inc.; ORNL/M-2751/V1/Pt. B; 9/93.
55. Oak Ridge National Laboratory Technology Logic Diagram, Volume 1, Technology Evaluation, Part C, Waste Management; Martin Marietta Energy Systems, Inc.; ORNL/M-2751/V1/Pt. C; 9/93.

Technology Summaries

56. Pollution Prevention Program, Technology Summary; U.S. Department of Energy; DOE/EM-0137P; 2/94.
57. Mixed Waste Integrated Program, Technology Summary; U.S. Department of Energy; DOE/EM-0125P; 2/94.
58. Mixed Waste Landfill Integrated Demonstration, Technology Summary; U.S. Department of Energy; DOE/EM-0128P; 2/94.
59. Uranium in Soils Integrated Demonstration, Technology Summary; U.S. Department of Energy; DOE/EM-0148P; 3/94.
60. Minimum Additive Waste Stabilization, Technology Summary; U.S. Department of Energy; DOE/EM-0124P; 2/94.
61. Buried Waste Integrated Demonstration, Technology Summary; U.S. Department of Energy; DOE/EM-0149P; 3/94.
62. Supercritical Water Oxidation Program, Technology Summary, U.S. Department of Energy; DOE/EM-0121P; 2/94.
63. Robotics Technology Development Program, Technology Summary; U.S. Department of Energy; DOE/EM-0127P; 2/94.
64. In Situ Remediation Integrated Program, Technology Summary; U.S. Department of Energy; DOE/EM-0134P; 2/94.
65. Underground Storage Tank Integrated Demonstration, Technology Summary; U.S. Department of Energy; DOE/EM-0122P; 2/94.
66. Efficient Separations and Processing Integrated Program, Technology Summary; U.S. Department of Energy; DOE/EM-0126P; 2/94.
67. Rocky Flats Compliance Program, Technology Summary; U.S. Department of Energy; DOE/EM-0123P; 2/94.
68. Characterization, Monitoring, and Sensor Technology Integrated Program, Technology Summary; U.S. Department of Energy; DOE/EM-0156T; 4/94.

Technology Compendiums/Overviews/Research Needs

69. Approaches for the Remediation of Federal Facility Sites Contaminated with Explosive or Radioactive Wastes; U.S. EPA; EPA/625/R-93/013; 9/93.
70. Technologies to Remediate Hazardous Waste Sites; Pacific Northwest Laboratory; PNL-SA—18030; 3/90.
71. Basic Research Needs for Management and Disposal of DOE Wastes; U.S. Department of Energy; DOE/ER-0492T; 4/91.
72. Technologies Applicable for the Remediation of Superfund Radiation Sites; Foster Wheeler Enviresponse, Inc. and U.S. EPA.
73. Assessment of Technologies for the Remediation of Radioactively Contaminated Superfund Sites; U.S. EPA; EPA/540/2-90/001; 1/90.
74. Synopses of Federal Demonstrations of Innovative Site Remediation Technologies, Third Edition; Member Agencies of the Federal Remediation Technologies Roundtable; EPA/542/B-93/009; 10/93.
75. Technological Approaches to the Cleanup of Radiologically Contaminated Superfund Sites; U.S. EPA; EPA/540/2-88/002. 8/88.
76. Technology Catalogue, First Edition; U.S. Department of Energy; DOE/EM-0138P; 2/94.
77. Innovation Investment Area, Technology Summary; U.S. Department of Energy; DOE/EM-0146P; 3/94.
78. FY 1995 Technology Development Needs Summary, U.S. Department of Energy; DOE/EM-0147P; 3/94.
79. FY 1994 Program Summary; U.S. Department of Energy; DOE/EM-0216; 10/94.
80. FY 1993 Program Summary; U.S. Department of Energy; DOE/EM-0109P; 2/94.
81. Model Conference on Waste Management and Environmental Restoration; U.S. Department of Energy; CONF-9010166; 10/90.

Technologies

82. High Level Radioactive Waste, Doing Something About It; E. I. du Pont de Nemours & Co.; DP-1777; 3/89.
83. Functions and Requirements for a Cesium Demonstration Unit; Westinghouse Hanford Company; WHC-SD-W236B-FRD-001; 4/94.
84. Treatment of Radioactive Wastes from DOE Underground Storage Tanks; Oak Ridge National Laboratory; CONF-940815-37; 8/94.
85. Hanford Tank Wastes, Salt Splitting, FY92 Activities; Lawrence Livermore National Laboratory; UCRL-ID—111983; 9/92.
86. Citrate Based 'TALSPEAK' Lanthanide-Actinide Separation Process; Oak Ridge National Laboratory; ORNL/TM-12785; 9/94.
87. Magnetic Separation for Environmental Remediation; Los Alamos National Laboratory; CONF-940301-44; 3/94.
88. Radionuclide Separations for the Reduction of High-Level Waste Volume; Battelle Pacific Northwest Labs; CONF-930906-22; 9/93.

89. Safety Evaluation of Interim Stabilization of Nonstabilized Single-Shell Watch List Tanks; Westinghouse Hanford Company; WHC-SD-WM-TI-656; 9/94.
90. Functions and Requirements for Single-Shell Tank Leakage Mitigation; Westinghouse Hanford Company; WHC-SD-WM-FRD-019; 9/94.
91. Efficient Separations and Processing Integrated Program; Battelle Pacific Northwest Labs; CONF-940815-78; 8/94.
92. Waste Partitioning and Transmutation as a Means Towards Long-Term Risk Reduction; Forschungszentrum Juelich G.m.b.H.; JUEL-2818; 9/93.
93. Extensive Separations (Clean) Processing Strategy Compared to TRUEX Strategy and Sludge Wash Ion Exchange; Westinghouse Hanford Company; WHC-EP-0791; 8/94.
94. Advantages of Separation of High-Level Nuclear Waste into Two Waste Categories; Technische University Delft; ORS-1987-1; 10/87.
95. High Level Waste Repository System; Massachusetts Institute of Technology; MITNE—281; 5/88.
96. Subsurface Barrier Demonstration Test Strategy and Performance Specification; Westinghouse Hanford Company; WHC-SD-WM-SP-001; 5/94.
97. Review of Potential Subsurface Permeable Barrier Emplacement and Monitoring Technologies; Pacific Northwest Laboratory; PNL-9335; 2/94.
98. An Overview of In Situ Waste Treatment Technologies; Idaho National Engineering Laboratory; EGG-M-92-342; 1992.
99. Land Containment System: Horizontal Grout Barrier: A Method for In Situ Waste Management; Fernald Environmental Restoration Management Corporation; CONF-940632-24; 6/94.
100. Uranium Removal from Soils, An Overview from the Uranium in Soils Integrated Demonstration Program; Oak Ridge National Laboratory; CONF-940225-121; 1994.
101. An Overview of In Situ Waste Treatment Technologies; EG&G Idaho, Inc.; EGG-M-92-342; Spectrum 1992.
102. Buried Waste Integrated Demonstration FY94 Deployment Plan; Idaho National Engineering Lab; EGG-WTD-11156; 5/94.
103. Handbook on In Situ Treatment of Hazardous Waste-Contaminated Soils; U.S. EPA; EPA/540/2-90/002; 1990.
104. An Overview of the Department of Energy's Soil Washington Workshop; U.S. Department of Energy; DOE/NV-348; 9/91.
105. Soil Washing and Radioactive Contamination; Westinghouse Idaho Nuclear Company, Inc.; WINCO-11812; 3/92.
106. Soil Washing: A Preliminary Assessment of Its Applicability to Hanford; Battelle Pacific Northwest Laboratories; PNL-7787; 9/91.
107. Technical Review of Molten Salt Oxidation; U.S. Department of Energy; DOE/EM-0139P; 12/93.

108. Vitrification Technologies for Treatment of Hazardous and Radioactive Waste; U.S. EPA; EPA/625/R-92/002; 5/92.
109. Radioactive Waste Vitrification: A Review; Oak Ridge National Laboratory; ORNL/TM—11292; 8/89.
110. Hanford Waste Vitrification Plant, Dangerous Waste Permit Application; U.S. Department of Energy; DOE/RL—89-02; 7/89.
111. ISV Technology Development Plan for Buried Waste; Idaho National Engineering Laboratory; EGG-WTD-10325; 7/92.
112. In Situ Vitrification Technology; Geosafe Corporation; EPA; EPA 540/R-94-520a; 11/94.
113. In Situ Vitrification Treatment; Science Applications International Corporation; EPA /540/S-94/504; 10/94.
114. In Situ Vitrification, Geosafe Corporation; EPA/540/MR-94/520; 8/94.
115. Development and Demonstration of Treatment Technologies for the Processing of U.S. Department of Energy Mixed Waste; Oak Ridge National Laboratory; ORNL/TM-12679; 1/94.
116. Assessing Mixed Waste Treatment Technologies; Oak Ridge National Laboratory; U.S. Department of Energy; CONF-9404160—1; 1994.
117. Proceedings of the International Topical Meeting on Nuclear and Hazardous Waste Management Spectrum '92; American Nuclear Society, Inc.; U.S. Department of Energy; Atomic Energy Society of Japan; 8/92.
118. Radiation and Mixed Waste Incineration, Background Information Document, Volume 1: Technology; U.S. EPA; EPA 520/1-91-010-1; 5/91.
119. Radiation and Mixed Waste Incineration, Background Information Document, Volume 2: Risk of Radiation Exposure; U.S. EPA; EPA 520/1-91-010-2; 5/91.
120. The Incineration of Low-Level Radioactive Waste; U.S. Nuclear Regulatory Commission; NUREG-1393; 6/90.
121. Quantum-Catalytic Extraction Process Application to Mixed Waste Processing; Molten Metal Technology; CONF-940815—59; 8/94.
122. Status of Activities Low-Level Radioactive Waste Management in the United States; Idaho National Engineering Laboratory; EGG-M-89195; 1990.
123. Foreign Trip Report; Oak Ridge National Laboratory; ORNL/FTR-2015; 12/89.
124. Analytical Support for a New Low Level-Radioactive Wastewater Treatment Plant (U); Westinghouse Hanford Corporation; WSRC-RP-89-1328; 1/90.
125. Laboratory Development of Methods for Centralized Treatment of Liquid Low-Level Waste at Oak Ridge National Laboratory; Oak Ridge National Laboratory; ORNL/TM-12786; 10/94.
126. Project Report for the Commercial Disposal of Mixed Low-Level Waste Debris; Idaho National Engineering Laboratory; EGG-WM-11342; 5/94.

127. Directions in Low-Level Radioactive Waste Management, A Brief History of Commercial Low-Level Radioactive Waste Disposal; EG&G Idaho, Inc.; DOE/LLW-103-Rev. 1; 8/94.
128. Advanced Technology for Disposal of Low-Level Radioactive/Waste; Chem-Nuclear Systems, Inc.
129. Impact of Technology Applications to the Management of Low-Level Radioactive Wastes; Argonne National Laboratory; CONF-8906240—1.
130. Low-Level Radioactive Waste Volume Reduction and Stabilization Technologies Resource Manual; Ebasco Services Incorporated; DOE/LLW—76T; 12/88.
131. Volume 1, Technology Evaluation, Part A, Decontamination and Decommissioning; Martin Marietta Energy Systems, Inc.; ORNL/M-2751/V1/Pt. A; 9/93.
132. Waste Recycling Workshop Proceedings; Ohio State University; FEMP/Sub-084; 4/94.
133. Demonstration of Technologies to Remove Contamination From Goundwater; Westinghouse Hanford Corporation; WHC-SA—0367; 3/89.
134. Colloid Polishing Filter Method; Filter Flow Technology, Inc.; EPA 540/R-94-501a; 7/94.
135. Landfill Stabilization Focus Area; U.S. Department of Energy; DOE/EM-0251; 6/95.
136. Contaminant Plumes Containment and Remediation Focus Area; U.S. Department of Energy; DOE/EM-0248; 6/95.
137. Decontamination and Decommissioning Focus Area; U.S. Department of Energy; DOE/EM-0253; 6/95.
138. Radioactive Tank Waste Remediation Focus Area; U.S. Department of Energy; DOE/EM-0255; 6/95.
139. Efficient Separations and Processing Crosscutting Program; U.S. Department of Energy; DOE/EM-0249; 6/95.
140. Mixed Waste Characterization, Treatment, and Disposal Focus Area; U.S. Department of Energy; DOE/EM-0252; 6/95.

Mill Tailings

141. EIA New Releases; Energy Information Administration; DOE/EIA-0204(95/01); 1-2/95.
142. Decommissioning of U.S. Uranium Production Facilities; Energy Information Administration; DOE-EIA-0592; 2/95.
143. Proceedings of Workshop on Uranium Production Environmental Restoration, An Exchange Between the United States and Germany; U.S. Department of Energy; CONF-9308217; 8/93.
144. Health and Environmental Standards for Uranium and Thorium Mill Tailings, Final Rule; U.S. EPA; 40 CFR Part 192; 11/93.

145. Information for Consideration in Reviewing Groundwater Protection Plans at Uranium Mill Tailings Sites; Pacific Northwest Laboratory; NUREG/CR-5858; 5/92.
146. Uranium In Situ Leach Mining in the United States; Energy Information Administration/Uranium Industry Annual 1993; 9/94.
147. Uranium Mill Tailings Remedial Action (UMTRA) Project; U.S. Department of Energy; DOE/UMTRA—400124-0167; 9/89.

Disposal/Storage

148. Nuclear Waste, Foreign Countries' Approaches to High-Level Waste Storage and Disposal; GAO; GAO/RCED-94-172; 8/94.
149. Nuclear Waste, DOE's Management and Organization of the Nevada Repository Project; GAO; GAO/RCED-95-27; 12/94.
150. Nuclear Waste, Yucca Mountain Project, Management Funding Issues; GAO; GAO/T-RCED-93-58; 7/93.
151. Nuclear Waste, Comprehensive Review of the Disposal Program Is Needed; GAO; GAO/RCED-94-299; 9/94.
152. Nuclear Waste, Yucca Mountain Project Behind Schedule and Facing Major Scientific Uncertainties; GAO; GAO/RCED-93-124; 5/93.
153. Nuclear Waste, DOE's Repository Site Investigations, a Long and Difficult Task; GAO; GAO/RCED-92-73; 5/92.
154. Nuclear Waste, Operation of Monitored Retrievable Storage Facility Is Unlikely by 1998; GAO; GAO/RCED-91-194; 9/91.
155. Nuclear Waste, Issues Affecting Land Withdrawal of DOE's Waste Isolation Pilot Project; GAO; GAO/T-RCED-91-38; 4/91.
156. Feasibility of Disposal of High-Level Radioactive Waste into the Seabed, Overview of Research and Conclusions, Volume 1; Nuclear Energy Agency; INIS-X1Y—787.
157. Depleted Uranium Disposal Options Evaluation; EG&G Idaho, Inc.; EGG-MS-11297; 5/94.
158. Use of Savannah River Site Facilities for Blend Down of Highly Enriched Uranium; Westinghouse Savannah River Company; WSRC-RP-93-1262; 2/94.
159. National Summary Report of Draft Site Treatment Plans; National Overview of Mixed Wastes and Treatment Options; U.S. Department of Energy; 11/94.

Index

Lightning Source UK Ltd.
Milton Keynes UK
UKHW022122311018
331536UK00003B/44/P